Lectures on Numerical Mathematics

Heinz Rutishauser

Lectures on Numerical Mathematics

Edited by
Martin Gutknecht

with the Assistance of
Peter Henrici
Peter Läuchli
Hans-Rudolf Schwarz

Translated by
Walter Gautschi

With 67 Figures

1990

Birkhäuser
Boston · Basel · Berlin

Translator
Walter Gautschi
Department of Computer Sciences
Purdue University
West Lafayette, IN 47907
USA

Library of Congress Cataloging-in-Publication Data
Rutishauser, Heinz, 1918–1970.
    [Vorlesungen über numerische Mathematik. English]
    Lectures on numerical mathematics/Heinz Rutishauser; edited by
Martin Gutknecht, with the assistance of Peter Henrici, Peter
Läuchli, Hans-Rudolf Schwarz; translated by Walter Gautschi.
        p.   cm.
    Translation of: Vorlesungen über numerische Mathematik; German
ed. issued in 2 vols.
    Includes bibliographical references.
    ISBN-13:978-1-4612-8035-4   e-ISBN-13:978-1-4612-3468-5
    DOI:10.1007/978-1-4612-3468-5
    1. Numerical analysis.   I. Gutknecht, Martin.   II. Tiue.
    QA297.R8713   1990
    512'.7—dc20   90-58

Printed on acid-free paper.

© Birkhäuser Boston, 1990
Softcover reprint of the hardcover 1st edition 1990

Camera-ready copy supplied by the translator using troff (UNIX).

9 8 7 6 5 4 3 2 1

Editor's Foreword

The present book is an edition of the manuscripts to the courses "Numerical Methods I" and "Numerical Mathematics I and II" which Professor H. Rutishauser held at the E.T.H. in Zurich. The first-named course was newly conceived in the spring semester of 1970, and intended for beginners, while the two others were given repeatedly as elective courses in the sixties. For an understanding of most chapters the fundamentals of linear algebra and calculus suffice. In some places a little complex variable theory is used in addition. However, the reader can get by without any knowledge of functional analysis.

The first seven chapters discuss the direct solution of systems of linear equations, the solution of nonlinear systems, least squares problems, interpolation by polynomials, numerical quadrature, and approximation by Chebyshev series and by Remez' algorithm. The remaining chapters include the treatment of ordinary and partial differential equations, the iterative solution of linear equations, and a discussion of eigenvalue problems. In addition, there is an appendix dealing with the qd-algorithm and with an axiomatic treatment of computer arithmetic.

For a few algorithms, also problems of programming are discussed and fragments of ALGOL-programs are given. It should be pointed out that a number of complete and safe procedures to the methods described here are to be published (also with Birkhäuser, as Vol. 33 of the International Series of Numerical Mathematics) by W. Gander, L. Molinari and H. Švecová under the title "Numerische Prozeduren aus Nachlass und Lehre von Prof. Heinz Rutishauser".

When Professor H. Rutishauser died on November 10, 1970, at the age of 52, he left behind, among other things, the manuscripts to the courses mentioned above. These notes, which Mrs. M. Rutishauser kindly made available to us, have in the course of the years been repeatedly revised and updated to reflect progress in research. His desire to publish them later as a text book was known; some sections were already finished in an almost ready-to-print form. Unfortunately, however, he did not live to see the book completed. In view of the quality of these manuscripts, and given the world-wide reputation of the author, we deemed it more than justified that these manuscripts be published, even though in form and volume they certainly did not yet fully come up to the high demands of their author. From the beginning it was our intention to change the text as little as possible and to make no extensive reorganizations or additions. We were equally intent on not altering the character of the work and, for example, on preserving the occasionally pictorial language, which facilitates reading and understanding. Nevertheless, much remained to be done, especially in those parts which were only drafted by the author. Also, individual chapters which in part were available in several versions had to be merged as smoothly as possible into a seamless whole. The numerical examples were almost all newly recomputed. Finally, the figures needed to be prepared; the drawing of mathematically defined curves was generally done by a plotter.

Throughout the editing of the text, I was assisted above all by the coeditors Profs. P. Henrici, P. Läuchli and H.R. Schwarz, who read the entire manuscript and who consulted with me during many hours on questions of principles and details. Many further colleagues also helped me through their criticism; to be mentioned are particularly Prof. R. Jeltsch, Dr. R. Bloch and Dr. J. Waldvogel. Thanks go also to Miss G. Bönzli and Mrs. L. Gutknecht, who typed large parts of the text, and to Dr. V. Šechovcov, who drew the figures in ink. I am also very pleased that Mr. Stutz and others agreed to help me with the correction of the galley proofs.

My editing work for the most part was financed by the Swiss National Science Foundation. Finally, I wish to thank the publisher for the very careful and speedy printing.

Vancouver, B.C., February 1976   M. GUTKNECHT

Preface

Heinz Rutishauser is one of the pioneers of modern numerical mathematics. Educated originally as a function theorist, he in 1950 joined as a collaborator the Institute of Applied Mathematics, founded shortly before at the Federal Institute of Technology. There, his extraordinary algorithmic talent soon became evident. With concisely written publications he introduced methods and directions of research into numerical mathematics which later on proved to be fundamental. The stability theory in the numerical solution of ordinary differential equations, "economization" of power series by the use of Chebyshev polynomials, the quotient-difference algorithm, the LR-method, the exact justification of the Romberg algorithm, and many other contributions all go back to Rutishauser. He was also one of the first to recognize that the computer itself could be used for the preparation of computer programs, and he played a leading role in the development of the programming language ALGOL. In the last years of his life, Rutishauser concerned himself with the axiomatization of numerical computation and as a result gave perhaps the most satisfactory treatment, from a theoretical point of view, of the propagation of rounding errors. His health-related aversion to travel and, no doubt, a touch of introversion, prevented all these achievements from becoming known and appreciated as they deserved to be.

After Rutishauser's death in 1970, his widow, Mrs. Margrit Rutishauser, asked the undersigned to sift through his unpublished scientific notes. It became immediately clear to us that Rutishauser's lectures on numerical mathematics constituted an important part of these notes. The lectures, which in quality and originality far excel the average presentations in this area, were already intended for publication by

Rutishauser himself, but have only partly been prepared in detail for publication. It so happened that Dr. Martin Gutknecht, who still heard these lectures as a student, and who also has the necessary technical knowledge, could be prevailed upon to successfully complete the preparation for publication. Commendably, the work of Dr. Gutknecht has been supported by the Swiss National Science Foundation. We are pleased, thanks to the cooperation of the Birkhäuser publishing house, to be able to present the outcome to the public.

Zurich, February 1976

P. HENRICI
P. LÄUCHLI
H.R. SCHWARZ

Translator's Preface

Rutishauser's *Vorlesungen über numerische Mathematik* appeared in 1976 in two volumes. Even though more than twelve years have elapsed since the work was first published, it has retained much of its freshness and timeliness. The material treated, though no longer entirely up-to-date in some areas, still provides a sound and stimulating introduction to the field of scientific computing. It was felt desirable, therefore, to make the work accessible to a wider audience by providing an English translation.

The undersigned was happy to undertake this task, as he has known Rutishauser personally and has great admiration for his scientific achievements. In preparing the translation, he has combined the original two volumes into a single volume. He has refrained from making any major changes to the text itself, other than correcting a fair number of typographical errors. However, following a suggestion already made by G.W. Stewart in his review of the German original (cf. *Bull. Amer. Math. Soc.*, v. 84, 1978, pp. 660–663), he has supplemented each chapter with notes designed to make the reader aware of significant developments in computational techniques that occurred since the original volumes have appeared, and to direct him to appropriate sources for further study. The preparation of these notes took considerably longer than anticipated, and in fact would never have been completed, were it not for the invaluable assistance he has received from a number of colleagues. John K. Reid helped with the notes to Chapters 2 and 3, Florian A. Potra and Hermann Brunner with those to Chapters 4 and 5, and Chapters 8 and 9, respectively. The notes to Chapters 10 and 11 were contributed entirely by Lars B. Wahlbin, those to Chapters 12 and 13 in large part by Beresford N. Parlett. Comments from Carl de Boor pertaining to the notes for Chapters

6 and 7, and from Hans J. Stetter on the notes for Chapter 8, were also incorporated. The help of all these colleagues is herewith gratefully acknowledged.

Thanks are also due to Ms. Connie Heer, who capably and unremittingly prepared the photo-ready copy of the manuscript on a computer of the Department of Computer Sciences at Purdue University, using UNIX's *troff* system. Finally, we thank the publisher for patiently waiting for the completion of this project and for assisting us in the production of this volume.

West Lafayette, Ind., November 1989   WALTER GAUTSCHI

Contents

CHAPTER 1

An Outline of the Problems

§1.1. Reliability of programs

The object of numerical mathematics is to devise a numerical approach for solving mathematically defined problems, i.e., to exhibit a detailed description of the computational process which eventually produces the solution of the problem in numerical form (for example, a numerical table). In so doing, one must, of course, be cognizant of the fact that a numerical computation almost never is entirely exact, but is more or less perturbed by the so-called rounding errors. The computing process, indeed, is executed in *finite arithmetic,* for example in floating-point arithmetic (number representation: $z = a \times 10^b$), where only a finite number of digits are at disposal both for the mantissa a and for the exponent b.

Depending on how well the effects of finite arithmetic are taken into consideration, a computational process is classified as:

(a) a formal algorithm

(b) a naive program

(c) a strict program.

By a *formal algorithm* we mean a description of the basic course of computation. It represents the first step towards the solution of a problem, which, however, need not yet consider any limitations in arithmetic.

For example, in

$$x_{k+1} := x_k - f(x_k)/f'(x_k)$$

one has a formal algorithm for Newton's method for determining zeros of a function $f(x)$. Note that this algorithm offers no protection against division by 0. (To forbid dividing by 0, of course, is also an arithmetic limitation.)

One speaks of a *naive program* when one has an unequivocal definition of the computational process. The word "naive" is meant to convey the notion that although the finiteness of arithmetic is taken into account, the provisions made are based more on empirical grounds than on solid theory. From a naive program, therefore, one can generally expect reasonable results, but this cannot be guaranteed with absolute certainty.

Of a *strict program* we require not only that it should run correctly in spite of the finiteness of arithmetic, but also that it should do so on the basis of a rigorous proof.

Now a strict program still offers only *sequential reliability,* that is, one guarantees only the correctness of execution – in particular, correct termination – with no assertions being made concerning the accuracy of the results. If, however, one can guarantee in addition that the errors of the results lie within certain bounds (which are either produced along with the results, or can be preimposed together with the initial data), then the program is said to be *numerically reliable.*

Obviously, numerical reliability presupposes sequential reliability; if a naive program is still claimed to be numerically reliable, then this can only be meant conditionally.

§1.2. The evolution of a program

Given an applied mathematics problem, it is one of the tasks of numerical mathematics to first of all set up a formal algorithm for the solution procedure, and then from this develop a naive or, if possible, a strict program. (Here we shall be satisfied, however, with naive programs.) Such a program, i.e., the detailed computational steps for the solution of a problem, is always written in an internationally standardized algorithmic language (e.g., IFIP-ALGOL, ASA-FORTRAN, etc.).

The whole process can be explained by the following scheme.

Basic scheme for the solution of a problem on a computer

	Point of departure	*Activity*	*Domain of relevance*
↑	mathematical problem		domain of analysis
		discretization	
Applied mathematics	discrete mathematical problem		domain of algebra
		development of a numerical method	
	formal algorithm		numerical computation in exact arithmetic
		consideration of finite arithmetic	
	naive program (quality of program ascertained only empirically)		numerical computation in finite arithmetic
Numerical mathematics	strict program (quality of program guaranteed by rigorous proofs)		sequential reliability
↓	strict program with a priori or a posteriori error estimates		numerical reliability

§1.3. Difficulties

Just what kind of difficulties we may encounter in constructing a program, i.e., in defining a computational process, will now be explained in the case of a few miniature problems.

A) To be solved is the *quadratic equation*

$$x^2 - 742\,x + 2 = 0.$$

Proceeding quite naively, one obtains with 6-digit computation

$$x = 371 \pm \sqrt{137639} = 371 \pm 370.997,$$
$$x_1 = 741.997, \quad x_2 = .003,$$

where x_2, as a small difference of large numbers, i.e., owing to cancellation, has poor relative accuracy. However, one can easily determine x_2 more accurately, namely according to

$$x_2 = 2/x_1 = 2\,/\,741.997 = .00269543.$$

For the solution of a quadratic equation

$$x^2 + px + q = 0$$

we thus note: *The absolutely largest root must be computed first; then the smaller one can be determined by Vieta's rule.*

This leads to the following piece of ALGOL program:

```
x1: = abs(p/2) + sqrt(p↑2/4 − q);
if p > 0 then x1: = − x1;
x2: = q/x1;
```

However, this is still a naive program; it can only be applied as long as

1) the roots are real,
2) one does not have $p = q = 0$ ($x_1 = x_2 = 0$),
3) p^2 is still representable in machine arithmetic.

The last cannot be taken for granted: in the example

$$x^2 - 10^{200}x + 10^{50} = 0$$

the coefficients and the two roots $x_1 = 10^{200}$, $x_2 \doteq 10^{-150}$ are representable on a CDC-6000 computer, but $p^2 = 10^{400}$ is not.

B) As a further example, we briefly touch on the solution of *linear systems of equations:* Suppose one has to solve

$$1002x + 1003y = 1000$$
$$1003x + 1005y = 1000.$$

In 4-digit arithmetic one obtains with Gauss elimination

$$x = 1.999, \quad y = -1.000,$$

where these values, however, are quite uncertain because of cancellation. Now, perhaps, the client persists on physical grounds that the solution is sharply defined. One can react in two ways:

(a) Compute with more digits, which in the case at hand leads to

$$x = 1.998000, \quad y = -.998999.$$

This is meaningful if one deals with a purely mathematical problem, that is, if the coefficients 1002, 1003, etc. are *exact* numbers. How absurd this easy expedient of higher precision can be, is shown by the other recourse:

(b) One returns to the origin of the problem. Perhaps it was

$$1000z + 2.2x + 2.9y = \ \ .2$$
$$1000z + 2.9x + 5.4y = -.2,$$

where $z = x + y - 1$. By substituting for z and rounding to four decimals, one recovers the system of equations mentioned in the beginning. However, if the above double precision result is inserted into the original system, one obtains, first of all, $z = x + y - 1 = -.000999$; but then, substitution into the left-hand side of the first equation yields .4995 instead of .2, and in the second equation one finds $-.5994$ instead of $-.2$.

It would have been far better, here, to work with three unknowns:

$$2.2x + 2.9y + 1000z = \ \ .2$$
$$2.9x + 5.4y + 1000z = -.2$$
$$x + \ \ \ y - \ \ \ \ \ z = \ \ 1.$$

From this system one obtains, even with slide rule precision,

$$x = 1.61, \quad y = -.61, \quad z = -.00157,$$

which is a much better solution, since substitution into the left-hand side of the first equation yields .203, and into that of the second equation, $-.195$.

The initially given system of equations also might have been the normal equations of a least squares problem. In this case one would do better to solve it by orthogonalization (see Chapter 5).

C) To be solved is a *differential equation with strong damping*:

$$y' = 5xy^3 - 1000y + \sin x, \quad y(0) = 0.$$

Here, one would first look around for available programs for the numerical integration of differential equations. Most computing centers have for this purpose a program for the so-called Runge-Kutta method. This ([1]) in fact produces with stepsize $h = .005$ the useless results given in the column y_A of Table 1.1. In a case like this, only an extreme reduction of the integration step will help – which entails an equally severe increase in computational effort – , or one develops completely new methods. One such method ([2]), indeed, yields the values in the column y_B and with double the stepsize $h = .01$ even the practically identical values in column y_C. The exact solution, incidentally, is close to the function

Table 1.1. *Numerical integration of a differential equation with strong damping*

x	y_A	y_B	y_C	y_D
0.000	0	0	0	0
0.005	$1.770830_{10}-5$	$4.006721_{10}-6$		$4.006726_{10}-6$
0.010	$1.969178_{10}-4$	$8.999908_{10}-6$	$8.999895_{10}-6$	$8.999920_{10}-6$
0.015	$2.590041_{10}-3$	$1.399952_{10}-5$		$1.399954_{10}-5$
0.020	$3.533222_{10}-2$	$1.899882_{10}-5$	$1.899878_{10}-5$	$1.899885_{10}-5$
0.025	$4.840539_{10}-1$	$2.399765_{10}-5$		$2.399768_{10}-5$
0.030	6.463935	$2.899588_{10}-5$	$2.899582_{10}-5$	$2.899592_{10}-5$
0.035	$-1.882310_{10}2$	$3.399338_{10}-5$		$3.399343_{10}-5$
0.040	$-3.437826_{10}49$	$3.899004_{10}-5$	$3.898995_{10}-5$	$3.899010_{10}-5$
0.045	overflow	$4.398572_{10}-5$		$4.398578_{10}-5$
0.050		$4.898030_{10}-5$	$4.898019_{10}-5$	$4.898037_{10}-5$

([1]) Procedure *rksstp* in the program library of the ALCOR users group.

([2]) Procedure *damint* in the program library of the ALCOR users group.

$$y(x) = \frac{1000 \sin x - \cos x + e^{-1000x}}{1000001},$$

which satisfies the differential equation $y' = -1000y + \sin x$. Its values are given in column y_D of Table 1.1.

D) When devising a computational process, one constantly has to keep in mind that something that is correct in pure mathematics can be totally absurd in a numerical context. For example, $(a - b)^2$ and $a^2 - 2ab + b^2$ are not the same at all, numerically; in 3-digit computation one has, say, for $a = 15.6$, $b = 15.7$,

$$(a - b)^2 = .1^2 = .01,$$
$$a^2 - 2ab + b^2 = 243 - 490 + 246 = -1,$$

that is, the expanded form not even guarantees a positive result.

Likewise, in the expression

$$s = \sum_{k=1}^{n} \sqrt{a_k^2 - 2a_k b_k \cos \gamma_k + b_k^2}$$

one cannot be sure that the root radicands turn out to be positive; even if this were the case, individual terms of the sum may become rather inaccurate because of cancellation. For example, with $a = 15.6$, $b = 15.7$, $\gamma = 5°$, and again 3-digit computation, we have

$$a^2 = 243, \quad b^2 = 246, \quad 2ab = 490,$$
$$\cos \gamma = .996, \quad 2ab \cos \gamma = 488,$$

thus

$$\sqrt{a^2 - 2ab \cos \gamma + b^2} = 1$$

instead of the more accurate value $\sqrt{1.87399} = 1.36894$.

One might of course argue that these inaccurate terms are relatively small, and hence in effect contribute little to the total error. This would be quite true if it weren't for the fact that through the square root the small terms (and their errors) are enhanced in an undesirable way.

How, then, should one remedy this obvious deficiency? We use the identity

$$a^2 - 2ab \cos \gamma + b^2 = (a - b)^2 + 4ab \sin^2(\tfrac{1}{2} \gamma)$$

and thus compute

$$s = \sum_{k=1}^{n} \sqrt{(a_k - b_k)^2 + 4a_k b_k \sin^2(\tfrac{1}{2} \gamma_k)}.$$

In this way, every cancellation is eliminated. One obtains for the above example, in 3-digit computation,

$$(a - b)^2 = .01, \quad 4ab = 980, \quad \sin^2(\tfrac{1}{2} \gamma) = .00190,$$

thus, in all, $\sqrt{.01 + 1.86} = 1.37$, which lies well within the computing precision.

In summary, we conclude that in numerical computation many ways of thinking that have become dear to us must be thrown overboard. In extreme situations, for each individual problem, a method especially appropriate for it must be developed from scratch. Under no circumstances is it advisable to copy formulas from books of pure mathematics and use them indiscriminately for programming.

Notes to Chapter 1

§1.3 A detailed and unusually thorough discussion of the floating-point number system and its implications can be found in Sterbenz [1974]. There, the reader will learn, for example, that computing the average of two floating-point numbers, or solving a quadratic equation, can be fairly intricate tasks, if they are to be made foolproof. The quadratic equations problem is also considered at some length in Young & Gregory [1972, §3.4], where further references are given to earlier work of W. Kahan and G.E. Forsythe.

The fact that thoughtless use of mathematical formulae and numerical methods, or inherent sensitivities in the problem, can lead to disastrous results, is illustrated by well-chosen examples in Stegun & Abramowitz [1956] and Forsythe [1970]. Sometimes, nearby singularities will also cause the accuracy to deteriorate, unless corrective measures are taken; Forsythe [1958] has an interesting discussion of this.

To assess the errors in the final answers of a long computation is still a formidable task. There are two general approaches that deserve to be briefly mentioned here – *backward error analysis* and *interval arithmetic*. In the first, one attempts to interpret the computed answers as the exact answers to a slightly perturbed problem and one seeks to estimate the perturbation involved. If one knows, then, how strongly the solution of the problem reacts to small perturbations, one can estimate the error in the computed solution. The reader is referred to Wilkinson [1963] for a systematic and skillful application of this idea to problems in algebra and linear algebra. The goal of interval arithmetic, on the other hand, is to produce intervals that are guaranteed to contain the desired answers. This is achieved (at a cost) by operating consistently on floating-point intervals, rather than floating-point numbers. Enclosing also the initial data in appropriate intervals allows one to study the effect of uncertainties in the data. Good accounts of interval analysis and some of its applications can be found in Moore [1966], [1979]. Interval analysis is basically an a posteriori approach, i.e., error bounds are produced only after the computation has been completed. For generating a priori bounds, a new version of error arithmetic, developed by Olver [1978], appears to be more promising.

References

Forsythe, G.E. [1958]: *Singularity and near singularity in numerical analysis*, Amer. Math. Monthly **65**, 229–240.

Forsythe, G.E. [1970]: *Pitfalls in computation, or why a math book isn't enough*, Amer. Math. Monthly **77**, 931–956.

Moore, R.E. [1966]: *Interval Analysis*, Prentice-Hall, Englewood Cliffs, N.J.

Moore, R.E. [1979]: *Methods and Applications of Interval Analysis*, SIAM, Philadelphia.

Olver, F.W.J. [1978]: *A new approach to error arithmetic*, SIAM J. Numer. Anal. **15**, 368–393.

Stegun, I.A. and Abramowitz, M. [1956]: *Pitfalls in computation*, J. Soc. Indust. Appl. Math. **4**, 207–219.

Sterbenz, P.H. [1974]: *Floating-Point Computation*, Prentice-Hall, Englewood Cliffs, N.J.

Wilkinson, J.H. [1963]: *Rounding Errors in Algebraic Processes*, Prentice-Hall, Englewood Cliffs, N.J.

Young, D.M. and Gregory, R.T. [1972]: *A Survey of Numerical Mathematics*, Vol. I, Addison-Wesley, Reading, Mass.

CHAPTER 2

Linear Equations and Inequalities

The solution of systems of linear equations (briefly called equations) is probably the most important type of numerical computer application, because countless problems in applied mathematics ultimately – if only approximately – can be reduced to linear equations. Not surprisingly, therefore, interest in this problem has grown enormously in the computer age; what previously was viewed as tedious work has since become a legitimate and actively pursued area of mathematical research.([1])

The problem itself is rather simple: Desired are n numbers, denoted by $x_1, x_2, \ldots, x_n$, which are subject to n conditions in which, however, they enter only linearly:

$$a_{11}x_1 + a_{12}x_2 + \cdots + a_{1n}x_n + a_{10} = 0$$
$$a_{21}x_1 + a_{22}x_2 + \cdots + a_{2n}x_n + a_{20} = 0$$

$$\vdots$$

$$\quad (1)$$

$$a_{n1}x_1 + a_{n2}x_2 + \cdots + a_{nn}x_n + a_{n0} = 0.$$

Here the coefficients a_{kl} have prescribed values and the x_l are to be determined numerically. The a_{k0} are the constant terms which are sometimes given a different name, say $b_1, b_2, \ldots, b_n$, or are sometimes appended to the coefficient matrix as $(n+1)$st column $a_{1,n+1}, a_{2,n+1}, \ldots, a_{n,n+1}$.

It is customary to write down such equations in a compact form, say:

---

[1] Compare, e.g., Forsythe G.E., Moler C.B.: *Computer Solution of Linear Algebraic Systems*, Prentice-Hall, Englewood Cliffs, N.J., 1967.

$$\sum_{\ell=1}^{n} a_{k\ell} x_\ell + a_{k0} = 0 \quad (k = 1, 2, \ldots, n), \tag{2}$$

which can also be written in matrix form as

$$\mathbf{Ax} + \mathbf{v} = \mathbf{0}. \tag{3}$$

Here,

$$\mathbf{x} = \begin{bmatrix} x_1 \\ \cdot \\ \cdot \\ \cdot \\ x_n \end{bmatrix}$$

denotes the desired *solution vector*,

$$\mathbf{A} = \begin{bmatrix} a_{11} & \cdot & \cdot & \cdot & a_{1n} \\ \cdot & & & & \cdot \\ \cdot & & & & \cdot \\ \cdot & & & & \cdot \\ a_{n1} & \cdot & \cdot & \cdot & a_{nn} \end{bmatrix}$$

is the *coefficient matrix*, and

$$\mathbf{v} = \begin{bmatrix} a_{10} \\ \cdot \\ \cdot \\ \cdot \\ a_{n0} \end{bmatrix}$$

the *constant vector*. From (3) one obtains the solution at once as

$$\mathbf{x} = -\mathbf{A}^{-1}\mathbf{v}. \tag{4}$$

From a purely mathematical point of view, the problem is solved by (4), but for numerical purposes nothing is gained by it; on the contrary, the formula (4) embodies a suggestive force that has often misled uncritical programmers in unpleasant ways. Indeed, the inverse matrix is inappropriate as a tool for the numerical solution of linear equations, and the computation of $\mathbf{A}^{-1}$ is more a detour than a help. Of course, also the numerical analyst often, and gladly, makes use of the inverse matrix as an aid for theoretical investigations; he may even compute it once in a while, but hardly ever to determine the solution of a large system of linear equations by means of (4)([1]).

§2.1. The classical algorithm of Gauss

The linear equations (1) to be solved are written as a tableau:

	x_1	x_2	$\cdots$	x_n	1
$0 =$	a_{11}	a_{12}	$\cdots$	a_{1n}	a_{10}
$0 =$	a_{21}	a_{22}		a_{2n}	a_{20}
$\vdots$	$\vdots$	$\vdots$			
$0 =$	a_{n1}	a_{n2}		a_{nn}	a_{n0}

$$(5)$$

Such a tableau – filled with concrete numbers when actually used – is to be understood in the following sense: The sum of the products of the entries in a row and the corresponding quantities on top of the tableau (the so-called header row) is always to yield the value at the left margin of the row. If the prescribed row values, as here, are equal to 0, they can also be omitted. According to this convention, the tableau (5) indeed means the same thing as the system of equations (1); the latter, however, is to be solved now with the help of the tableau.

Since the row values of the tableau are equal to 0, the rows can be permuted at will, multiplied by constants, and added to one another. We begin by dividing the first row by $-a_{11}$ (where we tacitly assume that $a_{11} \neq 0$):

	x_1	x_2	$\cdots$	x_n	1
	-1	c_{12}	$\cdots$	c_{1n}	c_{10}
	a_{21}	a_{22}		a_{2n}	a_{20}
$\vdots$	$\vdots$				
	a_{n1}	a_{n2}		a_{nn}	a_{n0}

$$(6)$$

with $c_{1\ell} = -a_{1\ell}/a_{11}$ $(\ell = 1, 2, \ldots, n, 0)$

and then add (for $k = 2, 3, \ldots, n$) a_{k1}-times the new first row to the kth row; we obtain:

----

[1] On some parallel computers there may be an advantage in computing A^{-1} explicitly when solutions for many vectors v are desired. (Translator's note)

$$
\begin{array}{c|ccccc|c}
 & x_1 & x_2 & \cdots & x_n & & 1 \\
\hline
 & -1 & c_{12} & \cdots & c_{1n} & & c_{10} \\
 & 0 & a_{22}^* & & a_{2n}^* & & a_{20}^* \\
 & \vdots & & & & & \\
 & 0 & a_{n2}^* & & a_{nn}^* & & a_{n0}^*
\end{array}
\tag{7}
$$

with $a_{kl}^* = a_{kl} + a_{k1}c_{1l}$ $(k = 2,3,\ldots, n; l = 2,3,\ldots, n, 0)$.

This tableau, which is equivalent to (6), contains:

a) a *terminal equation*, which can also be given the form

$$
x_1 = c_{10} + \sum_{l=2}^{n} c_{1l}x_l,
\tag{8}
$$

and

b) a *reduced tableau* which corresponds to $n-1$ equations in the $n-1$ unknowns $x_2, x_3, \ldots, x_n$. As soon as the latter have been solved, (8) immediately yields also the missing unknown x_1.

The reduced tableau is now treated in the same way: its first row (the second of (7)) is divided by $-a_{22}^*$, which produces another terminal equation with coefficients $c_{2l} = -a_{2l}^*/a_{22}^*$. Through addition of multiples of this terminal equation to the remaining rows, one obtains a further reduced tableau with $n-2$ unknowns and coefficients a_{kl}^{**}, etc. Eventually, one arrives at a scheme of n terminal equations

$$
\begin{array}{ccccccc|c}
x_1 & x_2 & x_3 & x_4 & \cdots & x_n & & 1 \\
\hline
-1 & c_{12} & c_{13} & c_{14} & \cdots & c_{1n} & & c_{10} \\
0 & -1 & c_{23} & c_{24} & & c_{2n} & & c_{20} \\
0 & 0 & -1 & c_{34} & & c_{3n} & & c_{30} \\
\vdots & & & & & & & \\
0 & 0 & 0 & 0 & & -1 & & c_{n0}
\end{array}
\tag{9}
$$

from which one successively determines $x_n, x_{n-1}, \ldots, x_2, x_1$ according to

$$
x_k = c_{k0} + \sum_{l=k+1}^{n} c_{kl}x_l \quad (k = n, n-1, \ldots, 1).
\tag{10}
$$

The procedure described by (10), which follows immediately from the equation (9), is called *back substitution*.

Example. A polynomial $a_0 + a_1 x + a_2 x^2$ of degree 2 is to be constructed in such a way that it agrees at $x = 1, 2, 3$ with $y(x) = 1/x$. This problem is solved by the following tableaus:

problem statement:

	a_0	a_1	a_2	1
$0 =$	1	1	1	-1
$0 =$	1	2	4	$-\frac{1}{2}$
$0 =$	1	3	9	$-\frac{1}{3}$

$$(11)$$

first reduced tableau:

a_0	a_1	a_2	1
-1	-1	-1	1
0	1	3	$\frac{1}{2}$
0	2	8	$\frac{2}{3}$

second reduced tableau:

a_0	a_1	a_2	1
-1	-1	-1	1
0	-1	-3	$-\frac{1}{2}$
0	0	2	$-\frac{1}{3}$

third reduced tableau:

a_0	a_1	a_2	1
-1	-1	-1	1
0	-1	-3	$-\frac{1}{2}$
0	0	-1	$\frac{1}{6}$
$\frac{11}{6}$	-1	$\frac{1}{6}$	

From the third reduced tableau (with the terminal equations) the a_2, a_1, a_0

can be determined one after another by (10); it is expedient to write them in turn at the bottom of the terminal tableau. As a result, we obtain the polynomial

$$\frac{1}{6}(x^2 - 6x + 11).$$

Row interchanges. Up until now, the possibility was ignored that one of the quantities by which one must divide ($a_{11}, a_{22}^*, a_{33}^{**}$, etc.) could be 0. In the example

x_1	x_2	x_3	x_4	1
1	2	3	4	0
1	2	4	6	−1
1	3	6	9	−1
1	4	9	16	0

(12)

this situation is encountered after the first step:

x_1	x_2	x_3	x_4	1
−1	−2	−3	−4	0
0	0	1	2	−1
0	1	3	5	−1
0	2	6	12	0

Since we now have $a_{22}^* = 0$, one interchanges the second and third equation, and then proceeds with the computation. The second reduced tableau (after the interchange) reads:

x_1	x_2	x_3	x_4	1
−1	−2	−3	−4	0
0	−1	−3	−5	1
0	0	1	2	−1
0	0	0	2	2

Since a_{43}^{**} accidentally became 0, the third and fourth step can be carried out together; one obtains directly the terminal equations:

$$
\begin{array}{ccccc}
x_1 & x_2 & x_3 & x_4 & 1
\end{array}
$$

x_1	x_2	x_3	x_4	1
−1	−2	−3	−4	0
0	−1	−3	−5	1
0	0	−1	−2	1
0	0	0	−1	−1

$$
\begin{array}{cccc}
1 & -3 & 3 & -1
\end{array}
$$

At the bottom of this terminal tableau we again have the solution.

Note: If the matrix **A** is indeed nonsingular, one always gets through with suitable row interchanges. Nevertheless, interchanges should be made not only when a divisor becomes 0, but already when it has become small. We shall return to this point in §2.4.

§2.2. The triangular decomposition

The scheme (9) of terminal equations contains all the information necessary for the calculation of the unknowns. Still, with a view towards the computer organization of the computation, one has to ask oneself whether filling in the lower half of the scheme with zeros is really meaningful. After all, one knows that there have to be zeros in those places and numbers −1 on the diagonal.

As a matter of fact, in passing from the scheme (5) to the scheme (7), one realizes that by inserting the −1 and the zeros into the first column one pushes away precisely those row factors $a_{11}, a_{21}, \ldots, a_{n1}$ which could provide information as to how the scheme (7) has been computed. Likewise, in the next elimination step, one displaces the row factors $a_{22}^*, a_{32}^*, \ldots, a_{n2}^*$ which have served for the calculation of the second terminal equation and the coefficients a_{kl}^{**} of the $n-2$ reduced equations.

Considering that this history of successive generation is of importance in many respects, it surely would be more appropriate not to replace these row factors $a_{11}, a_{21}, \ldots, a_{n1}, a_{22}^*, \ldots, a_{n2}^*, a_{33}^*, \ldots$ by 0 and −1, respectively. We rather leave them at their places, but henceforth denote them by b instead of a, and without asterisks (but with the same indices). In this way the terminal scheme takes on the form

x_1	x_2	x_3	$\cdots$	x_n	1	
b_{11}	c_{12}	c_{13}	$\cdots$	c_{1n}	c_{10}	
b_{21}	b_{22}	c_{23}		c_{2n}	c_{20}	
b_{31}	b_{32}	b_{33}		c_{3n}	c_{30}	(13)
$\vdots$						
b_{n1}	b_{n2}	b_{n3}		b_{nn}	c_{n0}	

It is called the *BC-scheme*([1]). The pth elimination step evidently consists in renaming the elements in the first column of the reduced equations by b_{pp}, $b_{p+1,p}, \ldots, b_{np}$, and letting them stay where they are, while the coefficients of the first of these reduced equations are divided by $-b_{pp}$, thus giving rise to the terminal equation coefficients $c_{p,p+1}$, $c_{p,p+2}, \ldots, c_{pn}, c_{p0}$:

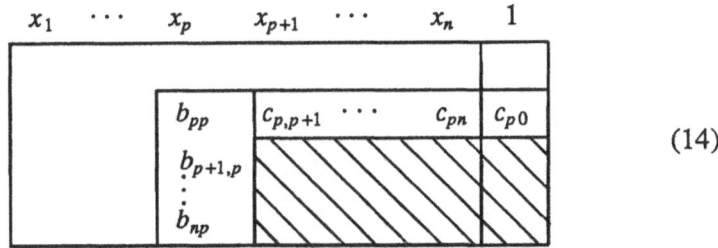

$$(14)$$

Thereafter, to each element in the hatched region one adds a product $b \times c$, namely $b_{kp}c_{p\ell}$ to the element in position $[k,\ell]$. With that, the pth system of reduced equations is completed.

The element in the position $[k,\ell]$ during the course of the complete elimination eventually will end up in the first row or column of a system of reduced equations, namely

(a)   if $k \geq \ell$, in the first column of the $(\ell - 1)$st system of reduced equations, and remains there unchanged as $b_{k\ell}$,

(b)   if $k < \ell$, in the first row of the $(k - 1)$st system of reduced equations, and then becomes after division by $-b_{kk}$ the terminal equation coefficient $c_{k\ell}$.

Consequently, for $k \geq \ell$,

$$b_{k\ell} = a_{k\ell} + \sum_{p=1}^{\ell-1} b_{kp}c_{p\ell} \tag{15}$$

---

[1] This scheme of course must not be read according to our convention of §2.1. (Editors' note)

or

$$a_{k\ell} = -\sum_{p=1}^{n} b_{kp}c_{p\ell},\tag{16}$$

provided one sets $c_{\ell\ell} = -1$ and $c_{p\ell} = 0$ for $p > \ell$. For $k < \ell$ one has

$$c_{k\ell} = -(a_{k\ell} + \sum_{p=1}^{k-1} b_{kp}c_{p\ell})/b_{kk}\tag{17}$$

or

$$a_{k\ell} = -\sum_{p=1}^{n} b_{kp}c_{p\ell},\tag{18}$$

if one defines $b_{kp} = 0$ for $p > k$. By (16) and (18), **A** is a matrix product,

$$\mathbf{A} = -\ \mathbf{BC},\tag{19}$$

where the matrices **B** and **C** are defined as follows:

$$\mathbf{B} = \begin{bmatrix} b_{11} & 0 & 0 & \cdot & \cdot & \cdot & 0 \\ b_{21} & b_{22} & 0 & & & & 0 \\ b_{31} & b_{32} & b_{33} & & & & 0 \\ \cdot & & & \cdot & & & \\ \cdot & & & & \cdot & & \\ \cdot & & & & & \cdot & \\ b_{n1} & b_{n2} & b_{n3} & & & & b_{nn} \end{bmatrix}, \quad \mathbf{C} = \begin{bmatrix} -1 & c_{12} & c_{13} & \cdot & \cdot & \cdot & c_{1n} \\ 0 & -1 & c_{23} & & & & c_{2n} \\ 0 & 0 & -1 & & & & c_{3n} \\ \cdot & & & \cdot & & & \\ \cdot & & & & \cdot & & \\ \cdot & & & & & \cdot & \\ 0 & 0 & 0 & & & & -1 \end{bmatrix}$$

One thus has:

Theorem 2.1. *The Gauss elimination algorithm, if it can be completed without row interchanges, achieves the decomposition of the coefficient matrix* **A** *into a product of two triangular matrices.*

The equations (17) and (18), of course, are valid also for $\ell = 0$:

$$c_{k0} = -(a_{k0} + \sum_{p=1}^{k-1} b_{kp}c_{p0})/b_{kk},\tag{20}$$

or

$$a_{k0} = -\sum_{p=1}^{n} b_{kp}c_{p0}.$$

This can be interpreted as

$$\mathbf{v} = -\mathbf{Bw}, \tag{21}$$

if in addition to the constant vector $\mathbf{v} = [a_{10}, a_{20}, \ldots, a_{n0}]^T$ one introduces also the vector $\mathbf{w} = [c_{10}, c_{20}, \ldots, c_{n0}]^T$. (The superscript T means "transposed".) Therefore, during the elimination process one also solves the additional system

$$\mathbf{Bw} + \mathbf{v} = \mathbf{0}. \tag{22}$$

If one now substitutes $\mathbf{A} = -\mathbf{BC}$ and (21) into the original system of equations, one obtains $-\mathbf{BCx} - \mathbf{Bw} = \mathbf{0}$ and thus

$$\mathbf{Cx} + \mathbf{w} = \mathbf{0}. \tag{23}$$

These, however, are precisely the terminal equations to which the elimination process has reduced the given system.

We now recognize how Gauss's algorithm works:

(1) The matrix $\mathbf{A}$ is decomposed into the factors $\mathbf{B}$ and $-\mathbf{C}$. (This operation takes place solely in the space of the matrix, and is called *triangular factorization*.)

(2) The system of equations $\mathbf{Bw} + \mathbf{v} = \mathbf{0}$ is solved. Owing to the triangular form of $\mathbf{B}$, one can give explicit formulae [cf. (20)] for this process, called *forward substitution*:

$$w_k = -\left(v_k + \sum_{p=1}^{k-1} b_{kp} w_p\right)/b_{kk} \quad (k = 1, 2, \ldots, n). \tag{24}$$

(3) The terminal equations $\mathbf{Cx} + \mathbf{w} = \mathbf{0}$ are solved, which is called *back substitution* and can also be described by explicit formulae [cf. (10)]:

$$x_k = w_k + \sum_{\ell=k+1}^{n} c_{k\ell} x_\ell \quad (k = n, n-1, \ldots, 2, 1). \tag{25}$$

One can carry out these three processes either separately, or, by including the constant vector in the tableau and subjecting it to the same transformation – as was done above – one can fuse the triangular decomposition and forward substitution into one process (the so-called *elimination*); then only back substitution remains to be done, for which the matrix $\mathbf{B}$ is not required.

Example. To the system of equations

	x_1	x_2	x_3	x_4	1
$0 =$	5	7	9	10	−1
$0 =$	6	8	10	9	−1
$0 =$	7	10	8	7	−1
$0 =$	5	7	6	5	−1

$$(26)$$

we first apply the second of the two variants, that is, the constant vector is carried along. We let the row factors stay in their places. After four elimination steps

	x_1	x_2	x_3	x_4	1
5	−1.4	−1.8	−2	.2	
6	−.4	−.8	−3	.2	
7	.2	−4.6	−7	.4	
5	0	−3	−5	0	

	x_1	x_2	x_3	x_4	1
5	−1.4	−1.8	−2	.2	
6	−.4	−2	−7.5	.5	
7	.2	−5	−8.5	.5	
5	0	−3	−5	0	

	x_1	x_2	x_3	x_4	1
5	−1.4	−1.8	−2	.2	
6	−.4	−2	−7.5	.5	
7	.2	−5	−1.7	.1	
5	0	−3	.1	−.3	

	x_1	x_2	x_3	x_4	1
5	−1.4	−1.8	−2	.2	
6	−.4	−2	−7.5	.5	
7	.2	−5	−1.7	.1	
5	0	−3	.1	3	

$$(27)$$

there results the *BC*-scheme, from which one infers the matrices

$$\mathbf{B} = \begin{bmatrix} 5 & 0 & 0 & 0 \\ 6 & -.4 & 0 & 0 \\ 7 & .2 & -5 & 0 \\ 5 & 0 & -3 & .1 \end{bmatrix}, \quad \mathbf{C} = \begin{bmatrix} -1 & -1.4 & -1.8 & -2 \\ 0 & -1 & -2 & -7.5 \\ 0 & 0 & -1 & -1.7 \\ 0 & 0 & 0 & -1 \end{bmatrix} \quad (28)$$

and the vector $\mathbf{w} = [.2, .5, .1, 3]^T$. Back substitution according to (25) then yields the solution:

$$x_4 = 3,$$
$$x_3 = .1 + (-1.7) \times 3 = -5,$$
$$x_2 = .5 + (-7.5) \times 3 + (-2) \times (-5) = -12,$$
$$x_1 = .2 + (-2) \times 3 + (-1.8) \times (-5) + (-1.4) \times (-12) = 20.$$

If the constant vector – according to the first variant – would not have participated in the transformations, that is, if in (27) the last column were missing, one could produce it by the forward substitution (24):

$$w_1 = - (-1)/5 = .2,$$
$$w_2 = - (-1 + 6 \times .2)/(-.4) = .5,$$
$$w_3 = - (-1 + 7 \times .2 + .2 \times .5)/(-5) = .1,$$
$$w_4 = - (-1 + 5 \times .2 + (-3) \times .1)/.1 = 3.$$

Subsequently, back substitution, as above, would again give the solution vector $\mathbf{x}$.

§2.3. Iterative refinement

The separate treatment of the three processes: triangular decomposition, forward substitution, and back substitution, is especially useful when, at some later time, a system of equations with the same coefficient matrix, but new constant terms, has to be solved again. Then the second solution in fact requires only forward and back substitution, for which the matrices $\mathbf{B}$ and $\mathbf{C}$ obtained in the first solution by decomposition of $\mathbf{A}$ can be reused without change. In this sense, the matrices $\mathbf{B}$ and $\mathbf{C}$ together are equivalent to the inverse matrix $\mathbf{A}^{-1}$.

As an example, we consider the so-called *iterative refinement*: suppose we test the computed solution vector $\mathbf{x}$ of the system $\mathbf{Ax} + \mathbf{v} = \mathbf{0}$, i.e., simply evaluate this expression through substitution. Because of the inaccuracies of the computation one obtains a *residual vector*

$$Ax + v = v_1, \tag{29}$$

which will be different from 0 in general. Thus, x is not the correct solution. One therefore tries a new corrected $x + x_1$, with the aim of making $A(x + x_1) + v = 0$. In view of (29), this is equivalent to

$$Ax_1 + v_1 = 0. \tag{30}$$

This system of equations for the correction x_1 indeed has the same coefficient matrix A and can be solved by forward and back substitution:

$$Bw_1 + v_1 = 0 \rightarrow w_1,$$

$$Cx_1 + w_1 = 0 \rightarrow x_1.$$

Numerical example. Suppose $x = [-.052, .2, .004, .184]^T$ has already been computed as a solution of the system of equations

	x_1	x_2	x_3	x_4	1
0 =	5	7	9	10	−3
0 =	6	8	10	9	−3
0 =	7	10	8	7	−3
0 =	5	7	6	5	−2

One has $Ax = [3.016, 2.984, 2.956, 2.084]^T$ and thus $v_1 = Ax + v = [.016, -.016, -.044, .084]^T$. The matrix A was already used in (26), and its triangular decomposition has been noted in (28). Forward and back substitution results in the values of $w_1 = [w_1^{(1)}, \ldots, w_4^{(1)}]^T$ and $x_1 = [x_1^{(1)}, \ldots, x_4^{(1)}]^T$ shown, respectively, at the bottom of the two following tableaus:

$w_1^{(1)}$	$w_2^{(1)}$	$w_3^{(1)}$	$w_4^{(1)}$	1
5	0	0	0	.016
6	−.4	0	0	−.016
7	.2	−5	0	−.044
5	0	−3	.1	.084
−.0032	−.088	−.0168	−1.184	

$x_1^{(1)}$	$x_2^{(1)}$	$x_3^{(1)}$	$x_4^{(1)}$	1
−1	−1.4	−1.8	−2	−.0032
0	−1	−2	−7.5	−.088
0	0	−1	−1.7	−.0168
0	0	0	−1	−1.184

−7.948	4.8	1.996	−1.184

We thus obtain the improved solution $x + x_1 = [-8, 5, 2, -1]^T$. As one easily checks, this solution satisfies the equation exactly.

Here the correction is substantially larger than the original solution x, and this in spite of the small residual vector $v_1 = Ax + v$. One sees that even for a rather inaccurate solution, the equations can be almost satisfied.[1]

§2.4. Pivoting strategies

Until now, the terminal equations have been obtained by always dividing the first of the reduced equations by its first coefficient. These divisors, which appear as diagonal elements $b_{11}, b_{22}, \ldots, b_{nn}$ in the BC-scheme, are called *pivot elements* and must of course be different from zero. If, however, in the pth elimination step it turns out that $b_{pp} = 0$, then the first reduced equation must be exchanged with another, whose first coefficient does not vanish. If, say $b_{kp} \neq 0$, then rows k and p of the scheme (14) are exchanged – the *whole rows,* of course. Therefore, the old b_{kp} will be used as pivot element; yet, it has been brought to the position of b_{pp}.

The question now arises as to what criteria are to be used for selecting the substitute pivot element b_{kp}. For numerical computation, it is indeed not only the case $b_{pp} = 0$ which is troublesome, but also the case where b_{pp} is very small in absolute value. We must not wait, therefore, until $b_{pp} = 0$, before we look for a substitute; rather, the question is always which of the elements $b_{pp}, b_{p+1,p}, \ldots, b_{np}$ in the first column of the reduced system is the best pivot element in the pth elimination step (and this for all p).

---

[1] To have such a large correction is not typical for practical computer computations where more significant figures are held, but the small residual is typical. (Translator's note)

Every selection criterion for deciding this question is called a *pivot-ing strategy*. Up until now, we used the *diagonal strategy*, i.e., we selected as pivot elements in turn the diagonal elements, without any interchanges whatsoever. The diagonal strategy of course is not generally applicable, since at any time it can trigger division by 0, even if in the original coefficient matrix all diagonal elements are $\neq 0$. Besides, also small diagonal elements are dangerous, as the following example (in four-digit computation) shows:

System of equations:

	x_1	x_2	1
0 =	.00031	1	−3
0 =	1	1	−7

BC-scheme:

x_1	x_2	1
.00031	−3226	9677
1	−3225	2.998

One obtains $x_2 = 2.998$, then $x_1 = 9677 - 3266 \times 2.998 = 5.000$, which, owing to cancellation, is totally unreliable.

Nevertheless, one has:

Theorem 2.2. *The diagonal strategy is always acceptable, if the coefficient matrix* **A** *is diagonally dominant, i.e., if in each row the diagonal element is larger in absolute value than the sum of the moduli of the off-diagonal elements.*([1]) (The proof by mathematical induction is straightforward.)

Partial pivoting strategy. In order to safely avoid the dangerous zero in the choice of pivots, most programmers select as pivot element in the pth elimination step simply the absolutely largest of the elements b_{pp}, $b_{p+1,p}, \ldots, b_{np}$ in question. If b_{kp} denotes this pivot element, one then

---

[1] It would suffice to assume that the matrix **A** is regular and *weakly* diagonally dominant (i.e., in each row the diagonal element is *not smaller* in absolute value than the sum of the moduli of the off-diagonal elements). (Editors' note)

Another instance in which the diagonal strategy is permissible is when **A** is symmetric and positive definite; see Chapter 3. (Translator's note)

interchanges the kth and pth row of the scheme, and carries out the elimination step. If, however, among those elements none is different from zero, the matrix $\mathbf{A}$ is singular and, at any rate, a unique resolution of the system of equations is not possible.

In the above example, this simple partial pivoting strategy yields the BC-scheme

x_1	x_2	1
1	−1	7
.00031	.9997	2.999

from which there follows $x_1 = 4.001$, $x_2 = 2.999$, exact to four decimals.

Complete pivoting strategy. This strategy consists in locating the pivot element not just in the first column of the matrix of reduced equations, but determining instead the absolutely largest element in the whole matrix of reduced equations. This maximum element is then brought into the position of a_{pp} by an interchange of rows *and* columns.

Now, while the partial as well as the complete pivoting strategies are much better than the diagonal strategy, they are not effective in all cases. For example, in

x_1	x_2	x_3	1
2	1	1	1
1	10^{-10}	0	0
1	0	10^{-10}	0

the element a_{11} is clearly the absolutely largest element in the first column as well as in the whole coefficient matrix. Both strategies therefore would select this element as pivot element; after one step one then obtains the scheme

x_1	x_2	x_3	1
2	−.5	−.5	−.5
1	−.5	−.5	−.5
1	−.5	−.5	−.5

since −.5, added to 10^{-10}, in 8-digit arithmetic, again gives −.5. Thus, the reduced equations have become linearly dependent (in fact identical); a unique solution is impossible.

After an interchange of rows 1 and 2, on the other hand, we obtain in the first step

	x_1	x_2	x_3	1
1	-10^{-10}	0	0
2	1	1	1
1	-10^{-10}	10^{-10}	0

and at the end the BC-scheme

	x_1	x_2	x_3	1
1	-10^{-10}	0	0
2	1	-1	-1
1	-10^{-10}	$2_{10}-10$	$-.5$

As solution one obtains $\mathbf{x} = [5_{10}-11, -.5, -.5]^T$.

Why is the pivot element 1 here better than the 2? Rather conspicuously, the 1 dominates the elements in the same row much more than is the case with the 2. This fact suggests the next strategy:

Relative partial pivoting strategy. In the first column of the matrix of reduced equations one selects the element as pivot element which, relative to the other elements in the same row, is the largest, i.e., one determines

$$\max_{p \le j \le n} \frac{|a_{jp}|}{\sum\limits_{k=p+1}^{n} |a_{jk}|},$$

where a_{jk} $(j,k = p, \ldots, n)$ are the coefficients of the reduced equations at the beginning of the pth elimination ([2]). (One has of course $a_{jp} = b_{jp}$ for $j = p, \ldots, n$.)

For the example above, the quotients in the first elimination step $(p=1)$ are 1 $(j=1)$, $_{10}10$ $(j=2)$, $_{10}10$ $(j=3)$. As first pivot one therefore has to take a_{21} or a_{31}.

---

[2] If $\sum\limits_{k=p+1}^{n} |a_{jk}| = 0$ for some j, it is true that the maximum becomes infinitely large, but then also, in this case, one has to select the jth row as pivot row. (Editors' note)

Relative complete pivoting strategy([3]). It is natural to seek a combination of the relative partial pivoting strategy and the complete pivoting strategy and to select among *all* elements of the matrix of reduced equations that one as pivot which, relative to the sum of the moduli of all elements in the same row, is the largest. It moreover turns out to be especially advantageous to include in the row sum also the elements b_{jk} with $k < p$, which, after all, are known at the beginning of the pth elimination step. For determining the pivot element, one thus selects an index pair $[j, \ell]$ for which the maximum

$$\max_{p \le j, \ell \le n} \frac{|a_{j\ell}|}{\sum_{k=1}^{p-1} |b_{jk}| + \sum_{k=p}^{n} |a_{jk}|}$$

is attained.

§2.5. Questions of programming

One has to keep in mind that the solution of the system of equations (1) is carried out on a computer and that, therefore, the coefficient matrix **A** together with the constant terms have first to be stored as **array** $a[1:n, 1:n+1]$. (The constant terms are now denoted by $a[k, n+1]$.)

The schemes derived from the initial tableau, placed in the **array** a, which always consist of terminal equations, row factors and reduced equations, are now stored in the same **array** a, and, naturally, this is true also for the *BC*-scheme obtained after n steps. At the end of the elimination, $a[k, \ell]$ therefore contains the element $b_{k\ell}$ or $c_{k\ell}$ of the *BC*-scheme, depending on whether $k \ge \ell$ or $k < \ell$, respectively.

At the beginning of the pth elimination step, on the other hand, one has [replace p by $p-1$ in (14)]

$$a[k, \ell] = \begin{cases} b_{k\ell} & \text{if } k \ge \ell \text{ and } \ell < p, \\ c_{k\ell} & \text{if } k < \ell \text{ and } k < p; \end{cases}$$

in all other cases ($k \ge p$ and $\ell \ge p$), $a[k, \ell]$ is a coefficient of a reduced

---

[3] Section added by the Editors. This strategy has been used by H. Rutishauser in the procedure *liglei*, which he has programmed for the computing center of the ETH, and which is published in: Gander W., Molinari L., Švecová H.: *Numerische Prozeduren aus Nachlass und Lehre von Prof. Heinz Rutishauser*, Birkhäuser Verlag, Basel, 1977.

equation.

In this way, the whole Gauss elimination process takes place in the **array** $a[1{:}n,\ 1{:}n+1]$, which means that one can get by with $n^2 + n$ storage cells.

However, the transition from the given scheme to the BC-scheme can be accomplished in different ways, quite apart from the fact that one can treat the constant terms either concurrently as $(n+1)$st column of the **array** a, or divorced from the coefficient matrix (now stored as **array** $a[1{:}n,\ 1{:}n]$) as vector **array** $v[1{:}n]$. Eventually, back substitution still has to be carried out.

The classical method of Gauss reduces the given equations step by step to reduced equations with less and less unknowns and at the same time builds up the terminal equations; the pth step has the form:

> **begin**
>   **for** $\ell := p+1$ **step** 1 **until** $n+1$ **do**
> $a[p,\ell] := -a[p,\ell]/a[p,p]$;
>   **for** $k := p+1$ **step** 1 **until** n **do**
>   **for** $\ell := p+1$ **step** 1 **until** $n+1$ **do**
> $a[k,\ell] := a[k,\ell] + a[k,p] \times a[p,\ell]$
> **end**;

The first **for**-loop here sets up the new terminal equation, while in the second part of this compound statement the reduced equations are being transformed. As to the row factors $a[k,p]$, we don't have to worry, since they remain unchanged at their places. Note also that the transformation of the constant terms is accomplished by always letting the index ℓ run up to $n+1$ [cf. (14)].

The complete elimination consists in executing this statement for $p = 1,2,\ldots, n$, where it is to be noted that for $p=n$ only the single operation $a[n,n+1] := -\,a[n,n+1]/a[n,n]$ occurs.

An important variant, the *columnwise elimination,* exploits the fact that by (15) and (17) each element of the BC-scheme can be built up directly from the corresponding coefficient $a_{k\ell}$ and certain products $b_{kj}c_{j\ell}$ of b- and c-elements. Since only c-elements above $a_{k\ell}$ and b-elements to the left of it are required, one can compute in turn, first by (17) with $k = 1,\ldots, \ell - 1$, and then by (15) with $k = \ell, \ell + 1, \ldots, n$, the quantities $c_{1\ell}, c_{2\ell}, \ldots, c_{\ell-1,\ell}, b_{\ell\ell}, b_{\ell+1,\ell}, \ldots, b_{n\ell}$, once all b-elements in columns 1 to $\ell - 1$ are known. This is accomplished, for example, by the program

```
begin
    for k := 1 step 1 until ℓ − 1 do
    begin
        s := a[k,ℓ];
        for j := 1 step 1 until k − 1 do  ·
            s := s+a[k,j] × a[j,ℓ];
        a[k,ℓ] := − s/a[k,k]
    end;
    for k := ℓ step 1 until n do
    begin
        s := a[k,ℓ];
        for j := 1 step 1 until ℓ − 1 do
            s := s+a[k,j] × a[j,ℓ];
        a[k,ℓ] := s;
    end
end;
```

For complete elimination, this is executed for $\ell = 1, 2, \ldots, n+1$. Since here, for $\ell = n+1$ (and only in this case), one treats precisely the constant terms, a possibility is indicated of computing separately the BC-scheme and the vector **w**. (Note that for $\ell = n+1$ the second k-loop is empty.)

This separation is achieved by executing the above statement only for $\ell = 1, \ldots, n$, and then, for $\ell = n+1$, by writing $v[j]$ in place of $a[j,\ell]$. This corresponds precisely to the forward substitution according to formula (24):

```
for k := 1 step 1 until n do
begin
    s := v[k];
    for j := 1 step 1 until k−1 do
        s := s+a[k,j] × v[j];
    v[k] := − s/a[k,k]
end;
```

Here also, one works "in place".

Finally, there comes the back substitution according to (25):

```
for k := n−1 step −1 until 1 do
begin
    s := v[k];
    for j := k+1 step 1 until n do
```

$$s := s + a[k,j] \times v[j];$$
$$v[k] := s$$
end;

(Here, **v** is already the solution vector; the original constant terms are destroyed.)

Interchanges. If one has to interchange two rows p and ℓ, one has to keep track of the indices. For that purpose one introduces an integer vector **integer array** $z[1:n]$ which initially is filled with $z[k]:=k$. Later, $z[k]=p$ is to signal that the original pth row resides in position k. In order that this always works, the interchange of the rows ℓ and p is done as follows (only in the matrix part, for the time being):

for $j := 1$ **step** 1 **until** n **do**
begin
    $h := a[\ell,j];$
    $a[\ell,j] := a[p,j];$
    $a[p,j] := h$
end;
$i := z[\ell];$
$z[\ell] := z[p];$
$z[p] := i;$

Now, given a constant vector **v**, one will first reload:

for $k := 1$ **step** 1 **until** n **do**
    $x[k] := v[z[k]];$

and then work with the constant vector **x**. In summary, we obtain the following procedure *gaukos* for the solution of linear systems of equations:

procedure *gaukos*$(n,a,v,x,z,sing)$;
    value n;
    integer n; **array** a,v,x; **integer array** z; **label** *sing*;
    comment *if $n > 0$: building - up of the bc-matrix columnwise*
        from left to right. pivot choice according to the partial
        pivoting strategy. row interchanges to bring the pivots into
        the diagonal. thereafter forward and back substitution.
        if $n < 0$: only forward and back substitution, it being
        assumed that the bc-matrix is already stored in a and the
        row interchange vector in z;
    begin
        real h,s,max;

```
integer i,j,k,ℓ,p;
boolean rep;
rep := (n <0);
n := abs (n);
if rep then goto con;
for k := 1 step 1 until n do z[k] := k;
comment triangular decomposition;
for ℓ := 1 step 1 until n do
begin
    comment first the coefficients of the terminal equations
        are computed in column ℓ;
    for k := 1 step 1 until ℓ−1 do
    begin
        s := a[k,ℓ];
        for j := 1 step 1 until k−1 do
            s := s+a[k,j] × a[j,ℓ];
        a[k,ℓ] := −s/a[k,k];
    end;
    comment the remaining coefficients of column ℓ are computed
        and at the same time their largest is determined as pivot;
    max := 0;
    for k := ℓ step 1 until n do
    begin
        s := a[k,ℓ];
        for j := 1 step 1 until ℓ−1 do
            s := s+a[k,j] × a[j,ℓ];
        a[k,ℓ] := s;
        if abs (s) > max then
            begin max := abs (s); p := k; end;
    end for k;
    if max = 0 then goto sing;
    comment if necessary, interchange rows ℓ and p;
    if p ≠ ℓ then
    begin
        for j := 1 step 1 until n do
        begin h := a[ℓ,j]; a[ℓ,j] := a[p,j]; a[p,j] := h; end;
        i := z[ℓ]; z[ℓ] := z[p]; z[p] := i;
    end if p;
end for ℓ;
comment forward substitution;
con:    for k := 1 step 1 until n do x[k] := v[z[k]];
```

```
for k := 1 step 1 until n do
begin
    s := x[k];
    for j := 1 step 1 until k−1 do s := s+a[k,j]×x[j];
    x[k] := −s/a[k,k];
end for k;
comment back substitution;
for k := n−1 step −1 until 1 do
begin
    s := x[k];
    for j := k+1 step 1 until n do s := s+a[k,j]×x[j];
    x[k] := s;
end for k;
end gaukos;
```

§2.6. The exchange algorithm

We consider s linear forms in t independent variables:

$$y_k = \sum_{l=1}^{t} a_{kl} x_l \quad (k = 1, \ldots, s). \tag{31}$$

For the time being, this should not be taken as a system of equations but merely as a fixed relationship between the $s+t$ variables $x_1, \ldots, x_t$, $y_1, \ldots, y_s$, by means of which the $y_1, \ldots, y_s$ can be computed from the $x_1, \ldots, x_t$. Written as a tableau:

$$
\begin{array}{c|cccc}
 & x_1 & x_2 & \cdots & x_t \\
\hline
y_1 = & a_{11} & a_{12} & \cdots & a_{1t} \\
y_2 = & a_{21} & a_{22} & & a_{2t} \\
\vdots & \vdots & & & \\
 & & & & \\
y_s = & a_{s1} & a_{s2} & & a_{st}
\end{array}
\tag{32}
$$

From this, one can now obtain a new tableau by solving, say, the pth linear form for the variable x_q:

$$x_q = \frac{1}{a_{pq}} \left[y_p - \sum_{l \neq q} a_{pl} x_l \right], \tag{33}$$

and subsequently substituting this expression in the remaining equations:

$$y_k = \sum_{\ell \neq q} a_{k\ell} x_\ell + \frac{a_{kq}}{a_{pq}} y_p - \frac{a_{kq}}{a_{pq}} \sum_{\ell \neq q} a_{p\ell} x_\ell \qquad (k \neq p)$$

or   (34)

$$y_k = \frac{a_{kq}}{a_{pq}} y_p + \sum_{\ell \neq q} \left[a_{k\ell} - \frac{a_{kq} a_{p\ell}}{a_{pq}} \right] x_\ell \qquad (k \neq p).$$

This exchange assumes that $a_{pq} \neq 0$.

Example. The tableau

	a	b	c
$d =$	3	$\boxed{2}$	1
$e =$	5	-1	-3

corresponds to the relations

$$d = 3a + 2b + c, \qquad e = 5a - b - 3c.$$

For the exchange of b and d, one obtains from the former relation first $b = -1.5a + .5d - .5c$, and then from the latter, $e = 6.5a - .5d - 2.5c$. Written as a tableau:

	a	d	c
$b =$	-1.5	$.5$	$-.5$
$e =$	6.5	$-.5$	-2.5

The above formulae for x_q and the y_k (excluding y_p) are again s linear forms in t variables, only the variables y_p and x_q have exchanged their roles. y_p is now an independent, x_q a dependent variable. This can be expressed in the form of a new tableau:

	x_1	$\cdots$	x_{q-1}	y_p	x_{q+1}	$\cdots$	x_t
$y_1 =$	a_{11}^*	$\cdots$		a_{1q}^*		$\cdots$	a_{1t}^*
$\vdots$	$\vdots$						
$y_{p-1} =$							
$x_q =$	a_{p1}^*			a_{pq}^*			a_{pt}^*
$y_{p+1} =$							
$\vdots$	$\vdots$						
$y_s =$	a_{s1}^*			a_{sq}^*			a_{st}^*

$$(35)$$

By examining the formulae (33) and (34), one sees that the coefficients a_{kl}^* of the new scheme (35) are defined as follows:

$$a_{pq}^* = 1/a_{pq}$$
$$a_{pl}^* = -\,a_{pl}/a_{pq} \qquad (l = 1, \ldots, q-1,\ q+1, \ldots, t)$$
$$a_{kq}^* = a_{kq}/a_{pq} \qquad (k = 1, \ldots, p-1,\ p+1, \ldots, s) \qquad (36)$$
$$a_{kl}^* = a_{kl} - \frac{a_{kq}a_{pl}}{a_{pq}} \qquad (k = 1, \ldots, p-1,\ p+1, \ldots, s;$$
$$l = 1, \ldots, q-1,\ q+1, \ldots, t).$$

It is true, though, that in practice one proceeds differently: Since a_{kl} and a_{kl}^* are always stored as $a[k,l]$, one must be careful to no longer use any a_{kl} for which the corresponding a_{kl}^* has already been formed. This is the case with the following arrangement:

$$a_{pq}^* = 1/a_{pq}$$
$$a_{pl}^* = -\,a_{pl}a_{pq}^* \qquad (l = 1, \ldots, q-1,\ q+1, \ldots, t)$$
$$a_{kl}^* = a_{kl} + a_{kq}a_{pl}^* \qquad (k = 1, \ldots, p-1,\ p+1, \ldots, s; \qquad (37)$$
$$l = 1, \ldots, q-1,\ q+1, \ldots, t)$$
$$a_{kq}^* = a_{kq}a_{pq}^* \qquad (k = 1, \ldots, p-1,\ p+1, \ldots, s).$$

The transition described by the formulae (36), resp. (37), from scheme (32) to the scheme (35) is called *exchange step with pivot element* a_{pq}.

Applications. The idea of the exchange step can be exploited in many different ways:

A) Let a tableau be given with $s = t = n$ dependent and independent variables:

	x_1	x_2	$\cdots$	x_n	
$y_1 =$	a_{11}	a_{12}	$\cdots$	a_{1n}	
$y_2 =$	a_{21}	a_{22}		a_{2n}	(38)
$\vdots$	$\vdots$	$\vdots$			
$y_n =$	a_{n1}	a_{n2}		a_{nn}	

This corresponds to the relation $\mathbf{y} = \mathbf{A}\mathbf{x}$ with the square matrix $\mathbf{A} = [a_{kl}]$.

Now after a variable x_{q_1} has been exchanged for y_{p_1}, one can apply to the resulting tableau $\{a_{kl}^*\}$ an additional exchange step, by further exchanging, say, x_{q_2} for y_{p_2}, provided $a_{p_2 q_2}^* \neq 0$.

Under appropriate conditions this process can be repeated so often until all variables x_l have turned into dependent, and all y_k into independent variables. In the final scheme we then have on top only y-variables, and on the left only x-variables, both, to be sure, in arbitrary order. However, by appropriately permuting the rows and columns of the final scheme, it will assume the following form:

	y_1	y_2	$\cdots$	y_n
$x_1 =$	α_{11}	α_{12}	$\cdots$	α_{1n}
$x_2 =$	α_{21}	α_{22}		α_{2n}
$\vdots$	$\vdots$	$\vdots$		
$x_n =$	α_{n1}	α_{n2}		α_{nn}

One thus has $x_l = \sum_{k=1}^{n} \alpha_{lk} y_k$, i.e., the matrix $\{\alpha_{lk}\}$ is the inverse $\mathbf{A}^{-1}$ of $\mathbf{A}$.
We have found a numerical method of *matrix inversion*.

Example. We compute the inverse of $\mathbf{A} = \begin{bmatrix} 1 & 10 \\ 1 & 5 \end{bmatrix}$. The pivot elements are put in boxes.

$$
\begin{array}{c}
\begin{array}{cc} x_1 & x_2 \end{array} \\
\begin{array}{c} y_1 = \\ y_2 = \end{array}
\boxed{\begin{array}{cc} 1 & \boxed{10} \\ 1 & 5 \end{array}}
\end{array}
\Rightarrow
\begin{array}{c}
\begin{array}{cc} x_1 & y_1 \end{array} \\
\begin{array}{c} x_2 = \\ y_2 = \end{array}
\boxed{\begin{array}{cc} -.1 & .1 \\ \boxed{.5} & .5 \end{array}}
\end{array}
\Rightarrow
\begin{array}{c}
\begin{array}{cc} y_2 & y_1 \end{array} \\
\begin{array}{c} x_2 = \\ x_1 = \end{array}
\boxed{\begin{array}{cc} -.2 & .2 \\ 2 & -1 \end{array}}
\end{array}
$$

Permuted:

$$
\begin{array}{c}
\begin{array}{cc} y_1 & y_2 \end{array} \\
\begin{array}{c} x_1 = \\ x_2 = \end{array}
\boxed{\begin{array}{cc} -1 & 2 \\ .2 & -.2 \end{array}}
\end{array}
\quad,\quad \text{i.e.,} \quad \mathbf{A}^{-1} = \begin{bmatrix} -1 & 2 \\ .2 & -.2 \end{bmatrix}.
$$

This inversion, naturally, is subject to certain conditions: In each of the n exchange steps one must be able to find a suitable pivot element, which can only be an element different from 0 at the intersection of an x-column and a y-row (one wants, after all, exchange an independent x for a dependent y).

This last condition, as the exchange proceeds, restricts the possible choices more and more, until in the last (nth) step one has no choice whatsoever, since there is only one column headed by an x and one row labeled on the left by a y; the element at the intersection therefore *must* be taken as pivot.

It goes without saying that also for the matrix inversion one must develop suitable pivot strategies. The points of view are the same as in Gauss's algorithm (diagonal strategy, partial pivoting strategy, etc.), although in practice one does not bring the pivot elements into the diagonal, through interchanges, but lets them stay in place. Only in the final tableau are rows and columns permuted to obtain the inverse.

In the following example we apply the complete pivoting strategy: in the tableau

$$
\begin{array}{c}
\begin{array}{ccc} x_1 & x_2 & x_3 \end{array} \\
\begin{array}{c} y_1 = \\ y_2 = \\ y_3 = \end{array}
\boxed{\begin{array}{ccc} 1 & 2 & 3 \\ 2 & 3 & 4 \\ 3 & 4 & \boxed{5} \end{array}}
\end{array}
$$

the 5 is the absolutely largest element. A first exchange step with this pivot yields:

	x_1	x_2	y_3
$y_1 =$	$-.8$	$-.4$	$.6$
$y_2 =$	$-.4$	$-.2$	$.8$
$x_3 =$	$-.6$	$-.8$	$.2$

Here, among the four elements located in x-columns and y-rows, $-.8$ is the absolutely largest. It becomes the second pivot:

	y_1	x_2	y_3
$x_1 =$	-1.25	$-.5$	$.75$
$y_2 =$	$.5$	0	$.5$
$x_3 =$	$.75$	$-.5$	$-.25$

Now only one x-column and one y-row remain; hence 0 at the intersection must be taken as pivot element. Since this is not possible, the process at this point breaks down; no inverse can be computed. The matrix is singular.

There is something, however, that can still be done. The 0 in question is obtained from 3 by subtraction, and therefore, in general, is subject to rounding errors, which in 6-digit computation, ought to be of the order of magnitude 10^{-6}. We therefore make things only a little worse if on top of this expected error we graft an additional error 10^{-8} and replace 0 by 10^{-8}. After that, one can go on with the computation and finds:

	y_1	y_2	y_3
$x_1 =$	$2.5_{10}7$	$-5_{10}7$	$2.5_{10}7$
$x_2 =$	$-5_{10}7$	10^8	$-5_{10}7$
$x_3 =$	$2.5_{10}7$	$-5_{10}7$	$2.5_{10}7$

Of course, this is not the actual inverse, which here does not even exist, but it is a matrix $\mathbf{B}$ for which $\mathbf{AB}-\mathbf{I}$ agrees with the zero matrix within the error bounds to be expected. (The elements are sums of products of the order of magnitude 10^8 in 6-digit computation.) The practical significance of the final tableau here lies in the fact that it makes the dependence of the columns and rows of the matrix $\mathbf{A}$ evident.

B) Let a tableau be given with $s=n$ dependent, and $t = n+1$ independent variables:

	x_1	x_2	$\cdots$	x_n	x_{n+1}
$y_1 =$	a_{11}	a_{12}	$\cdots$	a_{1n}	$a_{1,n+1}$
$y_2 =$	a_{21}	a_{22}		a_{2n}	$a_{2,n+1}$
$\vdots$	$\vdots$				
$y_n =$	a_{n1}	a_{n2}		a_{nn}	$a_{n,n+1}$

(39)

If now the variable x_{n+1} is given the fixed value 1 and in addition, one requires that the y all assume the value 0, then this means that

$$\sum_{l=1}^{n} a_{kl}x_l + a_{k,n+1} = 0 \quad (k = 1, \ldots, n),$$

i.e., we are dealing with a linear system of equations in which the unknowns $x_1, \ldots, x_n$ are to be determined in such a way that indeed $y_1 = y_2 = \cdots = y_n = 0$.

Now in order to obtain these x-values, we in turn exchange them all for the y, which requires n exchange steps. It must be observed, however, that x_{n+1} is not to be exchanged, i.e., that no pivots are selected from the last column.

There results a scheme which carries as labels on the left all the x, and on top all the y, in some arbitrary order; the last column, now as before, is labeled with $x_{n+1} = 1$; for example (with new a'_{kl}):

	y_3	y_7	$\cdots$	y_5	1
$x_7 =$	a'_{11}	a'_{12}	$\cdots$	a'_{1n}	$a'_{1,n+1}$
$x_1 =$	a'_{21}	a'_{22}		a'_{2n}	$a'_{2,n+1}$
$\vdots$	$\vdots$				
$x_4 =$	a'_{n1}	a'_{n2}		a'_{nn}	$a'_{n,n+1}$

True to our convention, this scheme is to be read as

$$x_7 = a'_{11}y_3 + a'_{12}y_7 + \cdots + a'_{1n}y_5 + a'_{1,n+1}$$
$$x_1 = a'_{21}y_3 + a'_{22}y_7 + \cdots + a'_{2n}y_5 + a'_{2,n+1}$$
$$\vdots$$
$$x_4 = a'_{n1}y_3 + a'_{n2}y_7 + \cdots + a'_{nn}y_5 + a'_{n,n+1} ;$$

since, however, all $y=0$, there thus follows $x_7 = a'_{1,n+1}$, $x_1 = a'_{2,n+1}$, etc.

Consequently: *The values in the last column of the final tableau represent the solution of the system of equations.*

The course of computation, however, still permits a reduction in work: with each exchange step there appears a new column, effectively labeled by 0; the elements of this column, therefore, are unimportant for the subsequent computation, since they are always multiplied by 0.

As a consequence, the exchange formulae need only be implemented for the elements of the *x*-columns; in the *y*-columns one can leave whatever numbers one wants. Inasmuch as fewer and fewer *x*-columns remain, the computational effort in this way is reduced by half. This procedure is known as the *Gauss-Jordan method*.

Example. In the tableau

	x_1	x_2	x_3	1
0 =	2.2	2.9	1000	−.2
0 =	2.9	5.4	1000	.2
0 =	1	1	−1	−1

using the relative partial pivoting strategy, the first step is carried out with the pivot $a_{31} = 1$ (4-digit computation):

	0	x_2	x_3	1
0 =		.7	1002	2
0 =		2.5	1003	3.1
x_1 =		−1	1	1

The second step with the pivot $a^*_{22} = 2.5$ yields:

	0	0	x_3	1
0 =			721.2	1.132
x_2 =			−401.2	−1.24
x_1 =			402.2	2.24

The third step with the pivot $a^{**}_{13} = 721.2$ leads to the result:

	0	0	0	1
$x_3 =$				−.001570
$x_2 =$				−.6101
$x_1 =$				1.609

C) There is one more possibility for savings: not only is it no longer necessary to compute the y-columns, but one can also freeze the newly-formed x-rows in the form in which they were created.

Thus, in the above example, one first freezes the x_1-row, i.e., no further exchange operations are applied to it:

	(x_1)	x_2	x_3	1
$0 =$		.7	1002	2
$0 =$		2.5	1003	3.1
$x_1 =$		−1	1	1

Further exchanges therefore take place only in the 0-rows; for the frozen rows the old labels (in parentheses) are still valid.

	(x_1)	(x_2)	x_3	1
$0 =$			721.2	1.132
$x_2 =$			−401.2	−1.24
$x_1 =$		−1	1	1

Third step:

	(x_1)	(x_2)	(x_3)	1
$x_3 =$				−.00157
$x_2 =$			−401.2	−1.24
$x_1 =$		−1	1	1

It is not difficult to see that with this last modification one has recovered Gauss's elimination. Since here the exchange formulae need only be applied to elements at the intersection of 0-rows and x-columns, it is evident that Gauss's elimination gets by with fewer operations than, say, the Gauss-Jordan method; in the former, one merely has to put up with the inconvenience of back substitution.

§2.7. Questions of programming

The exchange algorithm is very easy to program. The tableau is stored in the computer as **array** $a[1{:}s,\ 1{:}t]$. The notation $a[i,j]$ designates the element located in the position i,j of the tableau, and this regardless of the labeling on the left and on top and of the number of already completed exchange steps. Then for an exchange step with pivot element $a[p,q]$ the formulae (37) give rise to the following piece of program (s and t denote the number of rows and columns, respectively, of the tableau):

aa: $a[p,q] := 1/a[p,q]$;

bb: **for** $\ell := 1$ **step** 1 **until** $q-1$, $q+1$ **step** 1 **until** t **do**
  $a[p,\ell] := -a[p,\ell] \times a[p,q]$;

dd: **for** $k := 1$ **step** 1 **until** $p-1$, $p+1$ **step** 1 **until** s **do**
  for $\ell := 1$ **step** 1 **until** $q-1$, $q+1$ **step** 1 **until** t **do**
   $a[k,\ell] := a[k,\ell] + a[k,q] \times a[p,\ell]$;

cc: **for** $k := 1$ **step** 1 **until** $p-1$, $p+1$ **step** 1 **until** s **do**
  $a[k,q] := a[k,q] \times a[p,q]$;

The labels, here, serve only for the purpose of explanation: at aa: the pivot element, at bb: the pivot row, at cc: the pivot column, and at dd: the field of the $(s-1) \cdot (t-1)$ remaining elements are processed. Of course, these operations presuppose that the pivot element $a[p,q]$ has been selected appropriately.

The exchange step has been programmed here in such a way that the new tableau overwrites the old one; one therefore had to carefully make sure that no elements of the new tableau were computed as long as the corresponding elements of the old one were still needed. (Hence the order aa:, bb:, dd:, cc:.)

Now any such tableau is completely identified only together with the labeling on top and on the left. In order to represent this labeling in the computer, one introduces two integer vectors

integer array *left* $[1{:}s]$, *top* $[1{:}t]$

with the following meaning:

left $[k] = \ell > 0$: the kth row is labeled on the left by y_ℓ.

left $[k] = -\ell < 0$: the kth row is labeled on the left by x_ℓ.

top $[k] = \ell > 0$: the kth column is labeled on top by y_ℓ.

$top\ [k] = -\ell < 0$: the kth column is labeled on top by x_ℓ.

These conventions require additional operations, namely

a) *before the first* exchange step, since at this point, all rows are still labeled by y, and all columns by x:

for $k := 1$ **step** 1 **until** s **do** *left* $[k] := k$;

for $\ell := 1$ **step** 1 **until** t **do** *top* $[\ell] := -\ell$;

b) *after each* exchange step, if $a\ [p,q\ ]$ is the pivot element:

$k := left\ [p]$;
$left\ [p] := top\ [q]$;
$top\ [q] := k$;

(That is, *left* $[p]$ and *top* $[q]$ are exchanged.)

By means of the labeling now available in this form, one can also easily check the admissibility of an element as pivot element: in matrix inversion, for example, only an element at the intersection of a y-row and an x-column is permissible, which can be expressed by the condition

if *left* $[k] > 0 \wedge top\ [\ell] < 0$ **then.**

The labeling simulated in this way also permits, after n exchange steps, to transform the final array $\mathbf{A}_n$ by row- and column-permutation into the inverse $\mathbf{A}^{-1}$. Note, in this connection, that with *left* $[k] = -\ell$ one also must have *top* $[\ell] = k$. Therefore, if the kth row of $\mathbf{A}_n$ has to be put in place of the $-left\ [k]$th row, then this is equivalent to having to make the *top* $[\ell\ ]$th row of $\mathbf{A}_n$ the ℓth row of $\mathbf{A}^{-1}$.

In order that this permutation can be carried out within the matrix $\mathbf{A}$, one has first of all to lay away the ℓth row into a vector $\mathbf{b}$, so that the $top[\ell]$th row finds place in the ℓth one; then the $top[top[\ell]]$th row is reloaded into the $top[\ell]$th one, etc., until eventually $top[top[top[...[top[\ell]]...]]] = \ell$ and the vector $\mathbf{b}$ can be inserted exactly at the right place. Then one looks for further cycles, i.e., rows ℓ with $top[\ell] \neq \ell$. Subsequently, also the columns would have to be permuted.

For row permutation, one has the following program:

```
for ℓ := 1 step 1 until n do
if top [ℓ] ≠ ℓ then
begin
        comment lay away ℓth row in b;
```

```
    for j := 1 step 1 until n do b [j] := a [ℓ,j];
    q := ℓ;
    for p := top [q] while p ≠ ℓ do
    begin
        top [q] := q;
        for j := 1 step 1 until n do a [q,j] := a [p,j];
        q := p;
    end for p;
    top [q] := q;
    comment insert vector b, cycle completed;
    for j := 1 step 1 until n do a [q,j] := b [j];
end for ℓ;
```

§2.8. Linear inequalities (optimization)

A merchant has 4 lbs. of silver and 7 lbs. of gold, from which he can produce and sell the following alloys:

1) 50% gold, 50% silver at $3200/lb.

2) 75% gold, 25% silver at $6000/lb.

3) 100% gold at $5000/lb.

Which alloys should he produce in order to achieve a *maximum* return?

In view of the amounts of metal available one obtains certain inequalities. If the amounts of the 3 alloys produced are denoted by x_1, x_2, x_3, one must have:

$$.5x_1 + .75x_2 + x_3 \leq 7 \qquad \text{(supply of gold)},$$
$$.5x_1 + .25x_2 \qquad\quad \leq 4 \qquad \text{(supply of silver)}.$$

The return on the sale, i.e,

$$3200x_1 + 6000x_2 + 5000x_3,$$

then is to be made a maximum.

All this can be summarized in the tableau:

		x_1	x_2	x_3	1
y_1	=	−.5	−.75	−1	7
y_2	=	−.5	−.25	0	4
z	=	3200	6000	5000	0

(40)

where the value of z, under the constraints $x_1 \geq 0$, $x_2 \geq 0$, $x_3 \geq 0$, $y_1 \geq 0$, $y_2 \geq 0$ has to be maximized.

The problem is solved by arranging, through a sequence of exchange operations, that the maximum for z becomes explicitly evident. Under the stated conditions, this maximum is clearly achieved when the coefficients of z in the matrix part are all negative, e.g.

	x_1	y_1	x_3	1
$x_2 =$				
$y_2 =$				
$z =$	-800	-8000	-3000	56000

since in this case the maximum of z under the constraints $x_1 \geq 0$, $y_1 \geq 0$, $x_3 \geq 0$ obviously occurs for $x_1 = y_1 = x_3 = 0$ and has the value 56000. All other admissible x_1, y_1, x_3- values yield a smaller z-value. It is to be noted, however, that x_2 and y_2 are then equal to the values in the 1-column (i.e., the column of constants); these values must not become negative. If this condition were violated, one would have achieved the maximum 56000 for an inadmissible combination (namely $x_2 < 0$ or $y_2 < 0$). This would mean, that one would have to sell a negative amount of alloy 2 or that more than 4 lbs. of silver had been used, respectively.

The goal, therefore, is to achieve, by means of exchange steps, that the 1-column ends up with nonnegative elements and the z-row (disregarding the corner element at the lower right) with negative elements.

The exchange operations must therefore be chosen in such a way that the elements in the 1-column remain nonnegative, while at the same time the elements in the z-row are made negative. Pivot elements must not be selected either in the 1-column or in the z-row.

We now consider, more generally, $s-1$ linear inequalities

$$y_k = \sum_{l=1}^{t-1} a_{kl} x_l + a_{kt} \geq 0 \quad (k = 1, \ldots, s-1) \tag{41}$$

in $t-1$ nonnegative variables $x_l \geq 0$ ($l = 1, \ldots, t-1$), where the values of a_{kl} are assumed nonnegative. Subject to these inequalities, we wish to maximize the linear form

$$z = \sum_{l=1}^{t-1} a_{sl} x_l + a_{st}. \tag{42}$$

The associated tableau reads:

		x_1	$\cdots$	x_ℓ	$\cdots$	x_q	$\cdots$	x_{t-1}	1
y_1	$=$	a_{11}	$\cdots$	$a_{1\ell}$	$\cdots$	a_{1q}	$\cdots$	$a_{1,t-1}$	a_{1t}
$\vdots$		$\vdots$							
y_i	$=$	a_{i1}		$a_{i\ell}$		a_{iq}		$a_{i,t-1}$	a_{it}
$\vdots$		$\vdots$							
y_p	$=$	a_{p1}		$a_{p\ell}$		a_{pq}		$a_{p,t-1}$	a_{pt}
$\vdots$		$\vdots$							
y_{s-1}	$=$	$a_{s-1,1}$		$a_{s-1,\ell}$		$a_{s-1,q}$		$a_{s-1,t-1}$	$a_{s-1,t}$
z	$=$	a_{s1}		$a_{s\ell}$		a_{sq}		$a_{s,t-1}$	a_{st}

Observation 1: According to the exchange rules (36) the sign of the element at the intersection of the pivot row and the 1-column is preserved precisely if the pivot element a_{pq} is negative, since we have $a_{pt}^* = - a_{pt}/a_{pq}$.

Rule 1. *The pivot element must be negative.*

Observation 2: If the pivot element is negative, then by virtue of $a_{sq}^* = a_{sq}/a_{pq}$ the sign of the element in the z-row below the pivot element is reversed. Since one wants to make the z-row negative, there follows

Rule 2. *The pivot element must be chosen above a positive z-row element.*

Observation 3: The new element a_{it}^* of the 1-column, which again must be nonnegative, is obtained as

$$a_{it}^* = a_{it} - \frac{a_{pt}a_{iq}}{a_{pq}}.$$

If $a_{iq} \geq 0$, the condition is certainly fulfilled, since $a_{pt} \geq 0$ and $a_{pq} < 0$. If $a_{iq} < 0$, then for all such i ($1 \leq i \leq s-1$) one must have $a_{it} \geq \dfrac{a_{pt}a_{iq}}{a_{pq}}$,

hence $\dfrac{a_{it}}{a_{iq}} \leq \dfrac{a_{pt}}{a_{pq}}$ (¹).

---

¹ The argument here is a simplification of the argument given in the original. (Translator's remark)

Rule 3. *The pivot element a_{pq}, among all negative elements of the same column q, must have the property that a_{it}/a_{iq} (t is the index of the 1-column) is maximum for $i=p$ (i.e., minimum in absolute value). Having selected the pivot column q, one thus forms the characteristic quotients a_{it}/a_{iq} ($i = 1, \ldots, s - 1$; $a_{iq} < 0$) and determines their largest([2]).*

As a consequence of these rules one obtains:

Theorem 2.3. *As long as the pivot selection proceeds in accordance with the Rules 1, 2 and 3([3]), the value of the corner element at the lower right increases monotonically; if, for some exchange step, $a_{pt} = 0$, then, however, that value remains unchanged during this step([4]).*

Proof. For an exchange step with pivot element a_{pq} (where, as always, $p \neq s$, $q \neq t$) one obviously has $a_{st}^* = a_{st} - a_{sq}a_{pt}/a_{pq}$, but $a_{sq} > 0$ according to Rule 2, and $a_{pq} < 0$ according to Rule 1. Owing to the Rules 1 and 3, the conditions $a_{kt} \geq 0$ ($k = 1, \ldots, s - 1$) remain intact.

Example. The gold and silver problem, posed at the beginning, is already represented as tableau (40):

	x_1	x_2	x_3	1
$y_1 =$	$-.5$	$-.75$	-1	7
$y_2 =$	$-.5$	$-.25$	0	4
$z =$	3200	6000	5000	0

We choose the first pivot from the first column. The characteristic quotients here are $7/(-.5) = -14$, $4/(-.5) = -8$; the second row gives the absolutely smaller quotient, hence a_{21} is the pivot element:

---

[2] It can happen that in the chosen pivot column q (with $a_{sq} > 0$) no element a_{iq} ($i = 1, \ldots, s - 1$) is negative. Then the linear form z is unbounded on the set of admissible points $x = [x_1, \ldots, x_{t-1}]$ (i.e., points satisfying the constraints), and the problem has no finite solution. (Editors' remark)

[3] The exchange algorithm, resulting from these rules is called the *Simplex Algorithm*. (Editors' remark)

[4] This value of z would also remain stationary if one selected $a_{sq} = 0$. Such steps, however, can be disregarded. If in the z-row of the final tableau there would occur, next to negative coefficients, also coefficients that are 0 and in whose column there is a pivot element satisfying the Rules 1 and 3, this would mean that the maximum is not unique. (Editors' remark)

	y_2	x_2	x_3	1
$y_1 =$	1	−.5	−1	3
$x_1 =$	−2	−.5	0	8
$z =$	−6400	4400	5000	25600

Now we select the pivot from the second column (the third would also be possible). The characteristic quotients are $3/(-.5)$ and $8/(-.5)$; the first row gives the absolutely smaller quotient, hence one takes a_{12} as pivot element:

	y_2	y_1	x_3	1
$x_2 =$	2	−2	−2	6
$x_1 =$	−3	1	1	5
$z =$	2400	−8800	−3800	52000

There is still a positive element in the z-row. To remove it, a further exchange step with pivot in the first column is required, whereby only the element $a_{21} = -3$ qualifies as pivot:

	x_1	y_1	x_3	1
$x_2 =$	$-\frac{2}{3}$	$-\frac{4}{3}$	$-\frac{4}{3}$	$\frac{28}{3}$
$y_2 =$	$-\frac{1}{3}$	$\frac{1}{3}$	$\frac{1}{3}$	$\frac{5}{3}$
$z =$	−800	−8000	−3000	56000

Evidently, $z = 56000 - 800x_1 - 8000y_1 - 3000x_3$ is now maximum for $x_1 = y_1 = x_3 = 0$, and from this, one also gets $x_2 = \frac{28}{3} > 0$, $y_2 = \frac{5}{3} > 0$, so that the maximum of z under the given constraints is found. The solution means:

$x_1 = x_3 = 0,\ x_2 = \frac{28}{3}$:   Produce only $9\frac{1}{3}$ lbs. of the 75% gold-silver alloy.

$y_1 = 0$:   The gold supply is exhausted.

$y_2 = \frac{5}{3}$: $\frac{5}{3}$ lbs. of silver remain unused.

$z = 56000$:   The return on the sale is $56000.

On the basis of this result one also recognizes that one could have arrived from the initial to the final tableau in a single exchange step, namely with a_{12} as the pivot element.

Minimization (point of view of the consumer). If the problem, in contrast to the one posed at the beginning, has as objective the purchase of certain materials at a minimum cost, then a reformulation is necessary.

Let's suppose we go to our goldsmith and want to stock up on gold and silver in the amounts of at least 2 lbs. and 1 lb., respectively, by buying his alloys. If again x_1, x_2, x_3 denote the amounts of the three alloys, then the following conditions are to be satisfied:

$$.5x_1 + .75x_2 + x_3 \geq 2 \quad \text{(gold)}$$
$$.5x_1 + .25x_2 \qquad \geq 1 \quad \text{(silver)}$$
$$3200x_1 + 6000x_2 + 5000x_3 = \text{minimum}.$$

Schematically:

	x_1	x_2	x_3	1
y_1 =	.5	.75	1	-2
y_2 =	.5	.25	0	-1
z =	-3200	-6000	-5000	0

Here, z means the negative costs, which are to be made a maximum, subject to the constraints $x_i \geq 0$, $y_i \geq 0$.

While the coefficients of the z-row are already negative, the 1-column, contrary to the rules, contains negative elements, so that $x_1 = x_2 = x_3 = 0$ is not a solution. (We would have a deficit of 2 lbs. in gold and 1 lb. in silver.)

The normal process, therefore, must be prefaced by an extra step in which the 1-column can be made positive. For that purpose, we proceed according to the following recipe.

Select a pivot column (index q) in which all elements above the z-row are positive([5]), determine among all quotients a_{it}/a_{iq} the smallest

---

[5] If no such column exists, this recipe, which can be derived in the same way as Rule 3, is not applicable. There exist, then, more general and more complicated methods to make the 1-column positive. Compare for this, as also for the Simplex Algorithm in general, Collatz L., Wetterling W.: *Optimization Problems*, Springer, New York, 1975, Section 3.4, or Künzi H.P., Tzschach H.G., Zehnder C.A.: *Numerical Methods of Mathematical Optimization with ALGOL and FORTRAN Programs*, Academic Press, New York, 1971, Section 1.3, or Stiefel E.: *An Introduction to Numerical Mathematics*, Academic Press, New York, 1963, Section 2.41. (Editors' remarks)

(most negative) and – if a_{pt}/a_{pq} denotes this smallest quotient – make an exchange with pivot a_{pq}. Then the 1-column becomes positive, and one can continue normally.

In our example, we may take $q=1$; the quotients then are $-2/.5 = -4$, $-1/.5 = -2$, hence an exchange with pivot a_{11} must be made:

	y_1	x_2	x_3	1
$x_1 =$	2	-1.5	-2	+4
$y_2 =$	1	-.5	-1	+1
$z =$	-6400	-1200	1400	-12800

This scheme would indicate that one should buy 4 lbs. of alloy 1 and nothing else. Then one indeed has 2 lbs. of gold and 2 lbs. of silver, thus a surplus of 1 lb. in silver, which is also signaled by $y_2 = 1$. The costs, however, are not minimum, since $\partial z/\partial x_3 = a_{33} > 0$. Therefore, one now makes a normal simplex exchange step with pivot column 3. The characteristic quotients are $4/-2 = -2$ and $1/-1 = -1$; one thus must select a_{23} as pivot:

	y_1	x_2	y_2	1
$x_1 =$	0	-.5	2	2
$x_3 =$	1	-.5	-1	1
$z =$	-5000	-1900	-1400	-11400

Minimum cost, with $11400, is now achieved: one buys 2 lbs. of alloy 1 and 1 lb. of pure gold. Since the solution is obtained with $y_1 = y_2 = 0$, we don't have any surplus in gold or silver.

Notes to Chapter 2

The work of Wilkinson has had a profound effect on our understanding of the roundoff properties of Gaussian elimination; his book (Wilkinson [1965]) has become a classical reference work. Other useful reference books are Stewart [1973], Strang [1980], and Golub & Van Loan [1989].

§2.3 The residuals for iterative refinement are often calculated in double precision, while the rest of the calculation is performed in single precision. Typically, the iterates converge rapidly to a solution that is accurate to single precision; see Stewart [1973] or

Golub & Van Loan [1989]. Underlying this mode is the assumption that the matrix coefficients are exactly represented by single-precision values. If this is not the case, one has to be content with the residual $\mathbf{Ax} + \mathbf{v}$ being small. Skeel [1980] shows that iterative refinement, without double-precision computation of residuals, is very effective at producing a small relative residual

$$\max_i \frac{|\mathbf{Ax} + \mathbf{v}|_i}{(|\mathbf{A}||\mathbf{x}| + |\mathbf{v}|)_i},$$

where the modulus signs applied to a vector or matrix refer to the corresponding vector or matrix having each element replaced by its absolute value. Skeel shows that one iteration is sufficient under certain reasonable conditions.

§2.4 The backward error analysis of Wilkinson (cf. Notes to §1.3) gives us a better appreciation of the effects of pivoting. He has shown that the solution obtained is exact for a perturbed system, where the perturbations are small compared to the coefficients of the reduced matrices. Partial pivoting limits the growth in the size of the largest matrix coefficient to the factor 2 at each stage, and it is thought that complete pivoting limits it overall to the factor n. (As far as we know, this result has not been proved, though Wilkinson has demonstrated a slightly weaker result.) This is a satisfactory situation for a well-scaled matrix, and the first example gives a good illustration of its success. Unfortunately, we know of no totally satisfactory pivotal strategy for matrices whose coefficients vary widely in size and are all known with good relative accuracy. The second example illustrates the problem. Our recommendation is to rescale the problem, for instance by the algorithm of Curtis & Reid [1972]. This would rescale the second example to

$$\begin{bmatrix} 2 & \alpha & \alpha \\ \alpha & \alpha^{-1} & 0 \\ \alpha & 0 & \alpha^{-1} \end{bmatrix},$$

where $\alpha = 10^{10/3} \approx 2154$.

For large sparse problems, it is also desirable to choose pivots that preserve as many as possible of the zero entries. For a discussion of this aspect, see George & Liu [1981] for the symmetric and positive definite case, and Duff, Erisman & Reid [1986] for the general case. Fortran software for sparse problems is available in the Yale Sparse Matrix Package (Eisenstat, Gursky, Schultz & Sherman [1977]), SPARSPAK (Chu, George, Liu & Ng [1984]), and the Harwell Subroutine Library (Hopper [1989]).

§2.5 There are many good Algol 60 codes for linear equations in the handbook of Wilkinson & Reinsch [1971]. They have provided the basis for many of the Fortran subroutines in the IMSL and NAG libraries and in LINPACK (Dongarra, Moler, Bunch & Stewart [1979]). We strongly recommend the use of one of these sources of reliable and efficient codes. LINPACK is becoming a *de facto* standard; many vendors provide optimized versions of the most popular routines to exploit their particular hardware.

§2.8 This section provides an introduction to the solution of linear programs by the simplex method. For further reading, see Dantzig [1963] and Chvátal [1983]. We note here, especially, that the exploitation of sparsity is essential in many practical problems, and that many of the numbers a_{pt} are often zero, which leads to real problems with degeneracy (steps for which the objective value remains unchanged), mentioned in Theorem 2.3. There are several large commercial packages available for the linear programming problem.

Some versions of the simplex algorithm are known to have a worst-case running time which, in certain contrived examples, can be exponential in the number of variables and constraints. On most problems of practical interest, nevertheless, the simplex method behaves like a polynomial-time (in fact, quadratic-time) algorithm. Truly polynomial algorithms for solving the linear programming problem have only recently been discovered, the first by Khachiyan [1979], and another by Karmarkar [1984]. The latter, in particular, has the potential of becoming a serious competitor to the simplex algorithm. For these, and other interior-point methods, the reader is referred to Schrijver [1986, Chs. 13 and 15] and Goldfarb & Todd [1989].

References

Chu, E., George, A., Liu, J. and Ng, E. [1984]: *SPARSPAK: Waterloo Sparse Matrix Package User's Guide for SPARSPAK-A,* Report CS-84-36, Department of Computer Science, University of Waterloo, Ontario, Canada.

Chvátal, V. [1983]: *Linear Programming,* W.H. Freeman, New York.

Curtis, A.R. and Reid, J.K. [1972]: On the automatic scaling of matrices for Gaussian elimination, *J. Inst. Math. Appl.* **10**, 118–124.

Dantzig, G.B. [1963]: *Linear Programming and Extensions,* Princeton University Press, Princeton, N.J.

Dongarra, J.J., Moler, C.B., Bunch, J.R. and Stewart, G.W. [1979]: *LINPACK Users' Guide,* SIAM, Philadelphia.

Duff, I.S., Erisman, A.M. & Reid, J.K. [1986]: *Direct Methods for Sparse Matrices,* Clarendon Press, Oxford. [Paperback edition, 1989].

Eisenstat, S.C., Gursky, M.C., Schultz, M.H. and Sherman, A.H. [1977]: *Yale Sparse Matrix Package, I: The Symmetric Codes, II: The Nonsymmetric Codes,* Reports 112 and 114, Computer Science, Yale University.

George, A. and Liu, J.W.H. [1981]: *Computer Solution of Large Sparse Positive Definite Systems,* Prentice-Hall, Englewood Cliffs, N.J.

Goldfarb, D. and Todd, M.J. [1989]: Linear Programming, in *Handbooks in Operations Research and Management Science,* Vol. 1: *Optimization* (G.L. Nemhauser, A.H.G. Rinnooy Kan and M.J. Todd, eds.), pp. 73–170. North-Holland, Amsterdam.

Golub, G.H. and Van Loan, C.F. [1989]: *Matrix Computations,* 2nd ed., The Johns Hopkins University Press, Baltimore.

Hopper, M.J., ed. [1989]: *Harwell Subroutine Library: A Catalogue of Subroutines*, Report AERE R9185, Computer Science and Systems Division, Harwell Laboratory.

Karmarkar, N. [1984]: A new polynomial-time algorithm for linear programming, *Combinatorica* **4**, 373–395.

Khachiyan, L.G. [1979]: A polynomial algorithm in linear programming (Russian), *Dokl. Akad. Nauk SSSR* **244**, 1093–1096. [English translation in *Soviet Math. Dokl.* **20**, 1979, 191–194.]

Schrijver, A. [1986]: *Theory of Linear and Integer Programming*, Wiley, Chichester.

Skeel, R.D. [1980]: Iterative refinement implies numerical stability for Gaussian elimination, *Math. Comp.* **35**, 817–832.

Stewart, G.W. [1973]: *Introduction to Matrix Computations*, Academic Press, New York.

Strang, G. [1980]: *Linear Algebra and Its Applications*, 2nd ed., Academic Press, New York.

Wilkinson, J.H. [1965]: *The Algebraic Eigenvalue Problem*, Clarendon Press, Oxford. [Paperback edition, 1988].

Wilkinson, J.H. and Reinsch, C. [1971]: *Linear Algebra*, Handbook for Automatic Computation, Vol. II, Springer, New York.

CHAPTER 3

Systems of Equations With Positive Definite Symmetric Coefficient Matrix

We have seen that in the general case the solution of a linear system of equations may present difficulties because of pivot selection. These difficulties disappear when the coefficient matrix A of the system is *symmetric* and *positive definite*. We therefore wish to examine this class of matrices in more detail.

§3.1. Positive definite matrices

With a symmetric matrix $\mathbf{A}$ (satisfying $a_{ik} = a_{ki}$) one can associate in a one-to-one fashion a quadratic form([1])

$$Q(\mathbf{x}) = Q(x_1, x_2, \ldots, x_n) = (\mathbf{x}, \mathbf{A}\mathbf{x}) = \sum_{i=1}^{n} \sum_{k=1}^{n} a_{ik} x_i x_k \tag{1}$$

(i.e., a homogeneous quadratic function of the independent variables $x_1, x_2, \ldots, x_n$).

Definition. *The matrix* $\mathbf{A}$ *(and also the form Q) is called positive definite if the function $Q(\mathbf{x})$, with the sole exception $Q(0, 0, \ldots, 0) = 0$, can assume only positive values, i.e., if*

$$Q(\mathbf{x}) > 0 \quad for \quad \mathbf{x} \neq [0, 0, \ldots, 0]^T. \tag{2}$$

---

[1] $(\mathbf{x}, \mathbf{y}) = \mathbf{x}^T \mathbf{y} = \sum x_k y_k$ here and in the sequel denotes the Euclidean scalar product of the vectors $\mathbf{x}$ and $\mathbf{y}$. (Editors' remark)

For a positive definite matrix, therefore, $Q(\mathbf{x})$ is a function of the n variables $x_1, \ldots, x_n$ which at the point $x_1 = x_2 = \cdots = x_n = 0$ (and only there) assumes its minimum.

First the question arises whether positive definite matrices, according to this definition, indeed exist. This we can answer in the affirmative: for $\mathbf{A} = \mathbf{I}$ (unit matrix), for example, we have

$$Q(\mathbf{x}) = \sum_{i=1}^{n} \sum_{j=1}^{n} \delta_{ij} x_i x_j = \sum_{k=1}^{n} x_k^2,$$

and this is 0 only for $x_1 = x_2 = \cdots = x_n = 0$. But also

$$\mathbf{A} = \begin{bmatrix} 1 & 1 & 1 \\ 1 & 2 & 3 \\ 1 & 3 & 6 \end{bmatrix}$$

is positive definite, because here we have $Q(\mathbf{x}) = x_1^2 + 2x_1x_2 + 2x_1x_3 + 2x_2^2 + 6x_2x_3 + 6x_3^2 = (x_1 + x_2 + x_3)^2 + (x_2 + 2x_3)^2 + x_3^2$, and this cannot be < 0 and also not $= 0$, as long as one of the $x_i \neq 0$.

There are also symmetric matrices, however, which are not positive definite, for example

$$\mathbf{A} = \begin{bmatrix} 5 & 2 & 7 \\ 2 & 5 & 2 \\ 7 & 2 & 5 \end{bmatrix},$$

for, with $\mathbf{x} = [1,0,-1]^T$, we have here $Q = -4$. Also the zero matrix is not positive definite, as it yields $Q \equiv 0$, hence also, e.g., $Q(1,1,\ldots, 1) = 0$. On the other hand, the zero matrix is insofar a limit case as the associated $Q(\mathbf{x})$ at least cannot become negative.

Definition. *A symmetric matrix is called positive semidefinite if it is not positive definite, but the quadratic form associated with it satisfies* $Q(\mathbf{x}) \geq 0$ *for all* $\mathbf{x}$.

For example,

$$\mathbf{A} = \begin{bmatrix} 1 & 1 & 1 \\ 1 & 1 & 1 \\ 1 & 1 & 1 \end{bmatrix}$$

is positive semidefinite, for we have $Q(\mathbf{x}) = (x_1 + x_2 + x_3)^2$, and this is always ≥ 0, but $= 0$ for $\mathbf{x} = [1,-1,0]^T$.

Note: The concepts "positive definite" and "positive semidefinite" are only applicable for symmetric matrices.[2]

An important property is contained in the following

Theorem 3.1. *A positive definite matrix is nonsingular. A positive semidefinite matrix is singular.*

Proof. a) If for some $\mathbf{x} \neq \mathbf{0}$ we had $\mathbf{Ax} = \mathbf{0}$, then we would also have $(\mathbf{x}, \mathbf{Ax}) = \sum_i \sum_j a_{ij} x_i x_j = 0$, which for a positive definite matrix is impossible. b) If, on the other hand, $\mathbf{A}$ is positive semidefinite, then $(\mathbf{x}, \mathbf{Ax}) = 0$ for some $\mathbf{x} \neq \mathbf{0}$. Now either $\mathbf{Ax} = \mathbf{0}$, in which case $\mathbf{A}$ is singular, or $\mathbf{y} = \mathbf{Ax}$ is orthogonal to $\mathbf{x}$. We then consider $\mathbf{z}(t) = \mathbf{x} + t\mathbf{y}$ and find

$$Q(\mathbf{z}(t)) = (\mathbf{x} + t\mathbf{y}, \mathbf{Ax} + t\mathbf{Ay})$$
$$= (\mathbf{x}, \mathbf{Ax}) + t(\mathbf{y}, \mathbf{Ax}) + t(\mathbf{y}, \mathbf{Ax}) + t^2(\mathbf{y}, \mathbf{Ay})$$
$$= 2t(\mathbf{y}, \mathbf{y}) + t^2(\mathbf{y}, \mathbf{Ay}).$$

If now $\mathbf{y}$ were $\neq \mathbf{0}$, then $Q(\mathbf{z}(t))$ would be negative for $t < 0$ and $|t|$ sufficiently small, contrary to our assumption; q.e.d.

§3.2. Criteria for positive definiteness

There are a number of simple criteria for the positive definiteness of matrices, which, however, are only necessary or only sufficient, and therefore do not always permit a definitive settlement.

Criterion 3.1. *For a symmetric matrix* $\mathbf{A}$ *to be positive definite, all diagonal elements must necessarily be positive.*[1]

---

[2] In principle, one could use (1) and (2) also to define positive definite nonsymmetric matrices $\mathbf{A}$. The following would then be true: $\mathbf{A}$ is positive definite precisely if the symmetric part $\frac{1}{2}(\mathbf{A}^T + \mathbf{A})$ is positive definite (in the usual sense). (Editors' remark)

[1] Proof: From $a_{ii} \leq 0$, with $\mathbf{x} = i$th coordinate vector, there would follow $Q(\mathbf{x}) \leq 0$. (Editors' remark)

Criterion 3.2. *For a symmetric matrix* **A** *to be positive definite, the absolutely largest element must necessarily lie on the diagonal, more precisely*([2]),

$$\max_{i \neq j} |a_{ij}| < \max_k a_{kk}. \tag{3}$$

Examples. The matrix

$$\begin{bmatrix} 0 & 1 & 2 \\ 1 & 2 & 3 \\ 2 & 3 & 4 \end{bmatrix},$$

by virtue of Criterion 3.1, cannot be positive definite, even though Criterion 3.2 is fulfilled, while

$$\begin{bmatrix} 5 & 2 & 7 \\ 2 & 5 & 2 \\ 7 & 2 & 5 \end{bmatrix}$$

satisfies 3.1 but not 3.2. For

$$\begin{bmatrix} 1 & 1 & 1 \\ 1 & 2 & 4 \\ 1 & 4 & 5 \end{bmatrix} \tag{4}$$

both criteria are fulfilled, and still the matrix is not positive definite([3]); the criteria are simply not sufficient.

Criterion 3.3 (Strong row sum criterion). *If in each row of a symmetric matrix the diagonal element exceeds the sum of the absolute values of all other elements in the row, that is, if*

$$a_{kk} > \sum_{\substack{l=1 \\ l \neq k}}^{n} |a_{kl}| \quad (k = 1, 2, \dots, n), \tag{5}$$

then **A** *is positive definite.*

---

[2] A still sharper result is: $a_{ik}^2 < a_{ii}a_{kk}$ $(i \neq k)$; see, e.g., Schwarz H.R., Rutishauser H., Stiefel E.: *Numerical Analysis of Symmetric Matrices*, Prentice-Hall, Englewood Cliffs, N.J., 1973, Theorem 1.3. (Editors' remark)
[3] One has, e.g., $Q(2,-3,1) = -5$. (Editors' remark)

Proof(^{4}). We have by (5), for $\mathbf{x} \neq \mathbf{0}$,

$$Q(\mathbf{x}) = \sum_i \sum_j a_{ij} x_i x_j \geq \sum_i a_{ii} x_i^2 - \sum_i \sum_{j \neq i} |a_{ij}| \, |x_i| \, |x_j|$$

$$> \sum_i \left\{ \sum_{j \neq i} |a_{ij}| \right\} |x_i|^2 - \sum_i \sum_{j \neq i} |a_{ij}| \, |x_i| \, |x_j|$$

$$= \sum_i \sum_{j \neq i} |a_{ij}| \, |x_i| \{ |x_i| - |x_j| \} = \sum_i \sum_{j \neq i} |a_{ij}| \, |x_j| \{ |x_j| - |x_i| \}$$

$$= \frac{1}{2} \sum_i \sum_{j \neq i} |a_{ij}| \{ |x_i| - |x_j| \}^2 \geq 0, \quad \text{q.e.d.}$$

It is to be noted, however, that the row sum criterion is only sufficient, not necessary. Thus, for example, the matrix

$$\begin{bmatrix} 3 & 2 & 2 \\ 2 & 3 & 2 \\ 2 & 2 & 3 \end{bmatrix}$$

is positive definite, with

$$Q(\mathbf{x}) = x_1^2 + x_2^2 + x_3^2 + 2(x_1 + x_2 + x_3)^2,$$

although Criterion 3.3 is not satisfied.

Thus, there are numerous matrices for which the Criteria 3.1, 3.2, 3.3 do not bring about any conclusive answer. In such cases, one must reach for the methods of §3.3. One can, however, still weaken somewhat the Criterion 3.3, thereby extending its domain of applicability:

By examining the conditions under which $Q(\mathbf{x}) = 0$ can hold if in place of (5) one only requires $a_{kk} \geq \sum_{\ell \neq k} |a_{k\ell}|$, one finds that for every pair i, j ($i \neq j$) it would have to be true that

$$\begin{cases} \text{either} & |x_i| = |x_j| \\ \\ \text{or} & a_{ij} = 0 . \end{cases} \tag{6}$$

---

^{4} After N. Rauscher, personal communication.

Besides, according to the proof above, $x_k \neq 0$ can only hold if for this k-value $a_{kk} = \sum\limits_{\ell \neq k} |a_{k\ell}|$. From this, there follows:

Criterion 3.4 (Weak row sum criterion). *If the symmetric matrix* **A** *is irreducible, i.e., for each pair* i,j ($i \neq j$) *there is a sequence of nonvanishing elements*

$$a_{ik_1}, a_{k_1 k_2}, a_{k_2 k_3}, \ldots, a_{k_{p-1} j}, \tag{7}$$

and if

$$a_{kk} \geq \sum_{\substack{\ell = 1 \\ \ell \neq k}}^{n} |a_{k\ell}| \quad (k = 1, 2, \ldots, n), \tag{8}$$

where equality, however, is not permitted for all k, *then* **A** *is positive definite.*

Indeed, by (7) and (6), Q could only be $= 0$ if $|x_1| = |x_2| = \cdots = |x_n|$; since in (8), however, strict inequality holds for at least one k, the corresponding x_k must be 0, hence $x_1 = x_2 = \cdots = x_n = 0$, q.e.d.

On the basis of Criterion 3.4, the frequently used matrix

$$\begin{bmatrix} 2 & -1 & & & & & \\ -1 & 2 & -1 & & & 0 & \\ & -1 & 2 & -1 & & & \\ & & \cdot & \cdot & \cdot & & \\ & & & \cdot & \cdot & \cdot & \\ & & & & \cdot & \cdot & \cdot \\ & 0 & & & -1 & 2 & -1 \\ & & & & & -1 & 2 \end{bmatrix} \tag{9}$$

turns out to be positive definite, since (8) always holds, with strict inequality for $k=1$ and $k=n$. Furthermore, for each pair i,j ($j > i$) there exists the chain (7) of nonvanishing elements with $k_\ell = i + \ell$ ($\ell = 1, 2, \ldots, j - i - 1$).

§3.3. The Cholesky decomposition

We now wish to develop a necessary and sufficient criterion for the positive definiteness of a symmetric matrix **A**.

Suppose the matrix **A** is positive definite, hence $a_{11} > 0$. In the quadratic form $Q(\mathbf{x}) = \sum_i \sum_j a_{ij} x_i x_j$ all terms which depend on x_1, that is,

$$a_{11} x_1^2, \quad a_{1k} x_1 x_k, \quad a_{k1} x_k x_1 \quad (k = 2, \ldots, n),$$

can then be eliminated by subtracting the expression

$$\left[\sqrt{a_{11}}\, x_1 + \sum_{k=2}^{n} \frac{a_{1k}}{\sqrt{a_{11}}} x_k \right]^2 = \left[\sum_{k=1}^{n} r_{1k} x_k \right]^2, \tag{10}$$

where $r_{1k} = a_{1k}/\sqrt{a_{11}}$ $(k = 1, \ldots, n)$. (We choose $\sqrt{a_{11}}$ to be positive.) One so obtains

$$Q(\mathbf{x}) - \left[\sum_{k=1}^{n} r_{1k} x_k \right]^2 = \sum_{i=2}^{n} \sum_{j=2}^{n} (a_{ij} - r_{1i} r_{1j}) x_i x_j, \tag{11}$$

which is again a quadratic form,

$$Q_1(\mathbf{x}) = \sum_{i=2}^{n} \sum_{j=2}^{n} a_{ij}' x_i x_j, \tag{12}$$

in the variables $x_2, x_3, \ldots, x_n$, and in fact

$$a_{ij}' = a_{ij} - r_{1i} r_{1j} \quad (i, j = 2, 3, \ldots, n). \tag{13}$$

Theorem 3.2. *With $Q(\mathbf{x})$, also the quadratic form $Q_1(\mathbf{x})$ is positive definite.*

Proof. If for certain values $x_2, x_3, \ldots, x_n$ (not all $= 0$) we had $Q_1(\mathbf{x}) \leq 0$, then with $x_1 = -\dfrac{1}{r_{11}} \sum_{k=2}^{n} r_{1k} x_k$ we would also have

$$Q(\mathbf{x}) = Q_1(\mathbf{x}) + \left[\sum_{k=1}^{n} r_{1k} x_k \right]^2 \leq 0, \text{ contrary to the assumption; q.e.d.}$$

Since, therefore, $Q_1(\mathbf{x})$ is again positive definite, we have $a_{22}' > 0$, so that a further splitting off becomes possible: one subtracts

$$\left[\sqrt{a'_{22}}\, x_2 + \sum_{k=3}^{n} \frac{a'_{2k}}{\sqrt{a'_{22}}}\, x_k \right]^2 = \left[\sum_{k=2}^{n} r_{2k} x_k \right]^2,$$

where $r_{2k} = a'_{2k}/\sqrt{a'_{22}}$ $(k = 2, \ldots, n)$, which produces a new quadratic form Q_2 in the variables $x_3, x_4, \ldots, x_n$:

$$Q_2(\mathbf{x}) = Q_1(\mathbf{x}) - \left[\sum_{k=2}^{n} r_{2k} x_k \right]^2 = \sum_{i=3}^{n} \sum_{j=3}^{n} a''_{ij} x_i x_j . \qquad (14)$$

This form, again, must be positive definite. Similarly, one obtains the positive definite forms

$$Q_3(\mathbf{x}) = Q_2(\mathbf{x}) - \left[\sum_{k=3}^{n} r_{3k} x_k \right]^2,$$

$$Q_4(\mathbf{x}) = Q_3(\mathbf{x}) - \left[\sum_{k=4}^{n} r_{4k} x_k \right]^2,$$

$$(15)$$

$$\vdots$$

$$Q_{n-1}(\mathbf{x}) = Q_{n-2}(\mathbf{x}) - \left[\sum_{k=n-1}^{n} r_{n-1,k} x_k \right]^2 .$$

(Q_k is a quadratic form in the variables $x_{k+1}, x_{k+2}, \ldots, x_n$.) $Q_{n-1}(\mathbf{x})$ then depends only on the variable x_n and therefore (being homogeneous quadratic) must necessarily be of the form

$$Q_{n-1}(\mathbf{x}) = c x_n^2.$$

Since also this quadratic form must still be positive definite, we have $c > 0$, so that $r_{nn} = \sqrt{c} > 0$ can be computed. We thus have $Q_{n-1}(\mathbf{x}) = (r_{nn} x_n)^2$. Together with (11) to (15), there finally results the representation

$$Q(\mathbf{x}) = \sum_{j=1}^{n} \left[\sum_{k=j}^{n} r_{jk} x_k \right]^2 \qquad (16)$$

(where $r_{jj} > 0$ for $j = 1, 2, ... n$).

A positive definite quadratic form can thus be represented as a sum of pure squares, where in addition, $r_{jj} > 0$. On the other hand, one has:

Theorem 3.3. *If the representation* (16) *with positive coefficients* $r_{11}, r_{22}, \ldots, r_{nn}$ *is possible, then the form* $Q(\mathbf{x})$ *is positive definite.*

Proof. If not all $x_k = 0$, there is a last one, x_p, which is still different from 0 (usually, this will be x_n). Then

$$\sum_{k=p}^{n} r_{pk} x_k = r_{pp} x_p \neq 0,$$

and, according to (16), $Q(\mathbf{x})$ is a sum of nonnegative terms, among which is $(r_{pp} x_p)^2 > 0$; q.e.d.

From this it follows that for a form which is not positive definite the decomposition (16) is not possible. Yet, one can have $a_{11} > 0$, so that a splitting $Q(\mathbf{x}) = Q_1(\mathbf{x}) + (\ )^2$ (perhaps several such) can still be carried out. But it is not possible to carry out all n reduction steps; in other words, among the n quantities $a_{11}, a_{22}', a_{33}'', \ldots,$ from which roots are to be taken, necessarily one must be ≤ 0, at which point the process breaks down.

Matrix interpretation. We arrange the coefficients $r_{1k}, r_{2k}, \ldots, r_{nn}$ generated during the various splittings in the form of a matrix and fill the empty spaces with zeros. This matrix will be denoted by **R**. Thus,

$$\mathbf{R} = \begin{bmatrix} r_{11} & r_{12} & r_{13} & \cdots & r_{1n} \\ 0 & r_{22} & r_{23} & & r_{2n} \\ 0 & 0 & r_{33} & & r_{3n} \\ \cdot & & \cdot & & \\ \cdot & & & \cdot & \\ \cdot & & & & \cdot \\ 0 & 0 & 0 & & r_{nn} \end{bmatrix} \tag{17}$$

(The coefficients from the jth splitting lie in the jth row of **R**.)

The quantities $\sum_{k=j}^{n} r_{jk} x_k$ occurring in the relation (16) are obviously the components of the vector **Rx**; therefore, the value of the quadratic form, according to (16), becomes([1])

$$Q(\mathbf{x}) = \sum_{j=1}^{n} \left[\sum_{k=j}^{n} r_{jk} x_k \right]^2 = ||\mathbf{Rx}||^2. \tag{18}$$

But now, $||\mathbf{Rx}||^2 = (\mathbf{Rx}, \mathbf{Rx}) = (\mathbf{R}^T\mathbf{Rx}, \mathbf{x})$; on the other hand, $Q(\mathbf{x}) = \sum_i \sum_j a_{ij} x_i x_j = (\mathbf{Ax}, \mathbf{x})$, and therefore one has identically in $\mathbf{x}$: $(\mathbf{Ax}, \mathbf{x}) \equiv (\mathbf{R}^T\mathbf{Rx}, \mathbf{x})$. This identity, however, can only hold if

$$\mathbf{A} = \mathbf{R}^T\mathbf{R}. \tag{19}$$

This means: the representation (16) of a quadratic form as a sum of squares is equivalent to a decomposition of the matrix $\mathbf{A}$ into two factors which are transposed of each other. This decomposition is called *Cholesky decomposition* of the matrix $\mathbf{A}$. We thus have:

Theorem 3.4. *The Cholesky decomposition, i.e., the construction of the triangular matrix* $\mathbf{R}$ *(with positive diagonal elements) according to* (19), *is possible precisely if* $\mathbf{A}$ *is positive definite.*

The Cholesky decomposition is considered as having succeeded only if all r_{jj} are positive. There are also decompositions for positive semidefinite matrices, but then the r_{jj} can no longer be all positive. For example,

$$\mathbf{A} = \begin{bmatrix} 1 & 1 & 1 \\ 1 & 1 & 1 \\ 1 & 1 & 1 \end{bmatrix}$$

is positive semidefinite, but one has $\mathbf{A} = \mathbf{R}^T\mathbf{R}$ with

$$\mathbf{R} = \begin{bmatrix} 1 & 1 & 1 \\ 0 & 0 & 0 \\ 0 & 0 & 0 \end{bmatrix}.$$

---

[1] $||\mathbf{x}|| = \sqrt{(\mathbf{x},\mathbf{x})} = \sqrt{\mathbf{x}^T\mathbf{x}} = \sqrt{\sum x_k^2}$ denotes the Euclidean norm of the vector $\mathbf{x}$; cf. §10.7. (Editors' remark)

§3.4. **Programming the Cholesky decomposition**

The elimination of the variable x_p from the quadratic form

$$Q_{p-1}(\mathbf{x}) = \sum_{i=p}^{n} \sum_{j=p}^{n} a_{ij} x_i x_j \tag{20}$$

is described by[1]

$$Q_p(\mathbf{x}) = Q_{p-1}(\mathbf{x}) - \left[\sum_{k=p}^{n} r_{pk} x_k \right]^2. \tag{21}$$

According to our earlier discussion of splitting off $(\sum r_{1k} x_k)^2$ and $(\sum r_{2k} x_k)^2$, one clearly has $r_{pk} = a_{pk}/\sqrt{a_{pp}}$, and the splitting itself has the effect

$$a_{ij} := a_{ij} - r_{pi} r_{pj} \quad (i,j = p+1, \ldots, n), \tag{22}$$

where now the new a_{ij} are the coefficients of $Q_p(\mathbf{x})$. One thus obtains the following program for the splitting (21)[2]:

```
r[p,p] := sqrt(a[p,p]);
for k := p+1 step 1 until n do r[p,k] := a[p,k]/r[p,p];
for i := p+1 step 1 until n do                              (23)
    for j := i step 1 until n do
        a[i,j] := a[i,j] - r[p,i] × r[p,j];
```

This piece of program destroys the coefficients of $Q_{p-1}(\mathbf{x})$ and replaces them by those of $Q_p(\mathbf{x})$. This makes sense, since the form Q_{p-1} is no longer used in the subsequent course of the computation.

The Cholesky decomposition now proceeds as follows. The given quadratic form $Q(\mathbf{x}) = Q_0(\mathbf{x})$, each time through a splitting (23), is reduced to $Q_1(\mathbf{x})$, $Q_2(\mathbf{x})$, etc., until $Q_n(\mathbf{x}) \equiv 0$, which is evidently described by

---

[1] Actually, the a_{ij} would have to be distinguished by an upper index $p-1$, since they depend on p. It will transpire, however, that this is only a "conceptual" index, and for this reason we omit it here.

[2] As can be seen from the index range of j, only the matrix elements on and above the diagonal are processed, which is all that is needed because of symmetry.

> for p := 1 **step** 1 **until** n **do**
> **begin**
>   **if** $a[p,p] \leq 0$ **then goto** *indef*;   (24)
>   **comment** *insert here the piece of program* (23);
> **end** *for* p;

The test for $a[p,p] \leq 0$ is necessary in order to guarantee – through excluding matrices which are not positive definite – the safe progression of the computation. For this purpose there must be a label *indef* at the end of the program:

indef: **end** *of program*;

Even with this provision, the program is not yet strict, since this test is not completely adequate. Finite arithmetic, namely, entails that only numbers below a certain bound M are representable. The program (24), however, may well lead outside of this range, which will manifest itself in overflow. (For the CDC-6500 system, e.g., one has $M \sim 10^{320}$.)

Examples. For the matrix

$$
A = \begin{bmatrix} 10^{-250} & 10^{250} & . & . & . \\ 10^{250} & & & & \\ . & & & & \\ . & & & & \\ . & & & & \end{bmatrix}
$$

one gets $r_{11} = 10^{-125}$, $r_{12} = 10^{375}$, and thus already the computation of the r_{1k} yields overflow. For

$$
A = \begin{bmatrix} 10^{-180} & 10^{180} & . & . & . \\ 10^{180} & 10^{200} & & & \\ . & & & & \\ . & & & & \\ . & & & & \end{bmatrix}
$$

one finds $r_{11} = 10^{-90}$, $r_{12} = 10^{270}$, $a'_{22} = 10^{200} - 10^{540}$; thus, overflow first occurs during the attempt of computing a'_{22}.

Both phenomena are possible only for matrices which are not positive definite to begin with. (This follows from Criterion 3.2). Therefore, the irregular termination of the computing process, for once, can be tolerated.

Programming hints

1. *Remark.* Occasionally, one finds fault with the Cholesky decomposition because it requires the computation of n square roots. One could indeed avoid these square roots by means of a suitable rearrangement of the computing process ($L^T DL$-decomposition); however, the extra effort for the n square roots is not significant enough to be worthwhile to trade it for other disadvantages.

2. *Remark.* Since the original matrix elements a_{ij} are destroyed anyhow by the program (23), (24), one may just as well use the storage locations for other purposes. From (23) it indeed transpires that after the computation of $r[p,k]$, the variable $a[p,k]$ is never used again; one therefore can store the newly computed quantity $r[p,k]$ in the place of $a[p,k]$. This can be achieved in the program (23) by systematically replacing the name r by a. At the end, the **array** $a[1{:}n, 1{:}n]$ then contains the matrix **R** in place of **A**, i.e., the Cholesky decomposition is carried out "in place"(^{3}).

3. *Remark.* The program (23), (24) computes in turn the coefficients of the quadratic forms $Q_1, Q_2, \ldots, Q_{n-1}$, which actually are not needed at all. Indeed, one can improve the program by rearranging the run of the indices.

One observes that in (23), (24), for fixed i, j ($j \geq i$), in the course of computation one subtracts from $a[i,j]$ the products $r[p,i] \times r[p,j]$ ($p = 1,2, \ldots, i-1$) before one executes (for $p=i$) $r[i,i] := sqrt(a[i,i])$ or $r[i,j] := a[i,j]/r[i,i]$, respectively. However, one can also finish computing a $r[i,j]$ before one starts on the next one:

```
for i := 1 step 1 until n do
    for j := i step 1 until n do
    begin
        s := a[i,j];
        for p := 1 step 1 until i-1 do
            s := s-r[p,i] × r[p,j];
        comment here, s is the (i,j)-coefficient of the        (25)
            quadratic form Q_{i-1};
        if i=j then
        begin
            if s ≤ 0 then goto indef;
```

---

^{3} The triangular matrix **R** of course occupies only the storage locations $a[i,k]$ with $k \geq i$. Those with $k < i$ are neither used nor changed. (Editors' remark)

$$r[i,i] := sqrt(s);$$
end
    else $r[i,j] := s/r[i,i];$
end *for* $i,j;$

Note that in this arrangement the elements of the original matrix **A** are not destroyed. But here, too, one could dispense with **A** and replace r consistently by a.

§3.5. Solution of a linear system

The system of equations

$$\mathbf{Ax} = \mathbf{b} \qquad (26)$$

with positive definite symmetric matrix **A**, once the Cholesky decomposition has been completed, can be solved very easily. Indeed, with $\mathbf{A} = \mathbf{R}^T\mathbf{R}$ one has $(\mathbf{R}^T\mathbf{R})\mathbf{x} = \mathbf{R}^T(\mathbf{Rx}) = \mathbf{b}$, so that (26) can be replaced by the two systems

$$\mathbf{R}^T\mathbf{v} = \mathbf{b}, \quad \mathbf{Rx} = \mathbf{v}. \qquad (27)$$

One thus first solves

	v_1	v_2	$\cdots$	v_n	1	
$0 =$	r_{11}	0	$\cdots$	0	$-b_1$	
$0 =$	r_{12}	r_{22}		0	$-b_2$	(28)
.	.	.				
.	.	.				
.	.	.				
$0 =$	r_{1n}	r_{2n}		r_{nn}	$-b_n$	

in the order $v_1, v_2, \ldots, v_n$ (*forward substitution*). Thereafter, one solves

	x_1	x_2	$\cdots$	x_n	1	
$0 =$	r_{11}	r_{12}	$\cdots$	r_{1n}	$-v_1$	
$0 =$	0	r_{22}		r_{2n}	$-v_2$	(29)
.	.	.				
.	.	.				
.	.	.				
$0 =$	0	0		r_{nn}	$-v_n$	

in the order $x_n, x_{n-1}, \ldots, x_1$ (*back substitution*). Thanks to the triangular form of **R**, these systems of equations are fairly unproblematic([1]).

In computational practice, one usually exploits the fact that the three vectors **b**, **v**, **x** can be stored in the same **array** $s[1{:}n]$, which then yields the following program:

> **for** $i := 1$ **step** 1 **until** n **do**
> **begin**
>   **for** $j := 1$ **step** 1 **until** $i-1$ **do**
> $s[i] := s[i] - r[j,i] \times s[j]$;
> $s[i] := s[i]/r[i,i]$;
> **end** *for* i;
> **for** $i := n$ **step** -1 **until** 1 **do**   (30)
> **begin**
>   **for** $j := i+1$ **step** 1 **until** n **do**
> $s[i] := s[i] - r[i,j] \times s[j]$;
> $s[i] := s[i]/r[i,i]$;
> **end** *for* i;

If one enters this program with $s[i] = b_i$, then the $s[i]$ at the end are the desired x_i.

§3.6. Influence of rounding errors

It must not be concealed that the Cholesky decomposition can be significantly disturbed through rounding errors.

Example. The matrix

$$\mathbf{A} = \begin{bmatrix} 37 & 5 & 12 & 2 \\ & 62 & 58 & -1 \\ \text{sym-} & & 66 & 17 \\ \text{metric} & & & 30 \end{bmatrix} \qquad (31)$$

is positive definite; one has, indeed,

$$Q(\mathbf{x}) = (6x_1 + x_3)^2 + (6x_2 + 5x_3 - x_4)^2 + (x_1 + 5x_2 + 6x_3 + 2x_4)^2$$
$$+ (-x_2 + 2x_3 + 5x_4)^2.$$

---

[1] Compare with §2.2, where the same solution technique is used. (Editors' remark)

The computing process (23), if computations are carried out with 4 decimal digits after the decimal point, here yields for $p=1$ (A_1 is the matrix belonging to the form Q_1):

$$r_{1k} = \{6.0828, \; .8220, \; 1.9728, \; .3288\},$$

$$A_1 = \begin{bmatrix} 61.3243 & 56.3784 & -1.2703 \\ & 62.1081 & 16.3513 \\ \text{sym.} & & 29.8919 \end{bmatrix};$$

for $p=2$:

$$r_{2k} = \{7.8310, \; 7.1994, \; -.1622\},$$

$$A_2 = \begin{bmatrix} 10.2767 & 17.5190 \\ \text{sym.} & 29.8656 \end{bmatrix};$$

for $p=3$:

$$r_{3k} = \{3.2057, \; 5.4650\},$$

$$A_3 = [-.0006];$$

(32)

thus, the decomposition has failed.

How is this to be interpreted? In certain circumstances a positive definite matrix may be viewed by the Cholesky method as being not positive definite. When this happens, however, it means that the matrix A cannot be distinguished from a singular matrix within the computer precision, because the continuous transition from a positive definite to a indefinite matrix goes through a semidefinite matrix; such a matrix, however, is singular. In other words, the failure of the Cholesky decomposition owing to rounding errors is only possible if the solution of the system $Ax = b$ is threatened anyhow by rounding errors (which does not mean that in case of success the accuracy could not also be imperiled).

The following counter measures are available.

a) *Increasing the precision.* This is not always meaningful, but in the above example the Cholesky decomposition indeed succeeds with 7 decimals after the decimal point and yields

$$R = \begin{bmatrix} 6.0827625 & .8219949 & 1.9727879 & .3287980 \\ 0 & 7.8309849 & 7.1993982 & -.1622108 \\ 0 & 0 & 3.2057407 & 5.4649371 \\ 0 & 0 & 0 & .0064885 \end{bmatrix}. \quad (33)$$

Comparison with (32) shows that in 4-digit arithmetic, r_{34} in fact became slightly too large, which then led to $29.8656 - 5.4650^2 < 0$ ($29.8656 - 5.4649^2$ would have become positive).

b) *Search for the origin of the problem.* If it turns out that the problem can also be formulated as a least squares problem, one obtains better results with the methods of Chapter 5.

c) *Investigate the rounding errors.* In the numerical computation according to the program (25), the way the element r_{ij} ($i \leq j$) of the matrix **R** comes about, is by subtracting from a_{ij} (of the original matrix **A**) the products $r_{1i}r_{1j}, r_{2i}r_{2j}, \ldots, r_{i-1,i}r_{i-1,j}$; with the remainder s one forms

$$\text{for } i=j: \quad r_{ii} = \sqrt{s},$$
$$\text{for } j>i: \quad r_{ij} = s/r_{ii}, \tag{34}$$

so that in each case $s = r_{ii}r_{ij}$, i.e. (theoretically),

$$a_{ij} = \sum_{p=1}^{i} r_{pi}r_{pj} \quad (j \geq i), \tag{35}$$

which establishes (19) in a new way.

In practice, the two sides of (35) differ by the rounding errors which one commits in the arithmetic operations([1])

$$a_{ij}^{(p)} = a_{ij}^{(p-1)} - r_{pi}r_{pj} \quad (p = 1,2,\ldots, i-1) \tag{36}$$

and finally in the operations (34).

These rounding errors are (in floating-point arithmetic)([2]):

1) For the product in (36):

$$< \theta \, | r_{pi}r_{pj} |,$$

where θ is the smallest machine number for which $1 + \theta > 1$.

2) For the subtraction in (36)([3]):

---

[1] $A_p = [a_{ij}^{(p)}]$ here and in the sequel denotes the $(n-p) \times (n-p)$-matrix associated with the quadratic form Q_p (cf. the preceding example). In particular, $A_0 = [a_{ij}^{(0)}] = A$. (Editors' remark)

[2] Cf. Appendix, §A3.4, where, however, the rounding error bound for addition and subtraction is somewhat larger. Further literature: Wilkinson J.H.: *Rounding Errors in Algebraic Processes*, Prentice-Hall, Englewood Cliffs, N.J., 1963; Stoer J., Bulirsch R.: *Introduction to Numerical Analysis*, Springer, New York, 1980, Ch. 1. (Editors' remark)

[3] Terms of order $0(\theta^2)$, here and in the sequel, are neglected. (Editors' remark)

$$< \theta \{ |a_{ij}^{(p-1)}| + |r_{pi}r_{pj}| \} \quad (p = 1, 2, \ldots, i-1).$$

3) For the operations (34) the rounding errors have the effect that for the computed values r_{ii}, r_{ij}:

$$|r_{ii}^2 - s| < 2\theta s, \qquad |r_{ij}r_{ii} - s| < \theta s.$$

Altogether, one therefore obtains, in place of (35):

$$a_{ij} - \sum_{p=1}^{i} r_{pi}r_{pj} = -\Delta_{ij},$$

$$|\Delta_{ij}| < \theta \sum_{p=1}^{i} \{ |a_{ij}^{(p-1)}| + 2|r_{pi}r_{pj}| \}. \tag{37}$$

Thus, it is as if the Cholesky decomposition were applied *exactly* to a matrix $\mathbf{A} + \Delta$ with elements $a_{ij} + \Delta_{ij}$; in other words, not the matrix $\mathbf{A}$, but $\mathbf{A} + \Delta$ is tested for positive definiteness and decomposed in $\mathbf{R}^T\mathbf{R}$. This leads to wrong results, which are particularly disturbing when

$$\left\{ \begin{array}{c} \mathbf{A} \text{ is positive definite} \\ \mathbf{A} + \Delta \text{ is not positive definite} \end{array} \right\} \text{ or vice versa.} \tag{38}$$

Here, Δ is not known; what is available is only the estimate (37), with the help of which one must determine whether (38) can actually occur.

This determination can be made with the aid of the eigenvalues of $\mathbf{A}$: A symmetric matrix $\mathbf{A}$, as is well known, is characterized as positive definite by the fact that all its eigenvalues are not only real, but positive. Since, on the other hand, for symmetric matrices $\mathbf{A}$, Δ([4]),

$$\lambda_{\min}(\mathbf{A}) = \min_{||\mathbf{x}|| = 1} (\mathbf{x}, \mathbf{A}\mathbf{x}), \quad \lambda_{\min}(\mathbf{A} + \Delta) = \min_{||\mathbf{x}|| = 1} ((\mathbf{x}, \mathbf{A}\mathbf{x}) + (\mathbf{x}, \Delta\mathbf{x})),$$

$$\max_{||\mathbf{x}|| = 1} ||\Delta\mathbf{x}|| = \max_i |\lambda_i(\Delta)| = \max_{||\mathbf{x}|| = 1} |(\mathbf{x}, \Delta\mathbf{x})|,$$

one has

$$\lambda_{\min}(\mathbf{A}) - ||\Delta|| \leq \lambda_{\min}(\mathbf{A} + \Delta) \leq \lambda_{\min}(\mathbf{A}) + ||\Delta||, \tag{39}$$

---

[4] See, e.g., Schwarz H.R., Rutishauser H., Stiefel E.: *Numerical Analysis of Symmetric Matrices*, Prentice-Hall, Englewood Cliffs, N.J., 1973, Theorem 4.3 and Example 1.3. (Editors' remark)

if the spectral norm of Δ is defined by

$$||\Delta|| = \max_{||x|| = 1} ||\Delta x||. \tag{40}$$

It is clear, therefore, that (38) can only occur if

$$|\lambda_{min}(A)| \leq ||\Delta||. \tag{41}$$

To determine $||\Delta||$, we note that the norm is subadditive, i.e., one always has $||A + B|| \leq ||A|| + ||B||$, hence, according to (37),

$$||\Delta|| \leq \theta \sum_{p=1}^{n} \{||\overline{A}_{p-1}|| + 2||Z_p||\}, \tag{42}$$

where $\overline{A}_{p-1}$ is the matrix with elements $|a_{ij}^{(p-1)}|$ and Z_p the one with elements $|r_{pi}r_{pj}|$. As above, θ stands for the smallest machine number with $1 + \theta > 1$.

We now have, if the trace of A is denoted by t_A:

1) $||A|| \leq t_A$, if A is positive definite, and $||\overline{A}|| \leq t_{\overline{A}} = t_A$, even if $\overline{A}$ is no longer positive definite([5]).

2) The trace of A_p is smaller than the trace of A_{p-1} because, first, the element a_{pp} is no longer present and, secondly, the remaining diagonal elements a_{qq} are decreased by r_{pq}^2 (or at least not increased).

3) Z_p has a single eigenvalue([6]) different from zero, namely $\sum_{i=p}^{n} r_{pi}^2$.

---

[5] The first estimate holds because of

$$||A|| = \max_{||x||=1} ||Ax|| = \lambda_{max}(A) \leq t_A.$$

The second can be proved as follows:

$$||\overline{A}||^2 = \max_{||x||=1} (\overline{A}x, \overline{A}x) = \max_{||x||=1} (x, \overline{A}^2 x) = \lambda_{max}(\overline{A}^2)$$

$$\leq t_{\overline{A}}^2 = \sum_{i,k} |a_{ik}||a_{ki}| = \sum_{i,k} a_{ik}a_{ki}$$

$$= t_{A^2} = \sum_i \lambda_i(A^2) = \sum_i \lambda_i^2(A) \leq \left[\sum_i \lambda_i(A)\right]^2 = (t_A)^2.$$

(Editors' remark)

[6] Since z_p is a dyadic product, i.e., a product of a column vector times a row vector, z_p has rank 1, and the only eigenvalue different from 0 is equal to the trace. (Editors' remark)

Therefore, $||\mathbf{Z}_p|| = \sum_{i=p}^{n} r_{pi}^2$, and ([7])

$$\sum_{p=1}^{n} ||\mathbf{Z}_p|| = \sum_{p=1}^{n} \left[\sum_{i=p}^{n} r_{pi}^2 \right] = t_A.$$

By 1) and 2) we have $||\overline{\mathbf{A}}_p|| \leq t_A$, so that from (42) there finally follows:

$$||\Delta|| \leq \theta(n+2)t_A. \tag{43}$$

For a reliable determination of positive definiteness, one thus must have

$$\lambda_{\min}(\mathbf{A}) > \theta(n+2)t_A, \quad \text{i.e,} \quad \theta < \frac{\lambda_{\min}(\mathbf{A})}{(n+2)t_A}. \tag{44}$$

For the matrix (31) considered above as an example, one has $t_A = 195$, $\lambda_{\min}(\mathbf{A}) = 6.6_{10}-6$; therefore, a safe determination is possible only for

$$\theta < \frac{6.6_{10}-6}{6 \times 195} = 5.64_{10}-9, \tag{45}$$

that is, if the computation is carried out with at least 9 digits (floating-point). In fact, the Cholesky decomposition succeeded even with 9 fixed-point digits (7 of which after the decimal point).

The smallest eigenvalue of $\mathbf{A}$ in (44), however, is generally not known. In the subsequent discussion, we use only the quantity

$$\theta_A = \theta(n+2)t_A \tag{46}$$

which (for a positive definite matrix $\mathbf{A}$) by (43) is an upper bound for the falsification of the eigenvalues of $\mathbf{A}$ through the rounding errors of the Cholesky decomposition. If the Cholesky decomposition succeeds, we

---

[7] The second equality, by (37), is valid up to a term of $0(\theta)$, hence (43) up to a term of $0(\theta^2)$. (Translator's remark)

know that $\lambda_{\min} > -\theta_A$; if it fails, then $\lambda_{\min} < +\theta_A$. In other words, we have the

Theorem 3.5. *If the Cholesky decomposition* $\mathbf{A} - \theta_A\mathbf{I}$ *succeeds, then* $\mathbf{A}$ *is guaranteed to be positive definite*[8]; *if it fails for* $\mathbf{A} + \theta_A\mathbf{I}$, *then* $\mathbf{A}$ *is guaranteed to be not positive definite.*

Example. For the matrix (31) one has $t_A = 195$, thus with $\theta = 5_{10}-9$, i.e., with a 9-digit mantissa, $\theta_A = 6_{10}-6$. The Cholesky decomposition of $\mathbf{A} - 6_{10}-6\ \mathbf{I}$ (in floating-point arithmetic) yields

$$\mathbf{R} = \begin{bmatrix} 6.08276204 & .821995003 & 1.97278801 & .328798001 \\ 0 & 7.83098450 & 7.19939850 & -.162210806 \\ 0 & 0 & 3.20573903 & 5.46493998 \\ 0 & 0 & 0 & .002144761 \end{bmatrix} ; (33)$$

therefore, the matrix $\mathbf{A}$ is guaranteed to be positive definite.

§3.7. Linear systems of equations as a minimum problem

The system (26) is equivalent to a minimum problem:

Theorem 3.6. *For a linear system of equations* $\mathbf{A}\mathbf{x} + \mathbf{b} = 0$ *with positive definite symmetric matrix* $\mathbf{A}$ *the following is true. The system is uniquely solvable, and its solution is also the unique minimum of the quadratic function*

$$F(\mathbf{x}) = \frac{1}{2}(\mathbf{x}, \mathbf{A}\mathbf{x}) + (\mathbf{x}, \mathbf{b}) = \frac{1}{2}\sum_{i=1}^{n}\sum_{j=1}^{n} a_{ij}x_ix_j + \sum_{i=1}^{n} b_ix_i \qquad (47)$$

taken over all $\mathbf{x} = [x_1, x_2, \ldots, x_n]^T$.

---

[8] For the proof of the first assertion of the theorem one has to argue more precisely as follows: In deriving (43), it was assumed that $\mathbf{A}$ is positive definite. If, however, one only knows that the Cholesky decomposition of $\mathbf{A} - \theta_A\mathbf{I}$ succeeds, there follows at first only that $\mathbf{A} - \theta_A\mathbf{I} + \Delta$ is positive definite (where Δ is the rounding error matrix associated with this decomposition.) But from this it follows easily that (43) continues to hold up to terms of $0(\theta^2)$. The eigenvalues of $\mathbf{A} - \theta_A\mathbf{I}$ therefore are shifted at most by θ_A in first approximation. (Editors' remark)

Proof. This assertion will be proved without the use of previous knowledge and also without utilizing theorems on determinants and the like, solely on the basis of the definition of positive definiteness.

1) *Every solution of the linear system of equations is also a (relative and absolute) minimum of F* $(\mathbf{x})$: For a symmetric matrix $\mathbf{A}$ one has, in the notations of (1) and (47), the identity

$$F(\mathbf{x} + \mathbf{y}) = F(\mathbf{x}) + \frac{1}{2}Q(\mathbf{y}) + (\mathbf{A}\mathbf{x} + \mathbf{b},\mathbf{y}), \qquad (48)$$

because

$$\frac{1}{2}(\mathbf{x} + \mathbf{y}, \ \mathbf{A}\mathbf{x} + \mathbf{A}\mathbf{y}) + (\mathbf{b},\mathbf{x} + \mathbf{y})$$

$$= \frac{1}{2}(\mathbf{x},\mathbf{A}\mathbf{x}) + (\mathbf{A}\mathbf{x},\mathbf{y}) + \frac{1}{2}(\mathbf{y},\mathbf{A}\mathbf{y}) + (\mathbf{b},\mathbf{x}) + (\mathbf{b},\mathbf{y}).$$

Therefore, if $\mathbf{x}$ is a solution of the system $\mathbf{A}\mathbf{x} + \mathbf{b} = \mathbf{0}$, then for all $\mathbf{y}$ there holds

$$F(\mathbf{x} + \mathbf{y}) = F(\mathbf{x}) + \frac{1}{2}Q(\mathbf{y}), \qquad (49)$$

where by assumption $Q(\mathbf{y}) > 0$ for $\mathbf{y} \neq \mathbf{0}$, so that $F(\mathbf{x} + \mathbf{y}) > F(\mathbf{x})$.

2) *Every relative minimum of F* $(\mathbf{x})$ *is a solution of the linear system of equations*: If $\mathbf{c} = \mathbf{A}\mathbf{x} + \mathbf{b} \neq \mathbf{0}$, so that, say, the first component $c_1 \neq 0$, then by (48) one has for a vector $\mathbf{y} = [t, 0, 0, \ldots, 0]^T$

$$F(\mathbf{x} + \mathbf{y}) = F(\mathbf{x}) + \frac{1}{2}a_{11}t^2 + c_1 t.$$

This, however, is smaller than $F(\mathbf{x})$ for all t between 0 and $-2c_1/a_{11}$, so that $F(\mathbf{x})$ at the point $\mathbf{x}$ cannot have a relative minimum.

3) $F(\mathbf{x})$ *has at least one relative minimum*: Indeed, the quadratic form $Q(\mathbf{x}) = \sum \sum a_{ij}x_i x_j$ on the compact set $\sum x_i^2 = 1$ certainly assumes a minimum μ, which by virtue of the positive definiteness of $\mathbf{A}$ must be positive. Thus, for arbitrary $\mathbf{x} = [x_1, x_2, \ldots, x_n]^T$,

$$\sum_{i=1}^{n} \sum_{j=1}^{n} a_{ij}x_i x_j \geq \mu \sum_{i=1}^{n} x_i^2 = \mu ||\mathbf{x}||^2.$$

Furthermore, $|\sum_{i=1}^{n} b_i x_i| \leq ||\mathbf{b}|| \ ||\mathbf{x}||$, and thus

$$F(\mathbf{x}) \geq \frac{\mu}{2} ||\mathbf{x}||^2 - ||\mathbf{b}|| \ ||\mathbf{x}||,$$

hence

$$F(\mathbf{x}) \geq 4 \frac{||\mathbf{b}||^2}{\mu} \quad \text{for} \quad ||\mathbf{x}|| = \rho = 4 \frac{||\mathbf{b}||}{\mu} .$$

But since $F(\mathbf{x}) = 0$ for $\mathbf{x} = 0$, and $F(\mathbf{x})$ in the sphere $||\mathbf{x}|| \leq \rho$ is continuous, there follows the existence of at least one relative minimum of $F(\mathbf{x})$ in the interior of the sphere.

4) *There is only one relative minimum of* $F(\mathbf{x})$: If $\mathbf{x}$ is a relative minimum for $F(\mathbf{x})$, then by 2), $\mathbf{x}$ is a solution of $\mathbf{Ax} + \mathbf{b} = 0$. By 1), $\mathbf{x}$ is then an absolute minimum; more precisely, according to (49), there holds $F(\mathbf{z}) > F(\mathbf{x})$ for all $\mathbf{z} \neq \mathbf{x}$. In particular, for a second relative minimum $\mathbf{x}' \neq \mathbf{x}$, there of course would also have to be $F(\mathbf{z}) > F(\mathbf{x}')$ for all $\mathbf{z} \neq \mathbf{x}'$, which is not possible.

With this, the theorem is proved.

The minimum problem and the linear system of equations are thus equivalent; both have exactly one, and in fact the same, solution. We will see in Chapter 10 how this fact can be usefully exploited for the solution of a linear system of equations based on the minimum property.

Example. Let the system of equations be

$$137x - 100y - 11 = 0$$
$$-100x + 73y + 8 = 0.$$

The function to be minimized here is

$$F(x,y) = \frac{1}{2} (137x^2 - 200xy + 73y^2) - 11x + 8y .$$

The minimum is attained for $x=3$, $y=4$ and is equal to $-\frac{1}{2}$. Consider a few points in the neighborhood of the solution:

(a) $x = 1.9$, $y = 2.5$ (distance to the solution approx. 1.86). Here we get $F(x,y) = -.49$, which is .01 above the minimum.

(b) $x = 2.85$, $y = 4.11$ (distance to the solution approx. .186). Here we get $F(x,y) = 3.1329$, which is 3.63 above the minimum.

Therefore, as one moves away from the minimum in different directions, F increases with different speed, a fact that has something to do with the condition of the matrix $\mathbf{A}$ (cf. §10.7).

Note: If the matrix $\mathbf{A}$ is not positive definite, the function $F(\mathbf{x})$ defined in (47) has no minimum (or at least not a uniquely determined one).

Notes to Chapter 3

§3.6 This section takes a somewhat pessimistic view of the roundoff properties of Cholesky factorization, which is a very stable process. Apart from diagonal scaling, it is equivalent to Gaussian elimination (cf. Chapter 2, Eq. (19)) in the sense that the equations

$$\mathbf{B} = \mathbf{R}^T \text{diag}(r_{ii}), \quad \mathbf{C} = -\text{diag}(r_{ii}^{-1})\mathbf{R}$$

are satisfied. When regarded as Gaussian elimination, the process is very stable, since Wilkinson has shown (Wilkinson [1961]) that no entry in any reduced matrix ever exceeds in absolute value the largest absolute value of an entry of $\mathbf{A}$. Scaling is innocuous, since exactly the same computations (unless underflow or overflow occurs) are performed if the scaling is by powers of the radix, and other scalings perturb this result only by very minor roundoff effects.

Reference

Wilkinson, J.H. [1961]: Error analysis of direct methods of matrix inversion, *J. Assoc. Comput. Mach.* **8**, 281–330.

CHAPTER 4

Nonlinear Equations

To introduce the subject, we consider a few examples of nonlinear equations:

$$x^3 + x + 1 = 0$$

is an algebraic equation; there is only one unknown, but it occurs in the third power. There are three solutions, of which two are conjugate complex.

$$2x - \tan x = 0$$

is a transcendental equation. Again, only one unknown is present, but now in a transcendental function. There are denumerably many solutions.

$$\sin x + 3 \cos x = 2$$

is a transcendental equation only in an unessential way, since it can be transformed at once into a quadratic equation for e^{ix}. While there are infinitely many solutions, they can all be derived from two solutions through addition of multiples of 2π.

$$x^3 + y^2 + 5 = 0$$
$$2x + y^3 + 5y = 0$$

is a system of two nonlinear algebraic equations in two unknowns x and y. It can be reduced to *one* algebraic equation of degree 9 in only *one* unknown. This latter equation has nine solutions which generate nine pairs of numbers (x_i, y_i), $i = 1, \ldots, 9$, satisfying the given system. (There are fewer if only real x, y are admitted.)

In general, every system of n algebraic equations in n unknowns, according to a theory of Bézout([1]), can be reduced to one algebraic equation in one unknown, but the degree of this equation is often very high. The solution after Bézout, therefore, is usually not practicable, numerically.

A solution of a system of transcendental equations by similar means is even less practical. The only option left, as a rule, is linearization, which however leads to an infinite iterative process.

§4.1. The basic idea of linearization

We consider the following general problem: Given are n functions

$$f_j(x_1, x_2, \ldots, x_n) \quad (j = 1, 2, \ldots, n) \tag{1}$$

of n variables, and these variables $x_1, x_2, \ldots, x_n$ are to be determined in such a way that $f_1 = f_2 = \cdots = f_n = 0$. One has to distinguish, in this connection, between the case where only real, and the case where also complex values of the unknowns are admitted.

We start from some point $\mathbf{x} = [x_1, x_2, \ldots, x_n]^T$ and compute the vector $\mathbf{f} = [f_1, f_2, \ldots, f_n]^T$, where $f_j = f_j(x_1, x_2, \ldots, x_n)$. We now seek a correction $\Delta\mathbf{x}$ such that for $j = 1, 2, \ldots, n$

$$f_j^* = f_j(x_j + \Delta x_1, x_2 + \Delta x_2, \ldots, x_n + \Delta x_n) = 0.$$

In a first approximation, we have

$$f_j^* \approx f_j + \sum_{k=1}^{n} \frac{\partial f_j}{\partial x_k} \Delta x_k \,;$$

one can therefore determine approximate values for the correction by solving the system of equations

$$\sum_{k=1}^{n} \frac{\partial f_j}{\partial x_k} \Delta x_k + f_j = 0 \quad (j = 1, \ldots, n). \tag{2}$$

---

[1] See, for example, Walker R.J.: *Algebraic Curves*, Dover, New York, 1950, Ch. 3, §3.

This system for $\Delta x_1, \Delta x_2, \ldots, \Delta x_n$ is linear, and can therefore be solved by Gauss elimination. The corresponding tableau is given by:

	Δx_1	Δx_2	$\cdots$	Δx_n	1
$0 =$	$\dfrac{\partial f_1}{\partial x_1}$	$\dfrac{\partial f_1}{\partial x_2}$	$\cdots$	$\dfrac{\partial f_1}{\partial x_n}$	f_1
$0 =$	$\dfrac{\partial f_2}{\partial x_1}$	$\dfrac{\partial f_2}{\partial x_2}$		$\dfrac{\partial f_2}{\partial x_n}$	f_2
$\vdots$					
$0 =$	$\dfrac{\partial f_n}{\partial x_1}$	$\dfrac{\partial f_n}{\partial x_2}$		$\dfrac{\partial f_n}{\partial x_n}$	f_n

Introducing the matrix $\mathbf{F} = [\partial f_j / \partial x_k]$, which depends on $\mathbf{x}$, we can write the system in matrix form as

$$\mathbf{F}\,\Delta \mathbf{x} + \mathbf{f} = \mathbf{0}. \tag{3}$$

Once these equations are solved, the corrected point $\mathbf{x} + \Delta \mathbf{x}$ is again denoted by $\mathbf{x}$, and the process repeated with this new point, etc., until all f_j are sufficiently small.

In this way, the solution of a nonlinear system of equations is reduced to the solution of a sequence of linear systems of equations. This method is therefore called *linearization*; it represents an abundant source of linear systems of equations in computational practice.

Of course, there arises the question of convergence of the method, that is, whether the f_j indeed ever become sufficiently small. This, however, even in the case $n=1$, is a difficult question, and can be fully answered only in special cases. The difficulty, in fact, is that the matrix $\mathbf{F}$ may become singular, in which case the process falters.

Example. For the system

$$x^3 + y^2 + 5 = 0$$
$$2x + y^3 + 5y = 0,$$

the matrix $\mathbf{F}$ becomes

$$\mathbf{F} = \begin{bmatrix} 3x^2 & 2y \\ 2 & 3y^2 + 5 \end{bmatrix}.$$

Starting at the point $x = -2$, $y = 1$, we get $\mathbf{f} = [-2,2]^T$, and the system of equations for the first correction reads

Δx	Δy	1
12	2	-2
2	8	2

Its solution, rounded to 2 digits, is $\Delta x = .22$, $\Delta y = -.30$. As corrected point one thus obtains $x = -1.78$, $y = .70$, which leads to $\mathbf{f} = [-.1498, .2830]^T$. Second correction:

Δx	Δy	1
9.5052 1.4		$-.1498$
2	6.47	.2830

$$\Rightarrow \quad \begin{cases} \Delta x = .02326 \\ \Delta y = -.05093 \end{cases}$$

$$x = -1.75674, \quad y = .64907 \quad \Rightarrow \quad \mathbf{f} = \begin{bmatrix} -.00025 \\ .00532 \end{bmatrix}$$

One last, rather crude, correction with the same matrix $\mathbf{F}$ (actually, $\mathbf{F}$ should be computed anew)[1]:

Δx	Δy	1
9.5052 1.4		$-.00025$
2	6.47	.00532

$$\Rightarrow \quad \begin{cases} \Delta x = .00015 \\ \Delta y = -.00087 \end{cases}$$

$$x = -1.75659, \quad y = .64820 \quad \Rightarrow \quad \mathbf{f} = \begin{bmatrix} .000014 \\ .000170 \end{bmatrix}$$

---

[1] If the matrix $\mathbf{F}$ in a neighborhood of the desired solution changes only little – as is generally the case –, it makes no sense to compute it anew in each step. The system of equations for the corrections can then be solved simply by forward and back substitution. (Editors' remark)

Derivative-free linearization. For the elements of the coefficient matrix **F** in (3) one requires derivatives of the functions $f_j(x_1, \ldots, x_n)$, $j = 1, \ldots, n$. One must point out, however, that in certain cases the function values themselves are already extremely difficult to compute, so that derivatives can no longer be formed for all practical purposes.

An example for this is

$$f_1(x_1,x_2,x_3) = \int_0^\infty \frac{dt}{e^{x_1 t} + \dfrac{1}{e^{x_2 t} + \dfrac{1}{e^{x_3 t}}}}$$

In such cases one substitutes difference quotients for derivatives, thus for the (k,ℓ)-element of **F**, for example,

$$\frac{f_k(x_1,x_2, \ldots, x_\ell + h, \ldots, x_n) - f_k(x_1,x_2, \ldots, x_\ell, \ldots, x_n)}{h}$$

(in place of $\partial f_k/\partial x_\ell$). Here, whenever possible, h should be of the order of magnitude of the probable correction of the variable x_ℓ.[2]

Example. Let us again solve the simple system

$$f(x,y) = x^3 + y^2 + 5 = 0$$
$$g(x,y) = 2x + y^3 + 5y = 0,$$

but now without the use of derivatives of f and g. First, f and g must be evaluated in three points:

x	$= -2,$	y	$= 1$	$\Rightarrow$	$f = -2,$	$g = 2$
$x + h$	$= -1.5,$	y	$= 1$	$\Rightarrow$	$f = 2.625,$	$g = 3$
x	$= -2,$	$y + h$	$= 1.5$	$\Rightarrow$	$f = -.75,$	$g = 6.875.$

Then the difference quotients can be formed:

$$\frac{\Delta f}{\Delta x} = 9.25, \quad \frac{\Delta f}{\Delta y} = 2.5, \quad \frac{\Delta g}{\Delta x} = 2, \quad \frac{\Delta g}{\Delta y} = 9.75.$$

---

[2] Because of the danger of cancellation, however, h must not be chosen too small. (Editors' remark)

The system of equations for the first corrections therefore becomes:

Δx	Δy	1
9.25	2.5	−2
2	9.75	2

=> $\Delta x = .28760$
     $\Delta y = -.26412$

The corrected values are $x = -1.71240$, $y = .73588$ and give $f = .520$, $g = .653$. With these, the iteration would then be continued.

§4.2. Newton's method

For an equation $f(x) = 0$ in one unknown, the vector $\mathbf{f}$ reduces to a scalar f and the matrix $\mathbf{F}$ to a scalar $f'(x)$. The system of equations (3) consists only of one equation,

$$f'(x)\,\Delta x + f(x) = 0,$$

and has the solution $\Delta x = -f(x)/f'(x)$. This is *Newton's method*: starting with a suitable initial value x_0, one determines a sequence $x_1, x_2, x_3, \ldots$ in accordance with

$$x_{k+1} = x_k - \frac{f(x_k)}{f'(x_k)}, \quad k = 0,1,2, \ldots . \tag{4}$$

Example. In the case of the equation $x^2 - 2 = 0$, the recursion formula (4) reads

$$x_{k+1} = x_k - \frac{x_k^2 - 2}{2x_k}.$$

Starting with $x_0 = 1$, one obtains successively

$$x_1 = 1 - (-1)/2 = \underline{1}.5,$$
$$x_2 = 1.5 - .25/3 = \underline{1.41}66667,$$
$$x_3 = \underline{1.4142}157,$$
$$x_4 = \underline{1.4142135623747}.$$

(The correct digits are underlined.) The convergence, here, is obviously quite fast.

Geometrically, Newton's method can be interpreted very simply: linearization here means replacing the curve of $f(x)$ by its tangent at the point $(x_k, f(x_k))$; then, the "zero" of this tangent is determined (see Fig. 4.1). From the point x_{k+1} one proceeds in the same way, etc.

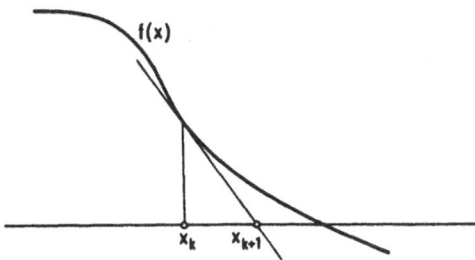

Figure 4.1. *Newton's method.*

We now examine the convergence of the method at least locally, that is, assuming that one is already near the zero. Let the zero be denoted by s; then (provided that $f(x)$ can be expanded in a Taylor series at s)[1]

$$f(x_k) = (x_k - s)f'(s) + \frac{(x_k - s)^2}{2} f''(s) + \ldots ,$$

$$f'(x_k) = f'(s) + (x_k - s)f''(s) + \ldots ,$$

hence

$$x_{k+1} \approx x_k - \frac{(x_k - s)f'(s) + \frac{1}{2}(x_k - s)^2 f''(s)}{f'(s) + (x_k - s)f''(s)} ,$$

and therefore

$$x_{k+1} - s \approx x_k - s - (x_k - s)\frac{f'(s) + \frac{1}{2}(x_k - s)f''(s)}{f'(s) + (x_k - s)f''(s)}$$

$$= \frac{\frac{1}{2}(x_k - s)^2 f''(s)}{f'(s) + (x_k - s)f''(s)} .$$

---

[1] A simpler derivation can be found in Björck Å., Dahlquist G: *Numerical Methods*, Prentice-Hall, Englewood-Cliffs, N.J., 1974. (Editors' remark)

Consequently, for small $x_k - s$, one has in first approximation

$$x_{k+1} - s \approx (x_k - s)^2 \; \frac{f''(s)}{2f'(s)} \; .$$

This is the asymptotic error law for Newton's method; it says that the error in each step is essentially *squared*, provided that $f'(s) \neq 0$, i.e., the zero s is simple. This is referred to as *quadratic convergence*.

In the preceding example we had $f(x) = x^2 - 2$, $f'(s) = 2s$, $f''(s) = 2$, $s = \sqrt{2}$; therefore,

$$x_{k+1} - \sqrt{2} \approx \frac{1}{2\sqrt{2}} \; (x_k - \sqrt{2})^2 \; .$$

In addition, we had $x_2 - \sqrt{2} \approx .0024531$, from which, by repeated application of this formula, one obtains the approximations

$$x_3 - \sqrt{2} \approx 2.1276_{10}-6,$$

$$x_4 - \sqrt{2} \approx 1.6_{10}-12,$$

$$x_5 - \sqrt{2} \approx .9_{10}-24$$

for the successive errors.

Naturally, if $f''(s)/2f'(s))$ is large, it takes a long time until quadratic convergence "takes hold", if it ever is achieved, which is by no means guaranteed.

On the other hand, it can happen that $f'(s) \neq 0$, $f''(s) = 0$ at the point s; then we obtain even cubic convergence, i.e., an error law

$$x_{k+1} - s \approx c(x_k - s)^3$$

(where c is a constant which depends on $f'(s)$ and $f'''(s)$).

§4.3. The regula falsi

If only real roots of the equation $f(x) = 0$ are desired, the *regula falsi* is quite suitable; for automatic computation, however, it must be modified([1]). The advantage of the method is that no derivatives need be computed and that a high reliability is achieved.

---

[1] The disadvantages of the classical regula falsi (in which at every step the secant is drawn between two function values with opposite signs) can be seen in the example $f(x) = 1 - x^x = 0$, $a = .1$, $b = 10$: In 100 steps the interval $[a,b]$ shrinks only to $[.10000002,10]$. The modified version given here, in contrast, produces the solution $x = 1$

For *initialization* one computes the values $f(x_k)$ at a sequence of points $x_0, x_1, x_2, \ldots$ (say, at $x_k = x_0 + kh$ with constant h). As soon as one finds a sign change in these function values, for example $f(x_k) > 0$, $f(x_{k+1}) < 0$, one switches over to a *bracketing procedure* for the zero:

Denote x_k by a, and x_{k+1} by b; then $f(a) > 0$, $f(b) < 0$, and we compute

$$c = \frac{af(b) - bf(a)}{f(b) - f(a)} \tag{5}$$

(the denominator is not 0), as well as $f(c)$. Geometrically, c can be interpreted as the "zero" of the secant from $(a, f(a))$ to $(b, f(b))$ (cf. Fig. 4.2). If now, for example, $f(c) > 0$, there lies a zero between b and c. However, we seek yet another point d between b and c with $f(d) < 0$. The first trial is made with $d = (b + c)/2$; if this still yields $f(d) > 0$, one chooses a new $c := d$, $d := (b + d)/2$ and repeats this (in the figure unnecessary) bisection until $f(d) < 0$. Then one sets $a := c$, $b := d$ and, as above, determines a new c according to formula (5), etc.

This algorithm can be programmed in the form of the following procedure *regfal*. It assumes that two points a,b are already known with $f(a) > 0, f(b) < 0$ or $f(a) < 0, f(b) > 0$.

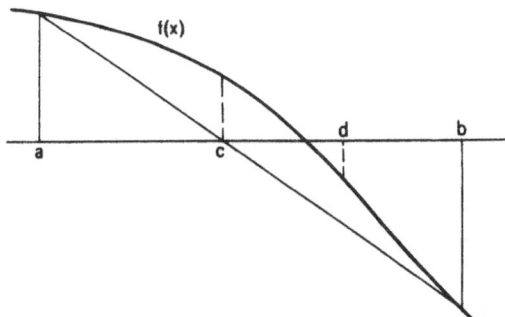

Figure 4.2. *The regula falsi.*

in 12 steps (24 evaluations of $f(x)$). A similar modification, converging more rapidly, asymptotically, is the Illinois algorithm; see Wilkes M.V., Wheeler D.J., Gill S.: *The Preparation of Programs for an Electronic Computer*, Addison-Wesley, Reading, Mass. 1951, and Anderson N., Björck Å.: A new high order method of regula falsi type for computing a root of an equation, *BIT* **13**, 253–264 (1973). (Editors' remark)

real procedure *regfal(a,b,f)*;
    value *a,b*;
    real *a,b*; **real procedure** *f*;
    begin
        real fa, fb, fc, fd, c, d;
        $fa := f(a)$; $fb := f(b)$;
        if $fa < 0$ **then**
        begin
            $c := a$; $fc := fa$; $a := b$; $fa := fb$; $b := c$; $fb := fc$
        end;
start:
        $c := (b \times fa - a \times fb)/(fa - fb)$;
        $fc := f(c)$;
        if $fc = 0 \wedge (c \geq a \wedge c \geq b) \wedge (c \leq a \wedge c \leq b)$ **then goto** *ex*;[2]
        if $fc > 0$ **then**
        begin
            $d := (b + c)/2$;
            $fd := f(d)$;
ld:
            if $fd > 0$ **then**
            begin
                if $c = d$ **then goto** *ex*;
                $c := d$; $fc := fd$;
                $d := (d + b)/2$;
                $fd := f(d)$;
                goto *ld*;
            end
        end
        else
        begin
            $d := c$;
            $fd := fc$;
            $c := (a + d)/2$;
            $fc := f(c)$;
lc:
            if $fc < 0$ **then**

---

[2] Note that the stopping rule used here is programmed machine-independently and leads automatically – except in extreme special cases – to the maximum attainable accuracy. (Editors' remark)

```
    begin
        if c = d then goto ex;
        d := c;  fd := fc;
        c := (a + c)/2;
        fc := f (c);
        goto lc;
    end
end;
a := c;  fa := fc;  b := d;  fb := fd;
goto start;
ex:
    regfal := c;
end regfal;
```

Example. We return to the equation $f(x) = x^2 - 2 = 0$, for which $a = 2, b = 1$ are admissible initial values. The subsequent course of computation is reproduced in Table 4.1. (The values of $c, d, f(c)$ are always rounded to 4 digits after the decimal point. Numbers in parentheses were obtained as a result of storage transfers.)

Table 4.1. *Regula falsi for* $x^2 - 2 = 0$

a	$f(a)$	b	$f(b)$	c	$f(c)$	d	$f(d)$
2.0	2.0	1.0	−1.0				
				1.3333	−.2223	(1.3333)	(−.2223)
				1.6667	.7779		
(1.6667)	(.7779)	(1.3333)	(−.2223)				
				1.4074	−.0192	(1.4074)	(−.0192)
				1.5370	.3624		
(1.5370)	(.3624)	(1.4074)	(−.0192)				
				1.4139	−.0009	(1.4139)	(−.0009)
				1.4754	.1768		
(1.4754)	(.1768)	(1.4139)	(−.0009)				
				1.4142	.0000		

§4.4. Algebraic equations

A particular class of equations in one unknown are algebraic equations,

$$\sum_{k=0}^{n} c_k z^k = 0, \quad \text{with} \quad c_0 \neq 0, \quad c_n \neq 0, \qquad (6)$$

as they have (in the domain of complex numbers) exactly n solutions z_1, $z_2, \ldots, z_n$, all of which are usually sought. It is necessary, then, to compute also in the complex domain.

There are, however, some questions that need to be raised in the case of algebraic equations which for more general equations are perhaps less relevant. These concern *the purpose which the solutions $z_1, \ldots, z_n$ are expected to serve, and the origin of the coefficients c_k.*

First of all, one must realize that the roots of an algebraic equation are poorly defined through the coefficients; small changes in the coefficients can cause large changes in the roots. This fact prompts many people to compute the roots in double precision. But then they only obtain accurate roots to an inaccurate equation, which is not of much help. A lot more important is to make sure that after the computation of all root approximations $s_1, s_2, \ldots, s_n$ the product $c_n(z - s_1)(z - s_2)$ $\cdots (z - s_n)$ agrees with $\sum c_k z^k$ as much as can be expected within the machine precision. To demand more is not reasonable, but that much at least can be achieved.

Example. In solving the equation

$$z^4 - 4z^3 + 6z^2 - 4z + 1 = 0,$$

the four-fold root at $z = 1$ certainly gives difficulties, inasmuch as only a root accuracy of one fourth of the total number of digits can be expected (thus 5-digit root accuracy in 20-digit computation).

We consider two computers:

A computes with 16 digits and thus obtains roots accurate to 4 digits:

$$1.00005, \quad 1, \quad .9999, \quad .99995.$$

B computes with only 8 digits and therefore must expect errors of the order of magnitude .01; he obtains

$$1.01, \quad .99, \quad 1+.01i, \quad 1-.01i.$$

Which are now the better results? The products of the linear factors in case A produces (up to errors which are smaller than 10^{-8})

$$z^4 - 3.9999\,z^3 + 5.9997z^2 - 3.9997z + .9999,$$

and in case B (exactly)

$$z^4 - 4z^3 + 6z^2 - 4z + .99999999.$$

We must therefore regard the results of B, in a certain sense, as the better ones; the way this was achieved, in spite of the lower computing precision, was that B had been mindful of suitably correlating the errors, while A computed the roots independently from each other, without attempting to do anything beyond that.

The crucial device for correlating the errors in the roots, which may result in the smallest possible reconstruction errors, is *deflation*: If z_1 is an exact root of the polynomial $f(z)$ in (6), then

$$f_1(z) = \frac{f(z)}{z - z_1}$$

is a polynomial of degree $n - 1$, from which the remaining $n - 1$ roots can be determined.

Now, however, one has computed only an approximate root $s_1 \approx z_1$, for which one knows only that

$$f(s_1) \approx \theta \sum_{k=0}^{n} |c_k s_1^k|,$$

where θ denotes a unit in the last position of the mantissa. To make $f(s_1)$ smaller is not possible, in general, since the sum of terms $c_k z^k$ is affected with an error which somehow is approximately proportional to $\sum |c_k z^k|$. The division $f(z)/(z - s_1)$, which, as is well known, is carried out by means of the Horner scheme([1]), thus produces not only an inaccurate $f_1(z)$, but also a remainder $f(s_1)$. Indeed, what the Horner scheme

---

[1] The (simple) Horner scheme allows one to compute the polynomial value $f(s_1)$ and the coefficients of the polynomial f_1 defined by (7). Putting

$$f(z) = \sum_{k=0}^{n} a_{n-k} z^k, \quad f_1(z) = \sum_{k=0}^{n-1} b_{n-1-k} z^k, \quad f(s_1) = b_n,$$

does is nothing but the decomposition

$$f(z) = f(s_1) + (z - s_1)f_1(z), \qquad (7)$$

while the reconstruction later produces

$$f^*(z) = (z - s_1)f_1^*(z)$$

(where f_1^* denotes the already reconstructed polynomial corresponding to f_1). Therefore, one obtains as reconstruction error

$$\delta f(z) = f(z) - f^*(z)$$
$$= f(s_1) + (z - s_1)\delta f_1(z) + \text{errors arising in the}$$
$$\text{performance of the multiplication } (z - s_1)f_1^*(z).$$

Apart from the third term, which has the order of magnitude of rounding errors, one commits during the first deflation the reconstruction error $f(s_1)$, which however falsifies only the constant term. If, in particular, s_1 is very small, then $|f(s_1)| \approx \theta \sum |c_k s_1^k| \approx \theta |c_0|$, that is, the reconstruction error, as far as it comes from $f(s_1)$, is small compared to c_0. (The first term, after all, affects only the constant term.) One has therefore established the rule that the absolutely smallest root of an algebraic equation should always be determined first; then one carries out the deflation, and afterwards the absolutely smallest root of $f_1(z)$ is computed, etc.

Example. Substituting $s_1 = .0026$ in the left-hand side of the equation (cf. §1.3)

$$z^2 - 742z + 2 = 0,$$

the Horner scheme becomes (in 6-digit computation):

---

it has the form

a_0	a_1	$\cdots$	a_{n-2}	a_{n-1}	a_n
b_0	b_1	$\cdots$	b_{n-2}	b_{n-1}	b_n

Computational rule for the usual construction from left to right (comparison of coefficients in (7)):

$$b_0 = a_0, \quad b_k = a_k + s_1 b_{k-1} \quad (k = 1, \ldots, n).$$

If $b_n = f(s_1)$ is already known (e.g. $f(s_1) = 0$), then the scheme can be built up from the right:

$$b_k = (b_{k+1} - a_{k+1})/s_1 \quad (k = n - 1, \ldots, 0).$$

(Editors' remark)

1	−742	2
1	−741.997	.07081

One therefore finds $f(s_1) = .07081$ and $s_2 = 741.997$, which is exact to 6 digits, even though we had $s_1 \ne z_1 = .00269543. \ldots$.

If, on the other hand, one starts with the deflation of $s_1 = 741.997$, one obtains the Horner scheme

1	−742	2
1	−.003	−.22599

and, with $f(s_1) = -.22599$, a substantially larger reconstruction error. Besides, $s_2 = .003$ is a poor approximation for the second root, even though s_1 was exact to 6 digits.

Unfortunately, the stated rule by no means guarantees that the reconstruction errors remain small, as is shown in the following example of an algebraic equation of degree 10, for which the Horner scheme for the deflation of $s_1 = .951$ in 3-digit computation looks as follows:

1	8.1	27.8	50.8	47.6	7	−36.4	−45.2	−26.2	−7.9	−1
1	9.05	36.4	85.4	129	130	87.6	38.1	10.0	1.61	.53

Thus the reconstruction error here amounts to .53, with the constant term being −1, which certainly lies no longer within the computing precision. In order to obtain better results, the Horner scheme must be built up also from right to left, putting first 0 in place of .53 and then running the computation backwards:

1	8.1	27.8	50.8	47.6	7	−36.4	−45.2	−26.2	−7.9	−1
−0.116	7.99	35.4	84.5	128	129	86.9	37.4	9.41	1.05	0

If one replaces 87.6 in the first scheme by 86.9 in the second, and further to the right uses the values of the second scheme, one obtains

$$f_1(z) = z^9 + 9.05z^8 + 36.4z^7 + 85.4.z^6 + 129z^5 + 130z^4$$
$$+ 86.9z^3 + 37.4z^2 + 9.41z + 1.05$$

and thereby commits a reconstruction error $.7z^4$, which however for the larger coefficient $c_4 = -36.4$ is tolerable (actually, .7 ought to be compared even with 86.9).

§4.5. Root squaring (Dandelin-Graeffe)

Let $f(z)$ be a polynomial of degree n. Then $f(z)f(-z)$ is a polynomial of degree $2n$ and moreover an even function, so that the odd powers of z cannot occur. For example, $f(z) = z^2 - 3z + 1$ produces the polynomial $f(z)f(-z) = z^4 - 7z^2 + 1$. Hence, $f_1(z^2) = f(z)f(-z)$ is a polynomial of degree n in the variable z^2. If s is a root of $f(z) = 0$, then $f_1(s^2) = 0$, thus s^2 is a zero of $f_1(z)$. If one next forms $f_2(z^2) = f_1(z)f_1(-z)$, one obtains a new polynomial $f_2(z)$, which again has degree n and whose zeros are the fourth powers of the zeros of $f(z)$, etc.

If, for example, we start with $f(z) = z^2 - z - 1$, we can form in this way successively:

		z^2	z	1
$f(z)$	$=$	1	-1	-1
$f(-z)$	$=$	1	1	-1
$f_1(z)$	$=$	1	-3	1
$f_1(-z)$	$=$	1	3	1
$f_2(z)$	$=$	1	-7	1
$f_2(-z)$	$=$	1	7	1
$f_3(z)$	$=$	1	-47	1
$f_3(-z)$	$=$	1	47	1
$f_4(z)$	$=$	1	-2207	1

$f_4(z)$ (in 6-digit computation) evidently has the zeros $s_1 = 2207$ and $s_2 = 1/2207$; the zeros are torn apart so much that they can be read off directly as quotients of the coefficients([1]).

Now the roots of the original equation are the 16th roots of 2207 and 1/2207, respectively (one has squared four times), thus 1.618034 and .618034; but there are 16 different 16th roots, and all would now have to be examined whether they also satisfy the original equation. In our case, the solutions are 1.618034 and $-.618034$.

---

[1] Plausibility argument: as is well known,

$$f(z) = z^n - \sigma_1 z^{n-1} + \sigma_2 z^{n-2} - \cdots + (-1)^n \sigma_n,$$

where $\sigma_1, \ldots, \sigma_n$ are the elementary symmetric functions of the zeros $z_1, \ldots, z_n$:

$$\sigma_k = \sum_{i_1 < i_2 < \cdots < i_k} z_{i_1} z_{i_2} \cdots z_{i_k}.$$

If now $|z_1| \gg |z_2| \gg \cdots \gg |z_n|$, then $\sigma_k \approx z_1 z_2 \cdots z_k$, $\sigma_k/\sigma_{k-1} \approx z_k$. (Editors' remark)

Let us consider a further

Example. In order to solve the algebraic equation $z^3 - 9z^2 - 8z + 2 = 0$, one forms:

		z^3	z^2	z	1
$f(z)$	=	1	−9	−8	2
$f(-z)$	=	−1	−9	8	2
$f_1(z)$	=	−1	97	−100	4
$f_1(-z)$	=	1	97	100	4
$f_2(z)$	=	−1	9209	−9224	16
$f_2(-z)$	=	1	9209	9224	16
$f_3(z)$	=	1	84787233	−84787488	256

For the polynomial $f_3(z)$ one now computes the negative quotients $-c_k/c_{k+1}$ $(k = 2,1,0)$ of successive coefficients and subsequently their (positive) 8th root. Here, these roots are

$$9.795832, \quad 1.0000005, \quad .2041684.$$

Provided one still supplies them with the correct argument, they agree very well with the exact roots

$$-1, \quad 5 + \sqrt{23} = 9.795832 \ldots, \quad 5 - \sqrt{23} = .2041685 \ldots.$$

This method of root squaring, however, suffers from serious drawbacks:

1) Repeated squaring produces such large (or extremely small) numbers that one has to worry about over- or underflow.

2) One obtains only the 2^pth powers of the zeros (if $f_p(z)$ is used) and must therefore still take roots. After that, one has to decide which of the 2^p root values is the correct one, i.e., a zero of $f(z)$.

3) Conjugate complex pairs of roots cause difficulties.

Only through extremely complicated programming can these drawbacks be overcome([2]). The method, therefore, is not used frequently. It can serve, however, as a stopgap for other methods.

---

[2] For a variant of the Graeffe method in which the occurrence of very large and very small numbers is avoided, see Grau A.A.: On the reduction of number range in the use of the Graeffe process, *J. Assoc. Comput. Mach.* **10**, 538–544 (1963).

§4.6.　Application of Newton's method to algebraic equations

First of all, we try to locate the roots of the given equation (6) in the complex plane. One has, in this connection:

Theorem 4.1. *All n solutions of* (6) *lie in the circle*

$$|z| \leq \rho = 2 \max_{1 \leq k \leq n} \left| \frac{c_{n-k}}{c_n} \right|^{1/k}.$$

Proof. If for all k

$$|z| > 2 \left| \frac{c_{n-k}}{c_n} \right|^{1/k},$$

then (also for all k)

$$2^{-k} |z|^k > \left| \frac{c_{n-k}}{c_n} \right|,$$

$$2^{-k} |c_n| \, |z|^n > |c_{n-k}| \, |z|^{n-k}.$$

There follows

$$\left| \sum_{k=0}^{n} c_k z^k \right| \geq |c_n z^n| - \sum_{k=1}^{n} |c_{n-k}| \, |z|^{n-k}$$

$$\geq |c_n z^n| - \sum_{k=1}^{n} 2^{-k} |c_n| \, |z|^n$$

$$\geq |c_n z^n| \left\{ 1 - \sum_{k=1}^{n} 2^{-k} \right\} > 0,$$

i.e., z cannot be a solution, q.e.d.

This theorem serves as a basis for a simple recipe for using Newton's method to at least come close to a root:

One chooses at random an initial point z_0 on the circle $|z| = \rho$ and generates a sequence of complex numbers $z_1, z_2, z_3, \ldots$ according to the formula of Newton's method:

$$z_{\ell+1} = z_\ell - \frac{f(z_\ell)}{f'(z_\ell)}, \quad \ell = 0, 1, 2, \ldots. \tag{8}$$

This is continued as long as the modulus $|f(z_\ell)|$ of the polynomial is reduced to less than half its value. As soon as this no longer holds, two possible cases are to be distinguished:

a) $|f(z_t)|$ is already so small, that it is seriously affected by rounding errors; this occurs when

$$\left| \sum_{k=0}^{n} c_k z^k \right| \left\{ \sum_{k=0}^{n} |c_k z^k| \right\}^{-1} \ll 1$$

(thus, for example, when the quantity on the left has the order of magnitude of 10 units in the last position of the mantissa).

b) z_t lies near a zero of the derivative $f'(z)$ of the polynomial.

In Case a) one stops, in Case b) one can try to start afresh with another point on the circle $|z| = \rho$.

A more reliable method, however, consists in re-expanding the polynomial $f(z)$ in a Taylor series about the last computed point (with the absolutely smallest $|f(z)|$):

$$f(z_t + w) = \sum_{k=0}^{n} d_k w^k = g(w), \tag{9}$$

and then in applying Newton's method, beginning with $w = 0$, to

$$g_1(w) = g(\sqrt{w})g(-\sqrt{w}), \tag{10}$$

for as long as the function value $|g(z)|$ is halved at each step([1]). One then has again the alternatives a) and b). In Case a), a zero has been found, and it can be removed (deflation). In Case b), one proceeds with

$$g_2(w) = g_1(\sqrt{w})g_1(-\sqrt{w}), \quad \text{etc. } ([2]).$$

---

[1] The transition from g to g_1 is Graeffe's root squaring (cf. §4.5). If $g'(0) = 0$ (or small), but $g''(0) \neq 0$, then $g_1'(0) \neq 0$, and one can apply Newton's method to g_1. (Editors' remark)

[2] The convergence of the procedure described here has not yet been investigated. Tests, however, show that global convergence is not achieved. (Editors' remark)

Example. For the equation

$$z^5 + 1000z^2 + 1000 = 0 \tag{11}$$

one finds

$$\left| \frac{c_4}{c_5} \right| = 0, \quad \left| \frac{c_3}{c_5} \right|^{1/2} = 0, \quad \left| \frac{c_2}{c_5} \right|^{1/3} = 0,$$

$$\left| \frac{c_1}{c_5} \right|^{1/4} = 0, \quad \left| \frac{c_0}{c_5} \right|^{1/5} \approx 4, \quad \text{thus} \quad \rho = 20.$$

(It is to be noted that one could almost always find a smaller ρ, here, for example, $\rho = 11$.) Starting with $z_0 = 20i$, Newton's method in the first 6 steps produces the points shown in Table 4.2. Thus, in the last step, $|f(z)|$ even grows, which is connected with the fact that $|f'(z_5)|$ is small ($f'(z) = 0$ for $z \approx 3.68 + 6.36i$). But already in the 5th step, $|f(z)|$ has no longer been halved, so that one could have saved oneself the last step. We have indeed argued, in this situation, to re-expand $f(z)$ in a Taylor series with origin at z_5. For simplicity we make the new development at the point $z = 4 + 6i$:

$$\begin{aligned} g(w) = {}& f(z+w) \\ = {}& -15096 + 28896i + (-1520 + 2400i)w \\ & + (-2680 + 720i)w^2 + (-200 + 480i)w^3 \\ & + (20 + 30i)w^4 + w^5. \end{aligned}$$

Then one gets

$$\begin{aligned} g(w) = {}& -607089600 - 872428032i + (42753920 - 169324800i)w \\ & + (6022400 - 1189920i)w^2 + (43040 + 55200i)w^3 \\ & + (-100 + 240i)w^4 - w^5. \end{aligned}$$

Table 4.2. *Application of Newton's method to the algebraic equation* (11)

z_ℓ	$|f(z_\ell)|$
20.000i	3224779
.298 + 15.985i	1061312
.704 + 12.760i	352628
1.298 + 10.146i	121198
2.245 + 7.956i	47382
4.250 + 5.740i	33618
−2.750 + 6.056i	56131

This yields immediately $w_1 = - g_1(0)/g_1'(0) = -3.992560 + 4.593463i$. If only this first approximation to $g_1(w) = 0$ is used, one already obtains $\sqrt{w_1} = \pm (1.023114 + 2.244844i)$ as Newton correction for $g(w)$, and with it the approximation $z = 5.023114 + 8.244844i$ for the desired solution of $f(z) = 0$. (The other root gives nothing useful.) Now already $|f(z)| \approx 10389$, which, compared with $|f(4 + 6i)| \approx 32602$, is reduced to less than a third. From here on, the method converges rapidly (in 4 steps, with 14-digit precision) towards the solution $z = 5.017003 + 8.631391i$.

Notes to Chapter 4

§4.1 The "method of linearization" discussed in this section is often referred to in the literature as Newton's method for systems of nonlinear equations. It is a natural generalization of Newton's method for a single equation (see §4.2) and, in fact, can be extended to equations in infinite-dimensional (function) spaces. The first such generalization was done in the context of nonlinear operator equations in Banach spaces by L.V. Kantorovich in 1948 (see Kantorovich & Akilov [1964]). This generalization is often called the *Newton-Kantorovich method*. Another important generalization was given by J. Moser [1961] for the case of operators acting on a continuous scale of Banach spaces with properties similar to the properties of Sobolev spaces. The above generalizations provide useful tools in the study of the solution of nonlinear differential and integral equations. For a recent analysis of the Newton-Kantorovich and Newton-Moser methods, see Potra & Ptak [1984].

The principal difficulty with Newton's method is its local character of convergence: the initial approximation has to be sufficiently close to the desired solution for convergence to take place. In practice, therefore, the initial phase of Newton's method, or indeed the entire iteration process, is modified to make sure that initially the approximations move closer to the solution. One might insist, for example, that the functions in question decrease in some suitable norm. There are many ways to do this, which guarantee convergence even if the initial approximation is far away from the desired solution. Some of these modified methods, in fact, automatically turn into Newton's method in the vicinity of the solution, thus sharing with Newton's method quadratic convergence, but unlike Newton's method, possess qualities of global convergence. It is also possible to dispense with derivative evaluations and build up the required matrix of derivatives gradually from information gained during the iteration. Such methods are usually described in the context of optimization problems; for example, to minimize $f^T(x)f(x)$, which is equivalent to $f(x) = 0$, if f and x are of the same dimension and solutions are known to exist. For a discussion of such "Newton-like" methods, and other methods that have proven effective in practice, the reader is referred to Gill, Murray & Wright [1981], Dennis & Schnabel [1983], Fletcher [1987]. Among software packages for solving systems of nonlinear equations we mention MINPACK-1 (Moré, Garbow & Hillstrom [1980]), which implements a modification of Powell's hybrid method (Powell [1970]). The *hybrid method* is a variation of Newton's method which takes precautions to avoid large steps or increasing residuals. The subroutine HYBRD1 of MINPACK uses a finite difference approximation of the Jacobian (a sort of "derivative-free linearization" like the one described in §4.1), while HYBRDJ employs a user-supplied Jacobian. The subroutine SNSQE described in Kahaner, Moler & Nash [1989] is an easy-to-use combination of both subroutines above. The IMSL subroutines NEQNF and NEQNJ are also based on the MINPACK routines (see IMSL [1987, Vol. 2]).

Other methods with global convergence properties have been developed by using *continuation algorithms*. In this approach one considers a family of equations depending continuously on a parameter t belonging to the interval [0,1]. The equation corresponding to $t = 0$ has a known solution, while the equation corresponding to $t = 1$ is the equation whose solution is sought. The problem then is to construct an increasing sequence of parameters so that the solution of the equation corresponding to a parameter of this sequence is a good starting point for an iterative method to solve the equation corresponding to the next parameter in the sequence. A portable software implementation of this approach is available (Watson, Billups & Morgan [1987]).

§4.2 Newton first applied his iterative method in 1669 for solving a cubic equation. The procedure was systematically discussed in print by J. Raphson as early as 1690. Therefore, the method is sometimes referred to as the *Newton-Raphson method*. For more details on the history of Newton's method, see Goldstine [1977] and Ostrowski [1973].

By contrast with the regula falsi described in §4.3, Newton's method does not produce a convergent sequence of nested intervals containing the solution. However, for convex functions, this can be accomplished by using Fourier's modification of Newton's method (see Ostrowski [1973, Ch. 9]). For nonconvex functions, one may use interval arithmetic and some interval variants of Newton's method in order to construct such a sequence of nested intervals (see Alefeld & Herzberger [1983]).

The notions of quadratic and cubic convergence introduced in §4.2 can be generalized as follows. Letting e_k denote the distance between the kth term of a convergent sequence and its limit, the *q-order of convergence* of the sequence is defined as the limit $m = \lim \inf [\log e_{k+1} / \log e_k]$, whenever this limit is greater than one. If e_{k+1} is proportional to the mth power of e_k, then the q-order of convergence is obviously equal to m, but the above definition uniquely defines the q-order of a sequence in much more general situations. One says also that the sequence is *q-superlinearly convergent* if lim sup $[e_{k+1} / e_k] = 0$. If $m > 1$, then the sequence is q-superlinearly convergent, but the converse is not true. Finally, one says that the sequence is q-linearly convergent if $0 < \lim \sup[e_{k+1} / e_k] < 1$. The speed of convergence of sequences can also be measured by their *r-orders of convergence*. For the definition of the r-order and its relationship with the q-order, see Ortega & Rheinboldt [1970], Potra [1989].

§4.3 The regula falsi originates in medieval Arabic mathematics, perhaps even earlier in China (see Maas [1985]). Leonardo Pisano, alias Fibonacci, in the early 13th century calls it "regula duarum falsarum positionum" (rule of two false positions). It received this strange name, since for linear equations (a problem in the forefront of medieval arithmetic!) the method produces from two approximations ("false positions") the exact root by linear interpolation. Peter Bienewitz (1527) explains it thus (cf. Maas [1985, pp. 312–313]): "Vnd heisst nit darum falsi dass sie falsch vnd vnrecht wehr, sunder, dass sie auss zweyen falschen vnd vnwahrhaftigen zalen, vnd zweyen lügen die wahrhaftige vnd begehrte zal finden lernt".

In its original form, in which at every step the secant is drawn between two function values of opposite signs, the regula falsi is only linearly convergent. By taking a and b in (5) to be the latest two iterates, even if f does not change sign at those points, one obtains the so-called *secant method*. The q-order of convergence of this method is $(1 + \sqrt{5})/2 = 1.618 \ldots$. Because it requires only one function evaluation per iteration, its numerical efficiency is ultimately higher than that of Newton's method (see Ostrowski [1973]).

There are a great number of methods that have been proposed for solving single equations in one unknown. Many of them combine bisection and interpolation devices with various safeguarding measures designed not only to guarantee convergence, but also to yield fast convergence in cases of well-behaved equations, and at least the speed of bisection in other more difficult cases. A thorough study of some such methods can be found in Brent [1973]. One of the first methods of this type, originally published by Dekker [1969], is incorporated in the subroutine FZERO described in Kahaner, Moler & Nash [1989]. The IMSL subroutine ZBREN (cf. IMSL [1987, Vol. 2]) is based on Brent's improvement of Dekker's algorithm (Brent [1973]), which is a combination of linear interpolation, inverse quadratic interpolation and bisection. A Fortran implementation of Brent's method, the real function ZEROIN, can be found in Forsythe, Malcolm & Moler [1977]. All subroutines above find a zero of a function in a given interval that has to be specified by the user. Some popular subroutines which do not require the prescription of such an interval are based on *Muller's method* (cf. Muller [1956]). Such is the IMSL subroutine ZREAL (IMSL [1987, Vol. 2]).

While there is basically a unique generalization of Newton's method for şolving systems of nonlinear equations, this is no longer the case for the secant method. For the nonlinear system $\mathbf{f}(\mathbf{x}) = \mathbf{0}$ of (1), the generalization of Newton's method described in §4.1 is based upon locally approximating the mapping $\mathbf{f}(\mathbf{x})$ by the affine mapping $\mathbf{f}(\mathbf{x}_k) + \mathbf{A}(\mathbf{x} - \mathbf{x}_k)$, where $\mathbf{A}$ is the Jacobian of $\mathbf{f}$ at $\mathbf{x}_k$. The secant method could be generalized by considering a similar affine approximation, but where this time the matrix $\mathbf{A}$ should satisfy the "secant condition" $\mathbf{A}(\mathbf{x}_k - \mathbf{x}_{k-1}) = \mathbf{f}(\mathbf{x}_k) - \mathbf{f}(\mathbf{x}_{k-1})$. This condition, however, does not uniquely determine the matrix $\mathbf{A}$ (except when $n=1$). One way of determining the matrix $\mathbf{A}$ was proposed by Schmidt [1963] and led to a generalization of the secant method that, in the general case, has the same r-order of convergence as in the one-dimensional case. Nevertheless, this method is rather expensive and sometimes computationally unstable. A more efficient generalization of the secant method has been proposed by Broyden, who computes the matrix A at each step via a rank-one update (see Dennis & Schnabel [1983, Ch. 8]). The nonlinear systems arising in convex optimization problems have symmetric positive definite Jacobians, and in such cases the matrix A should also be symmetric positive definite. This can be accomplished by various rank-two updates. One of the most successful generalizations of the secant method is based on the *BFGS update*, independently discovered by Broyden, Fletcher, Goldfarb and Shanno in 1970 (see Dennis & Schnabel [1983, Ch. 9]). Both Broyden's method and the BFGS method are q-superlinearly convergent.

§4.4 A quantitative discussion of the sensitivity of roots of algebraic equations to small perturbations in the coefficients is given in Wilkinson [1963, pp. 38ff]. One finds there, in particular, Wilkinson's famous example of an ill-conditioned equation, with roots at the integers 1, 2, . . . , 20. This is further discussed in Wilkinson [1984] and Gautschi [1984]. The cited book of Wilkinson is also a good source for the effects of rounding errors in polynomial evaluation, in Newton's method, and in polynomial deflation. For further practical remarks concerning the solution of polynomial equations, in particular for an analysis of forward and backward deflation, and a combination thereof, see Peters & Wilkinson [1971].

While Newton's method possesses some special properties when applied to algebraic equations (see, e.g., Stoer & Bulirsch [1980, §5.5]), it does not allow for the computation of complex roots from real starting values. A method that overcomes this deficiency is *Laguerre's method* (see, e.g., Fröberg [1985, §11.5]). This method has global convergence for real roots, local cubic convergence to a simple root, and local linear convergence to a multiple root. The IMSL subroutine ZPLRC is based on Laguerre's method, while the other IMSL subroutine (cf. IMSL [1987, Vol. 2]) for solving polynomial equations, ZPORC, is based on the *Jenkins-Traub three-stage algorithm* (cf. Jenkins & Traub [1970]).

§4.5 Wilkinson [1963, pp. 67ff] discusses stability aspects of the rootsquaring process in the presence of rounding errors. He makes the point that "squaring" a polynomial may in some cases result in a worsening of the condition of the polynomial (with respect to rootfinding), although, as a rule, one should expect the opposite to happen – a steady improvement of the condition.

References

Alefeld, G. and Herzberger, J. [1983]: *Introduction to Interval Computations*, Academic Press, New York.

Brent, R.P. [1973]: *Algorithms for Minimization Without Derivatives*, Prentice-Hall, Englewood Cliffs, N.J.

Dekker, T.J. [1969]: Finding a zero by means of successive linear interpolation, in *Constructive Aspects of the Fundamental Theorem of Algebra* (B. Dejon and P. Henrici, eds.), pp. 37–28. Wiley-Interscience, London.

Dennis, J.E., Jr. and Schnabel, R.B. [1983]: *Numerical Methods for Unconstrained Optimization and Nonlinear Equations*, Prentice-Hall, Englewood Cliffs, N. J.

Fletcher, R. [1987]: *Practical Methods of Optimization*, 2nd ed., Wiley, Chichester.

Forsythe, G.E., Malcolm, M.A. and Moler. C.B. [1977]: *Computer Methods for Mathematical Computations*, Prentice-Hall, Englewood Cliffs, N.J.

Fröberg, C.-E. [1985]: *Numerical Mathematics. Theory and Computer Applications*, Benjamin/Cummings, Menlo Park, California.

Gautschi, W. [1984]: Questions of numerical condition related to polynomials, in *Studies in Numerical Analysis* (G.H. Golub, ed.), pp. 140–177, Studies in Mathematics **24**, The Mathematical Association of America.

Gill, P.E., Murray, W. and Wright, M.H. [1981]: *Practical Optimization*, Academic Press, London.

Goldstine, H.H. [1977]: *A History of Numerical Analysis from the 16th Through the 19th Century*, Studies in the History of Mathematics and Physical Sciences **2**, Springer, New York.

IMSL [1987]: *Math/Library User's Manual*, Houston.

Jenkins, M.A. and Traub, J.F. [1970]: A three-stage variable-shift iteration for polynomial zeros and its relation to generalized Rayleigh iteration, *Numer. Math.* **14**, 252–263.

Kahaner, D., Moler, C. and Nash, S. [1989]: *Numerical Methods and Software*, Prentice-Hall, Englewood Cliffs, N.J.

Kantorovich, L. and Akilov, G.P. [1964]: *Functional Analysis in Normed Spaces*, International Series of Monographs in Pure and Applied Mathematics **46**, Macmillan, New York.

Maas, C. [1985]: Was ist das Falsche an der Regula Falsi? *Mitt. Math. Ges. Hamburg* **11**, H. 3, 311–317.

Moré, J.J., Garbow, B.S. and Hillstrom, K.E. [1980]: *User Guide for MINPACK-1*, Argonne National Laboratory, Report ANL-80-74.

Moser, J. [1961]: A new technique for the construction of solutions of nonlinear differential equations, *Proc. Nat. Acad. Sci. U.S.A.* **47**, 1824–1831.

Muller, D.E. [1956]: A method for solving algebraic equations using an automatic computer, *Math. Tables Aids Comput.* **10**, 208–215.

Ortega, J.M. and Rheinboldt, W.C. [1970]: *Iterative Solution of Nonlinear Equations in Several Variables* Academic Press, New York.

Ostrowski, A.M. [1973]: *Solution of Equations in Euclidean and Banach Spaces*, Pure and Applied Mathematics **9**, Academic Press, New York.

Peters, G. and Wilkinson, J.H. [1971]: Practical problems arising in the solution of polynomial equations, *J. Inst. Math. Appl.* **8**, 16–35.

Potra, F.A. [1989]: On q-order and r-order of convergence, *J. Optim. Theory Appl.* **63**, no. 3.

Potra, F.A. and Ptak, V. [1984]: *Nondiscrete Induction and Iterative Processes*, Pitman, Boston.

Powell, M.J.D. [1970]: A hybrid method for nonlinear equations, in *Numerical Methods for Nonlinear Algebraic Equations* (P. Rabinowitz, ed.), pp. 87–114. Gordon and Breach, London.

Schmidt, J.W. [1963]: Eine Übertragung der Regula Falsi auf Gleichungen in Banachräumen I, II, Z. *Angew. Math. Mech.* **43**, 1–8, 97–110.

Stoer, J. and Bulirsch, R. [1980]: *Introduction to Numerical Analysis*, Springer, New York.

Watson, L.T., Billups, S.C. and Morgan, A.P. [1987]: Algorithm 652 – HOMPACK: A suite of codes for globally convergent homotopy algorithms, *ACM Trans. Math. Software* **13**, 281–310.

Wilkinson, J.H. [1963]: *Rounding Errors in Algebraic Processes*, Prentice-Hall, Englewood Cliffs, N. J.

Wilkinson, J.H. [1984]: The perfidious polynomial, in *Studies in Numerical Analysis* (G.H. Golub, ed.), pp. 1–28, Studies in Mathematics **24**, The Mathematical Association of America.

CHAPTER 5

Least Squares Problems

§5.1. Nonlinear least squares problems

We consider once again a system of nonlinear equations

$$f_1(x_1, x_2, \ldots, x_p) = 0$$
$$f_2(x_1, x_2, \ldots, x_p) = 0$$

$$\cdot$$
$$\cdot$$
$$\cdot$$

$$f_n(x_1, x_2, \ldots, x_p) = 0,$$

but now assume that the number n of equations is larger than the number p of unknowns.

If, for example, the system

$$x \ + y \ = 1$$
$$x^2 + y^2 = .8$$
$$x^3 + y^3 = .68$$

is to be solved, one must note that this is an impossible task, since from the first two equations there follows immediately $xy = .1$, thus

$$x = \frac{1 \pm \sqrt{.6}}{2}, \quad y = \frac{1 \pm \sqrt{.6}}{2} .$$

But then,

$$x^3 + y^3 = (x + y)^3 - 3xy(x + y) = 1 - .3 = .7 \neq .68.$$

The way we have treated here an overdetermined system $f_k(x_1, \ldots, x_p)$ $= 0$ is to solve the first p equations $f_1 = \cdots = f_p = 0$, but completely ignore the others. Clearly, this is not the correct approach; rather, one ought to try to satisfy as many equations as possible, if only approximately.

In order to achieve this, we first recall the concept of *residual*: If in the left-hand side of the kth equation $f_k(x_1, \ldots, x_p) = 0$ one substitutes arbitrary, but fixed values $x_1, \ldots, x_p$, one does not obtain 0 in general, but a residual s_k; through substitution in all n equations one obtains the n residuals $s_1, s_2, \ldots, s_n$, which all depend on $x_1, \ldots, x_p$.

Ideally, one would like to make all residuals s_k equal to 0 by a suitable choice of the x_l. However, this cannot be done; one can only try to make the residuals as uniformly small as possible. But what should this mean? The residuals, indeed, can be made small with respect to several points of view:

a) make the sum of the absolute values, that is $\sum |s_k|$, as small as possible;

b) make the sum of squares $\sum s_k^2$ as small as possible (*method of least squares*);

c) make the absolutely largest, i.e., $\max |s_k|$, as small as possible (*Chebyshev approximation*).

In the following we shall deal with the method of least squares, and thus compute the minimum

$$\min_{x_1, \ldots, x_p} \sigma(x_1, \ldots, x_p)$$

and the corresponding values of $x_1, \ldots, x_p$, where

$$\sigma(x_1, \ldots, x_p) = \sum_{k=1}^{n} [f_k(x_1, \ldots, x_p)]^2 = \sum_{k=1}^{n} s_k^2.$$

For the example above, this would mean, e.g., that one determines $\min \sigma(x,y)$ with

$$\sigma(x,y) = (x + y - 1)^2 + (x^2 + y^2 - .8)^2 + (x^3 + y^3 - .68)^2.$$

There is a direct method to deal with this problem: In the "landscape" (in $(p + 1)$-dimensional space $\mathbf{R}^{p+1}$) defined by $z = \sigma(x_1, \ldots, x_p)$ one goes constantly downhill (cf. Fig. 5.1). To do this, one needs the gradient of the function $\sigma(x_1, \ldots, x_p)$,

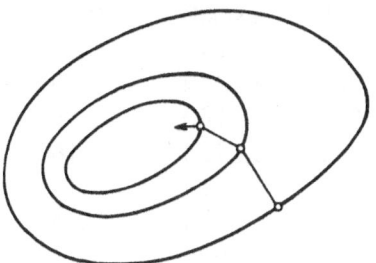

Figure 5.1. *Method of steepest descent*

$$(\text{grad } \sigma)_\ell = \frac{\partial \sigma}{\partial x_\ell} = 2 \sum_{k=1}^{n} f_k(x_1, \ldots, x_p) \frac{\partial f_k(x_1, \ldots, x_p)}{\partial x_\ell} \; ;$$

one then varies the x_ℓ according to

$$x_\ell := x_\ell + \Delta x_\ell, \quad \text{where} \quad \Delta x_\ell = -t \frac{\partial \sigma}{\partial x_\ell}.$$

(The choice of t is a problem in itself.)

In the above example one obtains

$$\frac{\partial \sigma}{\partial x} = 2s_1 + 4xs_2 + 6x^2 s_3,$$

$$\frac{\partial \sigma}{\partial y} = 2s_1 + 4ys_2 + 6y^2 s_3.$$

One can start at the point

$$x = \frac{1 + \sqrt{.6}}{2} = .88730 \ldots, \quad y = \frac{1 - \sqrt{.6}}{2} = .11270 \ldots,$$

and choose, say, $t = .05$. The resulting first ten steps are summarized in Table 5.1 (rounded results of the 14-digit computation).

Table 5.1. *Method of steepest descent for a nonlinear*
least squares problem

x	y	$\sigma \times 10^4$	$\dfrac{\partial \sigma}{\partial x} \times 10^2$	$\dfrac{\partial \sigma}{\partial y} \times 10^2$
.88730	.11270	4.0000	9.448	.152
.88257	.11263	1.7242	.242	−1.270
.88245	.11326	1.6440	.199	−1.173
.88235	.11385	1.5760	.183	−1.081
.88226	.11439	1.5183	.169	− .995
.88218	.11489	1.4693	.156	− .917
.88210	.11534	1.4278	.144	− .845
.88203	.11577	1.3925	.133	− .778
.88196	.11615	1.3626	.123	− .717
.88190	.11651	1.3372	.113	− .661
.88184	.11684	1.3156	.104	− .609

After 100 steps one would get

$$x = .88117, \quad y = .12073, \quad \sigma = 1.194018_{10}{-}4,$$

which, to the number of digits shown, agrees with the exact solution. The convergence, however, is very slow([1]).

This method of steepest descent is indeed not quite the right thing. Rather than just linearizing, we really ought to "quadratize";

$$\sigma(x_1 + \Delta x_1, \ldots, x_p + \Delta x_p) \approx$$

$$\sigma(x_1, \ldots, x_p) + \sum_{j=1}^{p} \Delta x_j \frac{\partial \sigma}{\partial x_j} + \sum_{i,j=1}^{p} \Delta x_i \Delta x_j \frac{\partial^2 \sigma}{\partial x_i \partial x_j}.$$

However, in the following, we shall turn our attention to the case of linear equations, which (apart from rounding errors) can be solved exactly. The given error equations then have the form

---

[1] Also a doubling of the stepsize to $t = .1$ would not bring the expected improvement in convergence. On the contrary, one quickly runs into an oscillatory regimen, and after 100 steps one is only as far as after 39 here. See also §10.3. (Editors' remark)

$$f_k(x_1, \ldots, x_p) = \sum_{\ell=1}^{p} f_{k\ell} x_\ell + g_k = s_k \quad (k = 1, \ldots, n).$$

But first, we describe this linear least squares problem in yet another way.

§5.2. Linear least squares problems and their classical solution

a) *Unconstrained least squares approximation* deals with the problem of approximating a vector **g** in $\mathbf{R}^n$ by means of m vectors $\mathbf{f}_1$, $\mathbf{f}_2, \ldots, \mathbf{f}_m$ $(m < n)$ in the sense of least squares, i.e., to find a vector

$$\mathbf{h} = \sum_{k=1}^{m} x_k \mathbf{f}_k$$

such that the Euclidean error norm $||\mathbf{h} - \mathbf{g}||$ becomes as small as possible. Desired, especially, are also the coefficients $x_1, \ldots, x_m$. This problem can be formulated also as

$$||\mathbf{Fx} - \mathbf{g}|| = \text{minimum},$$

or, after squaring, as

$$(\mathbf{Fx} - \mathbf{g}, \mathbf{Fx} - \mathbf{g}) = \text{minimum}, \tag{1}$$

if one collects the coefficients x_k into a vector **x** (in $\mathbf{R}^m$) and the vectors $\mathbf{f}_1, \mathbf{f}_2, \ldots, \mathbf{f}_m$ into a matrix **F** with m columns and n rows (cf. Fig. 5.2).

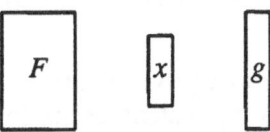

Figure 5.2. *Shape of* **F**, **x** *and* **g** *in unconstrained least squares approximation*

The problem can be phrased geometrically as follows: In the hyperplane of $\mathbf{R}^n$, spanned by $\mathbf{f}_1, \ldots, \mathbf{f}_m$, one seeks that vector **h** for which $\mathbf{h} - \mathbf{g}$ becomes shortest. This vector **h**, as is well known, can be constructed by dropping the perpendicular (from **g**) to the plane (cf. Fig. 5.3). One therefore has, for $i = 1, \ldots, m$, $(\mathbf{f}_i, \mathbf{Fx} - \mathbf{g}) = 0$, i.e., in matrix form, $\mathbf{F}^T(\mathbf{Fx} - \mathbf{g}) = 0$.

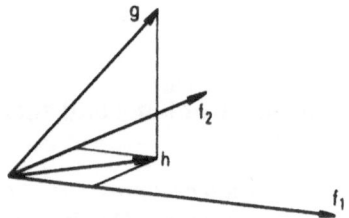

Figure 5.3. *Unconstrained least squares problem as approximation problem in* $\mathbf{R}^n$

Indeed, from (1) there first follows

$$(\mathbf{Fx},\mathbf{Fx}) - (\mathbf{Fx},\mathbf{g}) - (\mathbf{g},\mathbf{Fx}) + (\mathbf{g},\mathbf{g}) = \text{minimum},$$

thus

$$(\mathbf{x},\mathbf{F}^T\mathbf{Fx}) - 2(\mathbf{x},\mathbf{F}^T\mathbf{g}) = \text{minimum}, \tag{2}$$

where $\mathbf{F}^T\mathbf{F}$ is positive definite, provided the columns $\mathbf{f}_i$ of $\mathbf{F}$ are not linearly dependent. But now, according to §3.7, the minimum problem

$$\frac{1}{2}\,(\mathbf{x},\mathbf{Ax}) + (\mathbf{x},\mathbf{b}) = \text{minimum}$$

(with $\mathbf{A}$ positive definite) is equivalent to $\mathbf{Ax} + \mathbf{b} = \mathbf{0}$. Here, $\mathbf{A} = \mathbf{F}^T\mathbf{F}$, $\mathbf{b} = -\mathbf{F}^T\mathbf{g}$, and (2) is thus equivalent to the linear system (*normal equations*)

$$\mathbf{F}^T\mathbf{Fx} - \mathbf{F}^T\mathbf{g} = \mathbf{0}, \tag{3}$$

as we asserted above on geometrical grounds. $\mathbf{F}^T\mathbf{F}$ is a symmetric $m \times m$-matrix which, as mentioned, is positive definite in general, and $\mathbf{F}^T\mathbf{g}$ is an m-vector (cf. Fig. 5.4).

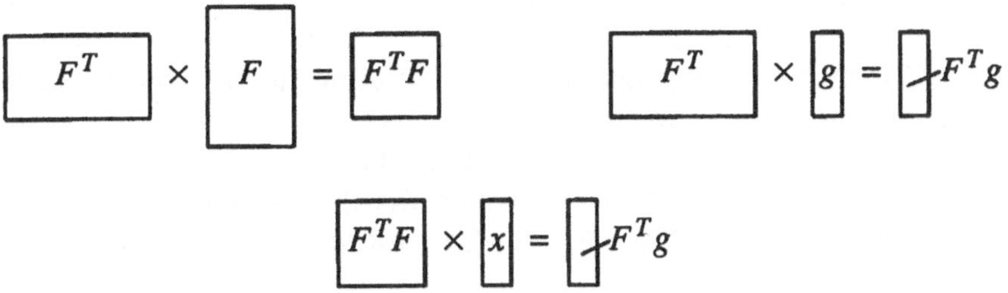

$$\boxed{F^T} \times \boxed{F} = \boxed{F^TF} \qquad\qquad \boxed{F^T} \times \boxed{g} = \boxed{F^Tg}$$

$$\boxed{F^TF} \times \boxed{x} = \boxed{F^Tg}$$

Figure 5.4. *Structure of the normal equations in unconstrained least squares approximation*

b) In *constrained least squares approximation* one deals with the following type of problem: Given m measurements $g_1, g_2, \ldots, g_m$, the "corrected" values $x_1, x_2, \ldots, x_m$ are to be determined such that

1) there are satisfied $p(<m)$ linear conditions $\displaystyle\sum_{j=1}^{m} c_{ij}x_j - d_i = 0$
   $(i = 1, 2, \ldots, p)$,

2) $\displaystyle\sum_{k=1}^{m} |x_k - g_k|^2$ becomes minimum.

In other words: the m measurements $g_1, \ldots, g_m$, through corrections which are as small as possible, are to be changed in such a way that the p conditions are satisfied.

In vector-matrix notation: *desired is a vector* $\mathbf{x}$ *such that* $||\mathbf{x} - \mathbf{g}||$ *is minimum subject to the constraint* $\mathbf{Cx} - \mathbf{d} = \mathbf{0}$. Here, $\mathbf{C}$ is a $p \times m$-matrix $(p < m)$, $\mathbf{d}$ a p-vector, and $\mathbf{x}, \mathbf{g}$ are vectors of dimension m (cf. Fig. 5.5).

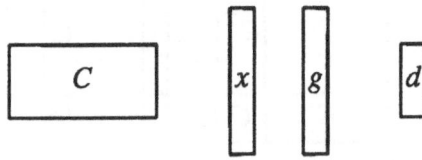

Figure 5.5. *Shape of* $\mathbf{C}, \mathbf{x}, \mathbf{g}$ *and* $\mathbf{d}$ *in constrained least squares approximation*

Example. If one measures the altitude $g(t)$ of a freely falling body at equal time intervals, that is, for $t_k = t_0 + k\Delta t$, $k = 1, \ldots, m$, the values $g_k = g(t_k)$ must lie on a parabola and therefore, in particular, the third differences of the numerical sequence $g_1, g_2, \ldots, g_m$ must vanish. Because of measurement errors, this is not the case exactly; one therefore determines adjusted values x_i for which the third differences are indeed equal to 0. This means $\mathbf{Cx} = \mathbf{0}$, with (say, for $m=7$, $p=4$)([1]):

$$\mathbf{C} = \begin{bmatrix} 1 & -3 & 3 & -1 & 0 & 0 & 0 \\ 0 & 1 & -3 & 3 & -1 & 0 & 0 \\ 0 & 0 & 1 & -3 & 3 & -1 & 0 \\ 0 & 0 & 0 & 1 & -3 & 3 & -1 \end{bmatrix}.$$

In constrained approximation, therefore, one has to determine a minimum of $||\mathbf{x} - \mathbf{g}||^2$ with side conditions; this is done, according to Lagrange, by computing the stationary values of

$$||\mathbf{x} - \mathbf{g}||^2 + \sum_{i=1}^{p} 2t_i \left[\sum_{j=1}^{m} c_{ij}x_j - d_i \right] = (\mathbf{x},\mathbf{x}) - 2(\mathbf{x},\mathbf{g}) + (\mathbf{g},\mathbf{g}) + (2\mathbf{t}, \mathbf{Cx} - \mathbf{d}),$$

where $\mathbf{t} = [t_1, \ldots, t_p]^T$ is the vector of the p Lagrange multipliers. Through partial differentiation with respect to all variables x_j, t_i one obtains from this immediately the system of equations in Fig. 5.6. The matrix is symmetric, but not positive definite (because of the Lagrange multipliers).

$$\begin{bmatrix} I & C^T \\ C & O \end{bmatrix} \times \begin{bmatrix} x \\ t \end{bmatrix} = \begin{bmatrix} g \\ d \end{bmatrix}$$

Figure 5.6. *System of equations for the "corrected" values* $\mathbf{x}$ *and the Lagrange multipliers* $\mathbf{t}$

[1] In the manuscript of this chapter all matrices are written as rectangular tableaus. Here we use instead the usual notation. (Editors' remark)

This can also be written as

$$
\begin{aligned}
\mathbf{x} + \mathbf{C}^T \mathbf{t} &= \mathbf{g} \\
\mathbf{C}\mathbf{x} &= \mathbf{d}
\end{aligned}
\tag{4}
$$

from which, by elimination of $\mathbf{x}$, there follows the system of normal equations

$$
\mathbf{C}\mathbf{C}^T \mathbf{t} - (\mathbf{C}\mathbf{g} - \mathbf{d}) = 0.
\tag{5}
$$

Here, $\mathbf{C}\mathbf{C}^T$ is a symmetric matrix of order p which, as a rule, is positive definite(^{2}) (cf. Fig. 5.7). From $\mathbf{t}$ one then obtains

$$
\mathbf{x} = \mathbf{g} - \mathbf{C}^T \mathbf{t}.
\tag{6}
$$

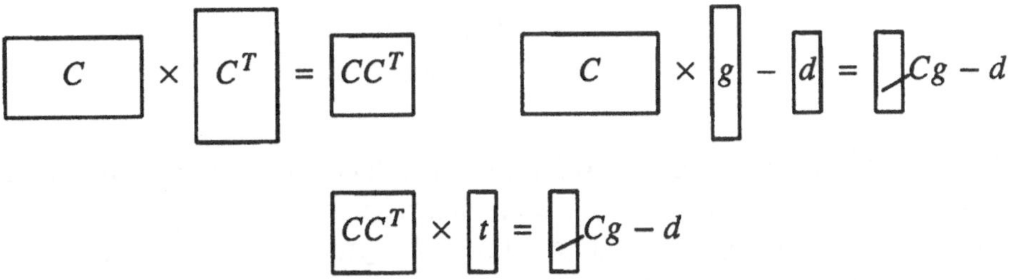

Figure 5.7. *Structure of the normal equations in constrained least squares approximation*

c) The *most general case: desired is a vector* $\mathbf{x}$ *such that* $||\mathbf{Fx} - \mathbf{g}||$ *becomes minimum subject to the side condition* $\mathbf{Cx} - \mathbf{d} = 0$.

This problem can be reduced to Case b); indeed, given the Cholesky decomposition $\mathbf{R}^T \mathbf{R}$ of $\mathbf{F}^T \mathbf{F}$, and introducing the vector $\mathbf{y} = \mathbf{R}\mathbf{x}$, one has

$$
(\mathbf{Fx} - \mathbf{g}, \mathbf{Fx} - \mathbf{g}) = (\mathbf{F}\mathbf{R}^{-1}\mathbf{y}, \mathbf{F}\mathbf{R}^{-1}\mathbf{y}) - 2(\mathbf{g}, \mathbf{F}\mathbf{R}^{-1}\mathbf{y}) + \text{const.}
$$

$$
= (\mathbf{y}, \mathbf{R}^{-1T}\mathbf{F}^T \mathbf{F}\mathbf{R}^{-1}\mathbf{y}) - 2(\mathbf{R}^{-1T}\mathbf{F}^T \mathbf{g}, \mathbf{y}) + \text{const.}
$$

---

^{2} Namely precisely in the case when the constraint equations are linearly independent. (Editors' remark)

Since

$$R^{-1T}F^TFR^{-1} = R^{-1T}R^TRR^{-1} = I,$$

one thus obtains

$$||Fx - g||^2 = (y,y) - 2(R^{-1T}F^Tg,y) + \text{const.}$$

$$= ||y - R^{-1T}F^Tg||^2 + \text{const.}$$

This is to be minimized under the requirement that $Cx - d = CR^{-1}y - d = 0$. The problem, therefore, is reduced to the case b), with

$$
\begin{array}{ll}
R^{-1T}F^Tg & \text{in place of } g, \\
Rx & \text{in place of } x, \\
CR^{-1} & \text{in place of } C.
\end{array}
\tag{7}
$$

d) *The curse of the classical methods.* The solution methods treated here all work with normal equation matrices, i.e., matrices of the form F^TF or CC^T, where the first matrix in the product is "wide", and the other "high". While, theoretically, such matrices are indeed positive definite, they nevertheless have often undesirable properties (ill-conditioning), so that one really should not use them in computational work.

The matrix

$$
F = \begin{bmatrix} 1.07 & 1.10 \\ 1.07 & 1.11 \\ 1.07 & 1.15 \end{bmatrix}
\tag{8}
$$

may serve as an example. Here the normal equations matrix, in strictly 3-digit computation, is

$$
F^TF = \begin{bmatrix} 3.42 & 3.60 \\ 3.60 & 3.76 \end{bmatrix};
\tag{9}
$$

however, this is not a positive definite matrix; already for $x = [-1,1]^T$ one finds that the value of the quadratic form is $-.02$.

We therefore propose to solve the problems a), b), c) with different methods, which we now discuss.

§5.3. Unconstrained least squares approximation through orthogonalization

The solution of the minimization problem can be simplified by subjecting the vectors $f_1, f_2, \ldots, f_m, g$ to a Schmidt orthogonalization process; this generates orthogonal vectors $u_1, \ldots, u_m, s$ with the following properties:

$$f_1 = r_{11}u_1$$
$$f_2 = r_{12}u_1 + r_{22}u_2$$
$$f_3 = r_{13}u_1 + r_{23}u_2 + r_{33}u_3$$

$$\cdot$$
$$\cdot$$
$$\cdot$$

(10)

$$f_m = r_{1m}u_1 + r_{2m}u_2 + \cdots + r_{mm}u_m$$
$$g = y_1u_1 + y_2u_2 + \cdots + y_mu_m - s,$$

where $r_{pp} > 0$ for $p = 1, \ldots, m$. The vectors $u_1, \ldots, u_m$ are also normalized, but s is not. The coefficients r_{pq} with $q > p$ and y_i are determined such that the vectors u_i, s become orthogonal, while the r_{pp} are normalization factors which are used to make the lengths of the u_i equal to 1.

Collecting the vectors $u_1, \ldots, u_m$ into a $n \times m$-matrix U, and the r_{pq} into an upper triangular[1] $m \times m$-matrix R, the relations (10) can be written as follows:

$$F = UR$$

(11)

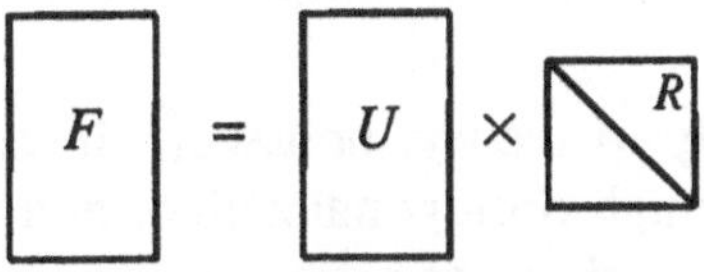

Figure 5.8. *UR-decomposition of* F

---

[1] Upper triangular matrix = matrix $[r_{pq}]$ with $r_{pq} = 0$ for $p > q$. (Editors' remark)

i.e., $\mathbf{F}$ is to be decomposed into a matrix $\mathbf{U}$ with orthonormal columns and an upper triangular matrix $\mathbf{R}$ (*UR*-decomposition; cf. Fig. 5.8). Furthermore,

$$\mathbf{g} = \mathbf{U}\mathbf{y} - \mathbf{s}, \quad \text{where} \quad \mathbf{y} = \mathbf{U}^T\mathbf{g}. \tag{12}$$

We then have identically in $\mathbf{x}$,

$$||\mathbf{F}\mathbf{x} - \mathbf{g}||^2 = ||\mathbf{U}\mathbf{R}\mathbf{x} - \mathbf{U}\mathbf{y} + \mathbf{s}||^2 = (\mathbf{U}(\mathbf{R}\mathbf{x} - \mathbf{y}) + \mathbf{s}, \mathbf{U}(\mathbf{R}\mathbf{x} - \mathbf{y}) + \mathbf{s})$$

$$= (\mathbf{U}^T\mathbf{U}(\mathbf{R}\mathbf{x} - \mathbf{y}), \mathbf{R}\mathbf{x} - \mathbf{y}) + 2(\mathbf{R}\mathbf{x} - \mathbf{y}, \mathbf{U}^T\mathbf{s}) + (\mathbf{s}, \mathbf{s}).$$

Since $\mathbf{U}^T\mathbf{s} = \mathbf{0}$ and $\mathbf{U}^T\mathbf{U} = \mathbf{I}_m$ (= unit matrix of order m), one obtains identically in $\mathbf{x}$,

$$||\mathbf{F}\mathbf{x} - \mathbf{g}||^2 = ||\mathbf{R}\mathbf{x} - \mathbf{y}||^2 + ||\mathbf{s}||^2,$$

where $\mathbf{s} = (\mathbf{U}\mathbf{U}^T - \mathbf{I})\mathbf{g}$ is constant, so that the minimum is obviously attained for $\mathbf{R}\mathbf{x} = \mathbf{y}$. One thus has to solve the system of equations

$$\mathbf{R}\mathbf{x} = \mathbf{y}, \tag{14}$$

which is quite easy, since $\mathbf{R}$ is a triangular matrix. For the solution $\mathbf{x}$, one has from (11), (12)

$$\mathbf{F}\mathbf{x} - \mathbf{g} = \mathbf{U}(\mathbf{R}\mathbf{x} - \mathbf{y}) + \mathbf{s} = \mathbf{s}, \tag{15}$$

i.e., $\mathbf{s}$ is precisely the residual vector, which is often more important than $\mathbf{x}$.

We note in passing that this matrix $\mathbf{R}$ is the same as the one that results from the Cholesky decomposition of the matrix $\mathbf{F}^T\mathbf{F}$. Indeed,

$$\mathbf{F}^T\mathbf{F} = \mathbf{R}^T\mathbf{U}^T\mathbf{U}\mathbf{R} = \mathbf{R}^T\mathbf{R}, \tag{16}$$

and the assertion follows from the uniqueness of the Cholesky decomposition.

This is valid only in theory, however. In computational work, the matrix $\mathbf{R}$ obtained through orthogonalization, and hence also the solution of the least squares approximation problem, is more accurate.

Example. The *UR*-decomposition (computed in 3-digit floating-point arithmetic) of the matrix (8) reads:

$$\begin{bmatrix} 1.07 & 1.10 \\ 1.07 & 1.11 \\ 1.07 & 1.15 \end{bmatrix} = \begin{bmatrix} .578 & -.535 \\ .578 & -.267 \\ .578 & .802 \end{bmatrix} \begin{bmatrix} 1.85 & 1.94 \\ 0 & .0374 \end{bmatrix}.$$

The values obtained are correct to 3 digits, while the triangular decomposition of $\mathbf{F}^T\mathbf{F}$ could not even have been executed.

The pseudoinverse. From $\mathbf{R}\mathbf{x} = \mathbf{y} = \mathbf{U}^T\mathbf{g}$ there follows

$$\mathbf{x} = \mathbf{R}^{-1}\mathbf{U}^T\mathbf{g}, \tag{17}$$

so that the matrix $\mathbf{R}^{-1}\mathbf{U}^T$ has the property that it yields, through multiplication into the vector $\mathbf{g}$, directly the solution $\mathbf{x}$, just like the solution of the linear system of equations $\mathbf{A}\mathbf{x} - \mathbf{b} = \mathbf{0}$ is obtained directly as $\mathbf{A}^{-1}\mathbf{b}$. Because of this analogy, the $m \times n$-matrix

$$\mathbf{Z} = \mathbf{R}^{-1}\mathbf{U}^T \tag{18}$$

is called the pseudoinverse of $\mathbf{F}$. It has the property

$$\mathbf{Z}\mathbf{F} = \mathbf{R}^{-1}\mathbf{U}^T\mathbf{U}\mathbf{R} = \mathbf{R}^{-1}\mathbf{R} = \mathbf{I}_m. \tag{19}$$

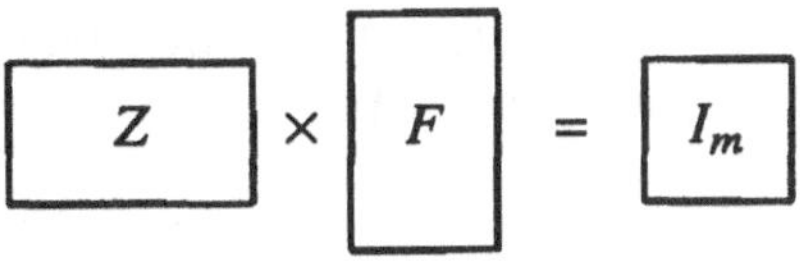

Figure 5.9. *The pseudoinverse* $\mathbf{Z}$ *of* $\mathbf{F}$

On the other hand, $\mathbf{F}\mathbf{Z}$ is a $n \times n$-matrix, but *not* the unit matrix.

Nevertheless, the pseudoinverse is more of theoretical interest than of practical significance for numerical computation. (There is also still a more general definition; see below.)

Orthogonalization without normalization. In smaller examples, computed by hand, the necessity of normalizing the vectors in the Schmidt orthogonalization process is annoying. One can indeed dispense with normalization: one determines orthogonal vectors $\mathbf{v}_1, \ldots, \mathbf{v}_m$ with

$$\mathbf{f}_1 = \mathbf{v}_1$$
$$\mathbf{f}_2 = s_{12}\mathbf{v}_1 + \mathbf{v}_2$$

$$\cdot$$
$$\cdot$$
$$\cdot$$

$$\mathbf{f}_m = s_{1m}\mathbf{v}_1 + s_{2m}\mathbf{v}_2 + \cdots + s_{m-1,m}\mathbf{v}_{m-1} + \mathbf{v}_m;$$

here,

$$s_{ij} = \frac{(f_j, v_i)}{(v_i, v_i)} \quad (i < j).$$

Thus, F is decomposed into $F = VS$, where V is a $n \times m$-matrix with the orthogonal columns $v_1, \ldots, v_m$ and S is an upper triangular $m \times m$ matrix with diagonal elements 1. Then $||Fx - g||$ is minimum when

$$V^T(VSx - g) = 0.$$

After the VS-decomposition, it remains therefore to solve the system of equations

$$Sx = (V^TV)^{-1}V^Tg$$

for x (back substitution). Note that V^TV is a diagonal matrix; the right-hand side of the system can therefore be computed very easily.

This approach has the advantage of requiring only rational operations. One trades this, however, for the disadvantage of uncontrolled growth in the elements of the matrix V.

The matrix

$$Z = S^{-1}(V^TV)^{-1}V^T$$

(generalizing the preceding definition) is also referred to as pseudoinverse of F. It again has the property

$$ZF = I_m, \quad x = Zg.$$

§5.4. Computational implementation of the orthogonalization

The orthonormalization process, starting from the vector f_1 proceeds to the vector f_m, whereby after the computation of u_k all remaining vectors $(f_{k+1}, f_{k+2}, \ldots, f_m)$ are *immediately* made orthogonal to u_k through the addition of suitable multiples of u_k. We therefore make the

Assumption: let $u_1, \ldots, u_{k-1}$ be already determined and orthonormal; let further $f_k, f_{k+1}, \ldots, f_m$ be orthogonal to $u_1, \ldots, u_{k-1}$ (but not among themselves). Then one computes:

$$\left. \begin{aligned} r_{kk} &:= ||f_k||, \\ u_k &:= f_k/r_{kk}, \\ r_{kj} &:= (u_k, f_j) \\ f_j &:= f_j - r_{kj}u_k \end{aligned} \right\} \quad j = k+1, \ldots, m. \tag{20}$$

In this way, the $\mathbf{u}_1$, $\mathbf{u}_2$, ..., $\mathbf{u}_k$ are now determined, and the new $\mathbf{f}_{k+1}$, ..., $\mathbf{f}_m$ are also orthogonal to $\mathbf{u}_k$. The step from $k - 1$ to k is thus completed. The whole UR-decomposition therefore requires the execution of the above equations (20) for $k = 1, 2, ..., m$.

Subsequently one computes

$$\left.\begin{array}{l} y_k := (\mathbf{u}_k, \mathbf{g}) \\ \mathbf{g} := \mathbf{g} - y_k \mathbf{u}_k \end{array}\right\} \quad k = 1, 2, ..., m, \tag{21}$$

and then solves the system $\mathbf{Rx} - \mathbf{y} = 0$. The latter is exactly the same computing process as the back substitution in the Cholesky method for the solution of the system $\mathbf{F}^T \mathbf{Fx} - \mathbf{F}^T \mathbf{g} = 0$.

Two difficulties now arise:

a) *Orthogonality of the* u_k. The orthonormalization process assumes that the generated vectors are orthogonal to machine precision. However, in orthogonalizing vectors which are almost parallel, rather oblique vectors $\mathbf{u}_k$ may be produced due to rounding errors (see Fig. 5.10). In other words: the inner products $(\mathbf{u}_i, \mathbf{u}_j)$ $(i \neq j)$ are substantially larger than a few units in the last position, which also adversely affects the accuracy of the solution $\mathbf{x}$. (The latter, to be sure, is still better than in direct solution of the normal equations.)

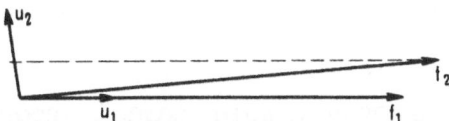

Figure 5.10. *Orthogonalization of two almost parallel vectors*

In order to guard against such inaccuracies in the orthogonalization process, it is advisable to repeat the orthogonalization of the vector $\mathbf{f}_k$ as soon as it becomes evident, during the computation of r_{kk}, that the length of $\mathbf{f}_k$ has been reduced by the orthogonalization process to less than 1/10 of its original value. One executes, in this case, the following additional operations:

$$\left.\begin{array}{l} d_j := (\mathbf{u}_j, \mathbf{f}_k) \\ \mathbf{f}_k := \mathbf{f}_k - d_j \mathbf{u}_j \end{array}\right\} \quad j = 1, 2, ..., k - 1, \tag{22}$$

$$r_{kk} := ||\mathbf{f}_k||.$$

Of course, this does not help if the vector $\mathbf{f}_k$ becomes exactly $\mathbf{0}$ (e.g., if the matrix $\mathbf{F}$ consists of all ones). This case will be treated later under b).

Numerical example (4-digit computation). For the matrix

$$\mathbf{F} = \begin{bmatrix} 8 & 21 \\ 13 & 34 \\ 21 & 55 \\ 34 & 89 \end{bmatrix}$$

one first obtains $r_{11} = ||\mathbf{f}_1|| = 42.78$ and

$$\mathbf{u}_1 = [.1870, .3039, .4909, .7948]^T ;$$

then $r_{12} = (\mathbf{u}_1, \mathbf{f}_2) = 112.0,$

$$\mathbf{f}_2 := \mathbf{f}_2 - 112.0 \, \mathbf{u}_1 = [.06, -.04, .02, -.02]^T .$$

The components of the new vector $\mathbf{f}_2$, owing to cancellation, have become 1-digit numbers (ca. 1000-fold reduction). The inner product $d_1 = (\mathbf{u}_1, \mathbf{f}_2)$ is $-.007022$. By adding $.007022 \, \mathbf{u}_1$ to $\mathbf{f}_2$, a change still occurs in 4-digit precision, because the first component, e.g., is stored in floating point arithmetic as $.06000$, to which is added $.007022 \times .1879 = .00131$. One so obtains a corrected vector:

$$\mathbf{f}_2^c = \mathbf{f}_2 + .007022 \, \mathbf{u}_1 = [.06131, -.03787, .02345, -.01442]^T .$$

Then, $r_{22} = ||\mathbf{f}_2^c|| = .07713,$

$$\mathbf{u}_2 = [.7949, -.4910, .3040, -.1879]^T .$$

Hence, altogether,

$$\mathbf{U} = \begin{bmatrix} .1870 & .7949 \\ .3039 & -.4910 \\ .4909 & .3040 \\ .7948 & -.1870 \end{bmatrix}, \quad \mathbf{R} = \begin{bmatrix} 42.78 & 112.0 \\ 0 & .07713 \end{bmatrix} .$$

One must not expect, however, that through reorthogonalization the exact values of $\mathbf{U}$ and $\mathbf{R}$ are obtained. Rather, one merely achieves that the columns of $\mathbf{U}$ are orthonormal to machine accuracy and that $\mathbf{UR}$ agrees with $\mathbf{F}$ to machine accuracy. Nevertheless, with these $\mathbf{U}$ and $\mathbf{R}$ one will obtain a vector $\mathbf{x}$ which almost yields the minimum value for $||\mathbf{Fx} - \mathbf{g}||$, even though $\mathbf{x}$ may be far from the theoretical value of the vector. For example, with $g = [13, 21, 34, 55]^T$, one gets (in exact computation) $U^T g = [69.2175, .0737]^T$; one thus has to solve the system

$$42.78 \quad x_1 + 112.0 \, x_2 = 69.2175$$

$$.07713 \, x_2 = .0737,$$

from which one obtains

$$x_2 = .9555, \quad x_1 = -.8836.$$

This gives $Fx = [12.9967, 21.0002, 33.9969, 54.9971]^T$, which, within the computing precision, agrees with g. Here, in fact, the exact solution is $x_1 = -1$, $x_2 = 1$, and this gives exactly $Fx = g$, i.e., the minimum of $||Fx - g||$ here is actually 0.

b) *Dependence of the columns of* F. In many cases, also reorthogonalization does not help, or does not help sufficiently, for example when the columns of F are linearly dependent, or become so during the course of the computation due to rounding errors. This is revealed – as for example in the case $f_k = 0$ – by the fact that reorthogonalization remains ineffective.

One can avoid such occurrences from the start by replacing the minimum problem by

$$||Fx - g||^2 + \varepsilon^2 ||x||^2 = \text{minimum}, \tag{23}$$

where ε is a sufficiently small number, so that the given problem is not changed in any practical sense.

For the computational implementation, this means that to the matrix F of the error equations one appends below the square matrix εI_m of order m, and to the vector g the zero vector in $\mathbf{R}^m$; the resulting vector in $\mathbf{R}^{n+m}$, depicted in Fig. 5.11, is to be made as short as possible. The appended matrix εI_m indeed corresponds to the term $\varepsilon^2 ||x||^2$ in (23).

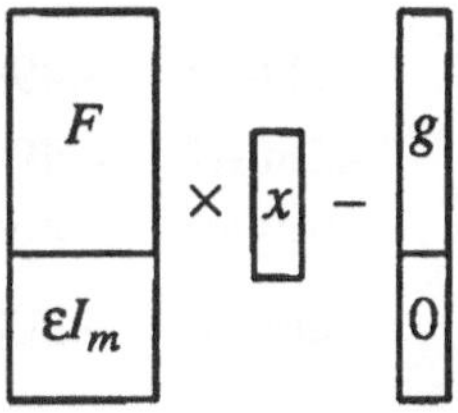

Figure 5.11. *Extended residual vector*

Note that in this case the occurrence of a linear dependence is impossible; for, even after orthogonalization, the extended vector $\mathbf{f}_k$ has still at least length ε. As a consequence, one also has $r_{jj} \geq \varepsilon$.

Numerical example (5-digit computation). To be orthogonalized is the matrix

$$\mathbf{F} = \begin{bmatrix} 1 & 1 & 1 \\ 1 & 1 & 1 \\ 1 & 1 & 1 \\ 1 & 1.01 & .99999 \end{bmatrix}.$$

The vectors $\mathbf{u}_k$, $\mathbf{f}_j$ and $\mathbf{f}^c_j$ determined according to (20) and – through reorthogonalization – by (22), are recorded in Table 5.2 in the order of their computation([1]). (At the bottom are appended the coefficients r_{11}, r_{12}, r_{13}, r_{22}, r_{23}, d_1, d_2, r_{33}.) Since the resulting $\mathbf{f}_2$, in spite of its reduction, becomes exactly orthogonal to $\mathbf{u}_1$, it does not need to be reorthogonalized.

Table 5.2. *Orthogonalization of a matrix with nearly linearly dependent columns*

u_1	f_2	$f_{3,1}$	u_2	$f_{3,2}$	$f^c_{3,1}$	$f^c_{3,2}$	u_3
0.5	-.0025	0	-.28867	$-2.4999_{10}-6$	0	$-4.9998_{10}-11$	-.49998
0.5	-.0025	0	-.28867	$-2.4999_{10}-6$	0	$-4.9998_{10}-11$	-.49998
0.5	-.0025	0	-.28867	$-2.4999_{10}-6$	0	$-4.9998_{10}-11$	-.49998
0.5	.0075	$-_{10}-5$	.86602	$-2.5001_{10}-6$	$-2_{10}-10$	$-5.0010_{10}-11$	-.50010
2	2.005	2	$8.6603_{10}-3$	$-8.6602_{10}-6$	$-4.9999_{10}-6$	$-1.732_{10}-10$	$_{10}-10$

One thus finds:

$$\mathbf{U} = \begin{bmatrix} .5 & -.28867 & -.49998 \\ .5 & -.28867 & -.49998 \\ .5 & -.28867 & -.49998 \\ .5 & .86602 & -.50010 \end{bmatrix},$$

---

[1] The additional second index in f_3 and f^c_3 refers to the respective value of k in (20), resp. j in (22). (Editors' remark)

$$\mathbf{R} = \begin{bmatrix} 2 & 2.005 & 2 \\ 0 & 8.6603_{10}-3 & -8.6602_{10}-6 \\ 0 & 0 & _{10}-10 \end{bmatrix} \; ;$$

but in spite of reorthogonalization, the first and last column of $\mathbf{U}$ are practically parallel. In addition, the small element r_{33} immediately causes difficulties in the solution of the least squares problem. For example, $\mathbf{g} = [1, 1, 1, 2]^T$ leads to the system of equations

	x_1	x_2	x_3	-1
$0 =$	2	2.005	2	2.5
$0 =$	0	$8.6603_{10}-3$	$-8.6602_{10}-6$	.86603
$0 =$	0	0	$_{10}-10$	-2.5001
	$2.5026_{10}10$	$-2.5_{10}7$	$-2.5001_{10}10$	

with the solution indicated at the bottom of the tableau. This solution, however, is totally meaningless, since in 5-digit computation the operation $\mathbf{Fx}$ results in complete cancellation.

On the other hand, the extended matrix $\mathbf{F}$, in the sense of Fig. 5.11 (with $\varepsilon = 10^{-3}$), has the following decomposition:

$$\mathbf{F} = \begin{bmatrix} 1 & 1 & 1 \\ 1 & 1 & 1 \\ 1 & 1 & 1 \\ 1 & 1.01 & .99999 \\ 10-3 & 0 & 0 \\ 0 & 10-3 & 0 \\ 0 & 0 & 10-3 \end{bmatrix}, \quad \mathbf{U} = \begin{bmatrix} .5 & -.28486 & .023311 \\ .5 & -.28486 & .023311 \\ .5 & -.28486 & .023311 \\ .5 & .85471 & -.069224 \\ 5_{10}-4 & -.11424 & -.70066 \\ 0 & .11396 & -.0085431 \\ 0 & 0 & .70897 \end{bmatrix},$$

$$\mathbf{R} = \begin{bmatrix} 2 & 2.005 & 2 \\ 0 & 8.7753_{10}-3 & 1.0569_{10}-4 \\ 0 & 0 & 1.4105_{10}-3 \end{bmatrix}.$$

Now, with $g = [1, 1, 1, 2]^T$, there results:

$$x = [-48.414, 97.997, -48.576]^T ,$$

$$Fx - g = [.007, .007, .007, -.013]^T .$$

§5.5. Constrained least squares approximation through orthogonalization

The constrained least squares problem

$$\min_{Cx - d = 0} ||x - g|| \tag{24}$$

can be reduced by means of an arbitrary vector x_0, with the property $Cx_0 - d = 0$, to

$$\min_{Cy = 0} ||y - h|| ; \tag{25}$$

simply put $y = x - x_0$, $h = g - x_0$.

The solution of the reduced problem, according to §5.2, is determined by the equations (5), (6), where now $d := 0$, $g := h$, $x := y$:

$$CC^T t - Ch = 0,$$
$$y = h - C^T t. \tag{26}$$

A comparison with the normal equations (3) and the relation (15) of unconstrained approximation shows the equivalence of the two problems, if one makes the following correspondences:

$$C^T \Leftrightarrow F \quad | \quad h \Leftrightarrow g$$
$$t \ \Leftrightarrow x \quad | \quad -y \Leftrightarrow s.$$

One can thus proceed as follows: first orthonormalize the columns of C^T and then make also $-h$ orthogonal to these columns, but no longer normalized. The resulting vector, according to the above correspondences, is $-y$. One obtains of course directly y by making h orthogonal to the columns of C^T. The procedure is summarized in Fig. 5.12.

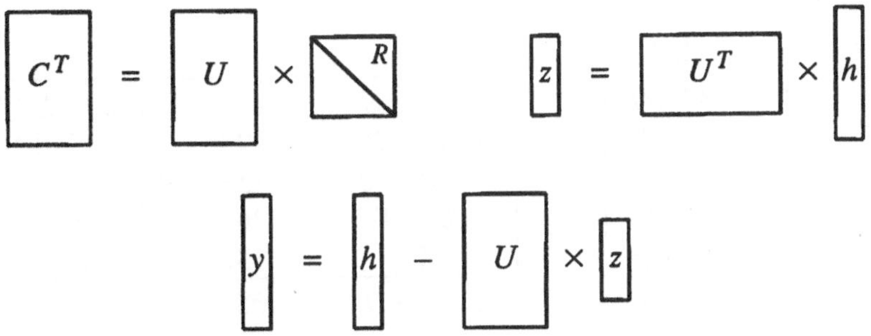

Figure 5.12. *Constrained least squares approximation through orthogonalization*

Numerical example. We are given the values of $f(t) = 10^5 \ln(t)$, rounded to integers, for $t = 10, 11, 12, 13, 14$:

t	$f(t)$
10	230259
11	239790
12	248491
13	256495
14	263906

These values are to be modified in such a way that they come to lie on a parabola (polynomial of degree 2 in the variable t). This simply means that the third differences of the corrected values must vanish:

$$y_k - 3y_{k+1} + 3y_{k+2} - y_{k+3} = 0 \quad \text{for} \quad k = 1, 2, \ldots, m - 3 .$$

In our example, $m = 5$, so that this condition needs to be written down only for $k = 1, 2$. Thus, $p = 2$, and

$$C = \begin{bmatrix} 1 & -3 & 3 & -1 & 0 \\ 0 & 1 & -3 & 3 & -1 \end{bmatrix} .$$

Orthogonalization:

C^T		h		U		y
1	0	230259		.223607	.253546	230283.11
−3	1	239790		−.670820	−.422577	239740.94
3	−3	248491	=>	.670820	−.253546	248493.49
−1	3	256495		−.223607	.760639	256540.74
0	−1	263906		0	.338062	263882.71

$$z^T = 29.74 \qquad 68.88$$

The last column contains the adjusted values.

Still a few words about the general problem

$$\min_{Bx = 0} \ ||Fx - g||, \tag{27}$$

where F is a $n \times m$-matrix, and B a $p \times m$-matrix. According to §5.2, this is reduced to the preceding problem (25) by means of the substitution [cf. (7)]

$$h = R^{-1T}F^Tg, \quad C = BR^{-1}, \quad y = Rx.$$

Since after the UR-decomposition of F there holds: $R^{-1T}F^T = U^T$, one obtains C as in Fig. 5.13.

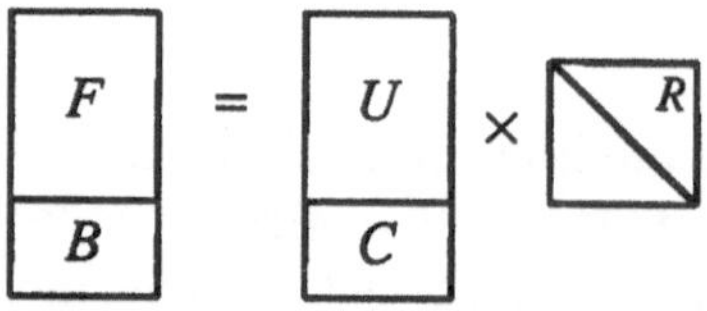

Figure 5.13. *Computation of* C *during the orthogonalization of* F

One thus makes a UR-decomposition of F, but executes the same operations with the columns of B as with the columns of F. After that, $h = U^Tg$; once y is computed as above, one finally obtains x from the equation

$$Rx - y = 0$$

Notes to Chapter 5

§5.1 Minimizing the sum of the squares of nonlinear functions is a special optimization problem which can be solved by methods applicable to general optimization problems, such as the methods discussed in Gill, Murray & Wright [1981], Dennis & Schnabel [1983], Fletcher [1987]. Methods tailored especially to least squares problems, however, are preferable. Among the more popular ones are the *Gauss-Newton method* and the *Levenberg-Marquardt algorithm*; see Dennis & Schnabel [1983, Ch. 10] or Fletcher [1987, Ch. 6]. The Gauss-Newton method is derived by considering a local affine model of the objective function around the current point, and by choosing the next point as the solution of the corresponding linear least squares problem. The Gauss-Newton method is locally q-quadratically convergent (cf. Notes to §4.2) on zero-residual problems, i.e., problems for which the function σ vanishes at the solution. For problems which have a small positive residual value at and around the solution, and which are not highly nonlinear, the Gauss-Newton method converges fast q-linearly. However, it may fail to converge on problems that are highly nonlinear and/or have large residuals. Nevertheless, it can be shown that the Gauss-Newton direction is always a descent direction, so that the method can be "globalized" by incorporating into the algorithm either a *line search* or a *trust region strategy*. The necessity of introducing such strategies is due to the fact that the model of the objective function is only locally valid. For a given descent direction, a line search algorithm will produce a shorter step in the same direction, while the trust region strategy first determines a shorter step length, and then produces a new step direction which gives the optimum of the model within a ball centered at the current point and having radius equal to the determined step length. The first approach leads to the so-called "damped Gauss-Newton" method, while the second underlies different variants of the Levenberg-Marquardt method. In particular, the modification of the Levenberg-Marquardt method due to Moré [1977] uses a scaled trust region strategy. The MINPACK subroutines LMDIF and LMDER are based on this modification (cf. Moré, Garbow & Hillstrom [1980]). The IMSL versions of these subroutines are named UNLSF, UNLSJ, respectively (cf. IMSL [1987, Vol. 3]). The subroutine NL2SOL developed by Dennis, Gay and Welsch (see Dennis & Schnabel [1983]) is a more sophisticated algorithm for solving nonlinear least squares problems. It is based on a local quadratic model of the objective function. The explicit quadratic model requires the Hessians of the residuals, which may be very expensive in general. In NL2SOL the second-order information is accumulated by a secant update approximation. NL2SOL is an adaptive procedure which uses either the Gauss-Newton or the Levenberg-Marquardt steps until sufficient information is obtained via secant updates, and then switches to the quadratic model, thus ensuring q-superlinear convergence on a large class of problems. Another very successful method has been recently developed by Fletcher and collaborators (see Fletcher [1987, Ch. 5]). Theirs is a hybrid method between the Gauss-Newton and the BFGS method. It uses a line search descent method defined by a positive definite approximate Hessian matrix. This matrix is either the Gauss-Newton matrix or a matrix obtained by using the BFGS update formula to the approximate Hessian matrix obtained in the previous step.

§5.2 One reason why the normal equations (3) are unsuitable for solving the least squares problem is the fact that the (Euclidean) condition number $\kappa(F^T F)$ of the matrix $F^T F$ is the square of the condition number $\kappa(F)$, where for any rectangular matrix A one defines $\kappa(A) = \max||Ax||/\min||Ax||$, the maximum and minimum being taken over all vectors x with (Euclidean) length $||x|| = 1$. Indeed, $\kappa(F)$ can be considered, under certain restrictions, to represent the condition of the least squares problem; see Björck [1967]. In spite of this, the normal equations method is almost universally used by statisticians. This is in part due to the lower computational complexity (see the floating-point operations count for different methods in the notes to §5.3), and in part to the fact that in most problems solved by statisticians the elements of the regression matrix are contaminated by errors of measurement which are substantially larger than the rounding errors contemplated by numerical analysts (cf. Higham & Stewart [1987]).

§5.3 Alternative methods for solving linear least squares problems use orthogonal matrix decomposition methods, based on Householder transformations (see §12.8), Givens rotations (called Jacobi rotations in §12.3) or singular value decomposition. A detailed discussion of such methods, including perturbation and rounding error analyses, as well as computer programs, can be found in Lawson & Hanson [1974]. For more recent developments, see Golub & Van Loan [1989]. The Householder and the Givens orthogonal transformations are used to factorize the matrix F into a product of an $n \times n$ orthogonal matrix and an $n \times m$ upper triangular matrix. By taking only the first m columns from the first matrix, and the first m rows of the second, one obtains a factorization of the form (11), and the solution of the least squares problem is then solved as indicated in (12) – (15). This yields the unique solution of the least squares problem whenever F has full rank. In case F is rank-deficient, one could use Householder transformations with column pivoting. However, in this case the solution is not unique, and additional work is needed to find the solution of minimal Euclidean norm (see Golub & Van Loan [1989, Ch. 6]). The method above, while working well on most rank-deficient problems, fails to detect near rank deficiency. The only fully reliable methods for handling near rank deficiency are based on the singular value decomposition of the matrix F, such as the *Golub-Reinsch method* and *Chan's method* (see Golub & Van Loan [1989, Ch.6]).

§5.4 Algorithm (20) is known as the *modified Gram-Schmidt algorithm*. Its superior stability properties, compared to classical Gram-Schmidt orthogonalization (10), have been noted experimentally by Rice [1966] and established theoretically by Björck [1967]. Nevertheless, both the classical and the modified Gram-Schmidt methods are considered of less practical importance nowadays, and mainly of historical interest (cf. Higham & Stewart [1987]). The reason is that the classical Gram-Schmidt algorithm is numerically unstable and modified Gram-Schmidt is slightly more expensive than Householder orthogonalization. The respective floating-point operations counts, indeed, are as follows: normal equations $nm^2/2 + m^3/6$; Householder orthogonalization $nm^2 - m^3/3$; modified Gram-Schmidt nm^2; Givens $2nm^2 - (2/3)m^3$; Golub-Reinsch $2nm^2 + 4m^3$; Chan $nm^2 + (17/3)m^3$ (cf. Golub & Van Loan [1989, Ch.6]). While Givens orthogonalization is twice as expensive as Householder orthogonalization for dense matrices, it is often more efficient in treating sparse matrix problems. Dense matrix least squares solvers can easily be assembled by calling the corresponding factorization subroutines and triangular systems solvers from LINPACK (cf. Dongarra et al. [1979]). The SQRLS subroutine from Kahaner, Moler & Nash [1989, Ch. 6] is such a program, which can solve overdetermined, underdetermined or singular systems of equations in the least squares sense. The IMSL subroutines LSQRR and LQRSL are also based on LINPACK. The LSBRR subroutine uses the iterative refinement of Björck [1967], described also in Golub & Van Loan [1989, Ch. 6].

§5.5 Linearly constrained least squares problems, including problems involving linear inequality constraints, are treated in Lawson & Hanson [1974] by orthogonal decomposition methods.

References

Björck, Å. [1967]: Solving linear least squares problems by Gram-Schmidt orthogonalization, *BIT* 7, 1–21.

Dennis, J.E. and Schnabel, R.B. [1983]: *Numerical Methods for Unconstrained Optimization and Nonlinear Equations*, Prentice-Hall, Englewood Cliffs, N.J.

Dongarra, J.J., Moler, C.B., Bunch, J.R. and Stewart, G.W. [1979]: *LINPACK Users' Guide*, SIAM, Philadelphia.

Fletcher, R. [1987]: *Practical Methods of Optimization*, 2nd ed., Wiley, Chichester.

Gill, P.E., Murray, W. and Wright, M.H. [1981]: *Practical Optimization*, Academic Press, London.

Golub, G.H. and Van Loan, C.F. [1989]: *Matrix Computations*, 2nd ed., The Johns Hopkins University Press, Baltimore.

Higham, N.J. and Stewart, G.W. [1987]: Numerical linear algebra in statistical computing, in *State of the Art in Numerical Analysis* (A. Iserles and M.J.D. Powell, eds.), pp. 41–57. Clarendon Press, Oxford.

IMSL [1987]: *Math/Library User's Manual*, Houston.

Kahaner, D., Moler, C. and Nash, S. [1989]: *Numerical Methods and Software*, Prentice-Hall, Englewood Cliffs, N.J.

Lawson, C.L. and Hanson, R.J. [1974]: *Solving Least Squares Problems*, Prentice-Hall, Englewood Cliffs, N.J.

Moré, J.J. [1977]: The Levenberg-Marquardt algorithm: implementation and theory, in *Numerical Analysis* (G.A. Watson, ed.), Lecture Notes Math. **630**, pp. 105–116, Springer, New York.

Moré, J.J., Garbow, B.S. and Hillstrom, K.E. [1980]: *User Guide for MINPACK-1*, Argonne National Laboratory, Report ANL-80-74.

Rice, J.R. [1966]: Experiments on Gram-Schmidt orthogonalization, *Math. Comp.* **20**, 325–328.

CHAPTER 6

Interpolation

Interpolation is the art of reading between the lines of a mathematical table. It can be used to express nonelementary functions approximately in terms of the four basic arithmetic operations, thus making them accessible to computer evaluation.

There is, however, a new point of view that emerges here: while an ordinary table of logarithms provides a dense enough tabulation so that one can interpolate linearly between two tabular values, it is our endeavor to tabulate as loosely as possible in order to avoid storing an excessive amount of numbers, for example:

x	10^x
0	1
0.01	1.023293
0.02	1.047129
.	.
.	.
.	.
1	10

These 101 values ought to be sufficient to interpolate to 10^x, and thus also, indirectly, to $\log x$. True, we then have to compute a little more until a function value $f(x)$ at an intermediate point x is determined with sufficient accuracy, but this surely can be entrusted to a computer.

The actual interpolation always proceeds as follows: the function $f(x)$ to be interpolated (ad hoc, or for permanent use) is replaced by another function which

a) deviates as little as possible from $f(x)$,

b) can be easily evaluated.

Normally, one replaces $f(x)$ by polynomials which agree with $f(x)$ at certain points; but rational functions or trigonometric polynomials are also used for interpolation.

§6.1. The interpolation polynomial

If at $n + 1$ pairwise distinct points $x_0, x_1, \ldots, x_n$ (*nodes*) we are given the function values $y_0, y_1, \ldots, y_n$ (*ordinates*), then, as is well known, there is exactly one polynomial $P(x)$ of degree $\leq n$ such that

$$P(x_k) = y_k \quad (k = 0, 1, \ldots, n) \tag{1}$$

(cf. Fig. 6.1). In general, $P(x)$ has degree n, but in the case $y_0 = y_1 = \cdots = y_n = 1$, for example, the uniquely determined polynomial is the constant 1.

For this existence and uniqueness theorem there is a constructive proof:

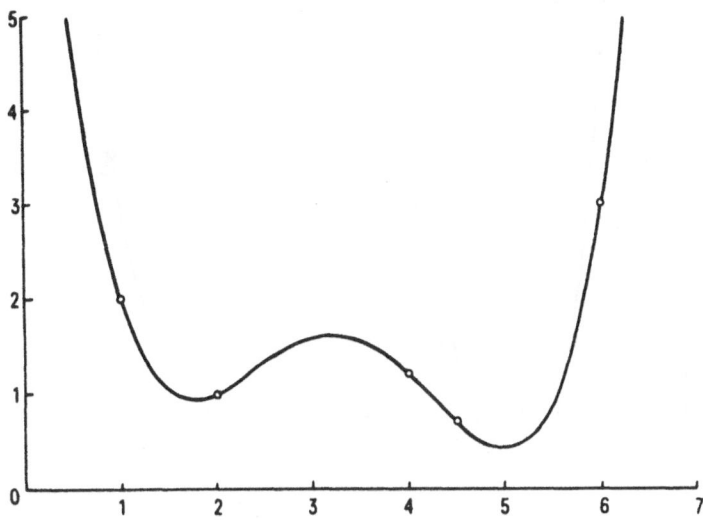

Figure 6.1. *An interpolation polynomial of degree 4*

We consider the Lagrange polynomial

$$\ell_k(x) = \prod_{\substack{j=0 \\ j \neq k}}^{n} \left[\frac{x - x_j}{x_k - x_j} \right] . \tag{2}$$

As a product of n linear factors, this is a polynomial of exact degree n; furthermore,

$$\ell_k(x_k) = 1,$$
$$\ell_k(x_i) = 0, \quad i \neq k. \tag{3}$$

For $n = 4$, for example, the graph of $\ell_2(x)$ looks as in Fig. 6.2.

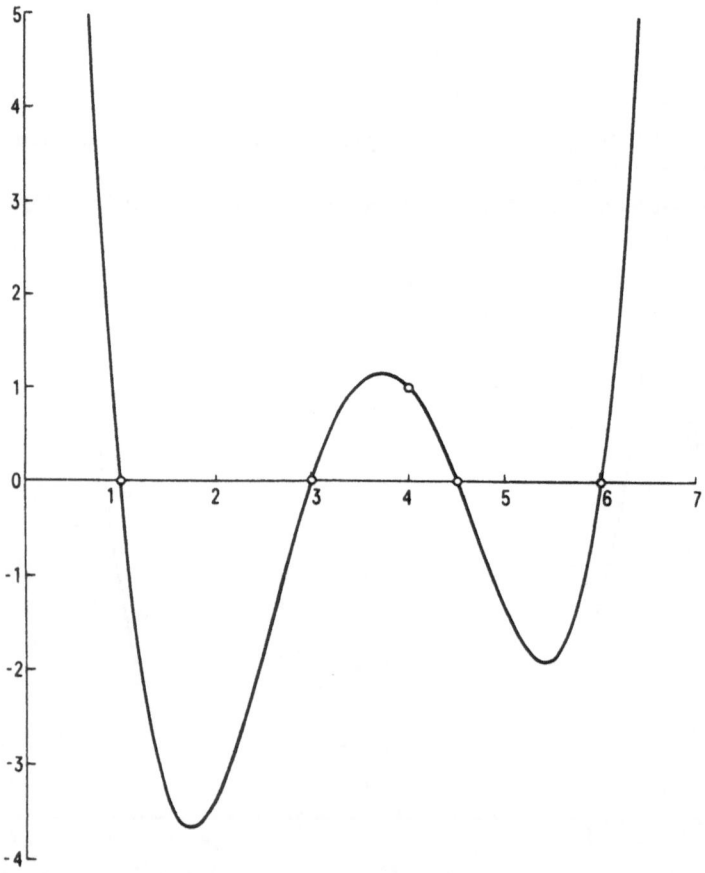

Figure 6.2. A *Lagrange polynomial* $\ell_2(x)$ *of degree 4*
$(x_0 = 1, x_1 = 3, x_2 = 4, x_3 = 4.5, x_4 = 6)$

Altogether, $n + 1$ such Lagrange polynomials can be constructed, namely for $k = 0, 1, \ldots, n$. Using them to form the linear combination

$$P(x) = \sum_{k=0}^{n} y_k \ell_k(x), \tag{4}$$

the following then holds:

1) $P(x)$ again is a polynomial of degree $\leq n$.

2) $P(x_j) = \sum_{k=0}^{n} y_k \ell_k(x_j)$, where now, among all factors $\ell_k(x_j)$ (j fixed), only $\ell_j(x_j) \neq 0$ (namely $= 1$), so that $P(x_j) = y_j$ ($j = 0, \ldots, n$). [$P(x)$ thus satisfies the conditions (1).]

3) If there is another polynomial $Q(x)$ of degree $\leq n$, which satisfies (1), then $P(x) - Q(x)$ is also a polynomial of degree $\leq n$, which vanishes at the $n + 1$ points $x_0, x_1, \ldots, x_n$; therefore, necessarily, $P(x) - Q(x) \equiv 0$.

Hence (4), or, what is the same, the so-called *Lagrange interpolation formula*

$$P(x) = \sum_{k=0}^{n} y_k \left\{ \prod_{\substack{i=0 \\ i \neq k}}^{n} \frac{x - x_i}{x_k - x_i} \right\}, \tag{5}$$

yields the uniquely determined interpolation polynomial of degree n corresponding to the given nodes and ordinates.

Example. For $x_0 = 1$, $x_1 = 2$, $x_2 = 4$, the Lagrange polynomials (2) are:

$$\ell_0(x) = \frac{x-2}{1-2} \cdot \frac{x-4}{1-4} = \frac{1}{3}(x^2 - 6x + 8),$$

$$\ell_1(x) = \frac{x-1}{2-1} \cdot \frac{x-4}{2-4} = -\frac{1}{2}(x^2 - 5x + 4),$$

$$\ell_2(x) = \frac{x-1}{4-1} \cdot \frac{x-2}{4-2} = \frac{1}{6}(x^2 - 3x + 2).$$

Therefore,

$$P(x) = \frac{y_0}{3}(x^2 - 6x + 8) - \frac{y_1}{2}(x^2 - 5x + 4) + \frac{y_2}{6}(x^2 - 3x + 2).$$

Unfortunately, for large n this formula becomes rather involved; not only do we have $n + 1$ terms, but each is a product of n linear factors. A simplified form will be discussed later.

Vandermonde matrix. It may be tempting to seek the polynomial $P(x)$ in the form $\sum c_k x^k$ and to determine the c_k by means of a linear system of equations

$$\sum_{k=0}^{n} c_k x_j^k - y_j = 0 \quad (j = 0, 1, \ldots, n) \tag{6}$$

in the $n + 1$ unknowns $c_0, c_1, \ldots, c_n$:

	c_0	c_1	c_2	$\cdots$	c_n	1
$0 =$	1	x_0	x_0^2	$\cdots$	x_0^n	$-y_0$
$0 =$	1	x_1	x_1^2		x_1^n	$-y_1$
$\vdots$						
$0 =$	1	x_n	x_n^2		x_n^n	$-y_n$

The coefficient matrix $\mathbf{V}$ here is a Vandermonde matrix, and therefore, as is well known, nonsingular as long as $x_i \neq x_j$ for $i \neq j$. The system (6), however, has only theoretical significance, since its solution by numerical methods is ill-advised on all counts (computational effort, storage requirement, accuracy). The solution of (6), indeed, is already accomplished by the Lagrange formula (4); even the following is true: the coefficients of the Lagrange polynomial $\ell_k(x)$ are the elements in the kth column of $\mathbf{V}^{-1}$. The latter means that, say in the preceding example (where $x_0 = 1$, $x_1 = 2$, $x_2 = 4$), one can read off

$$\mathbf{V} = \begin{bmatrix} 1 & 1 & 1 \\ 1 & 2 & 4 \\ 1 & 4 & 16 \end{bmatrix}, \quad \mathbf{V}^{-1} = \begin{bmatrix} \dfrac{8}{3} & -2 & \dfrac{1}{3} \\ -2 & \dfrac{5}{2} & -\dfrac{1}{2} \\ \dfrac{1}{3} & -\dfrac{1}{2} & \dfrac{1}{6} \end{bmatrix}.$$

§6.2. The barycentric formula

In order to make the Lagrange formula (5) more palatable for numerical computation, one divides it through by $\prod\limits_{i=0}^{n} (x - x_i)$ (this is the product of all $n + 1$ linear factors),

$$\frac{P(x)}{\prod\limits_{\substack{i=0}}^{n} (x - x_i)} = \sum_{k=0}^{n} \left\{ \frac{y_k}{\prod\limits_{\substack{i=0 \\ i \neq k}}^{n} (x_k - x_i)} \cdot \frac{\prod\limits_{\substack{i=0 \\ i \neq k}}^{n} (x - x_i)}{\prod\limits_{i=0}^{n} (x - x_i)} \right\} .$$

Observing that the second factor reduces to $1/(x - x_k)$, and introducing the constants

$$w_k = \frac{1}{\prod\limits_{\substack{i=0 \\ i \neq k}}^{n} (x_k - x_i)} \qquad (k = 0, 1, \ldots, n), \tag{7}$$

one gets

$$P(x) = \left\{ \prod_{i=0}^{n} (x - x_i) \right\} \left\{ \sum_{k=0}^{n} \frac{w_k y_k}{x - x_k} \right\} \tag{8}$$

(*first form of the barycentric interpolation formula*). This is certainly true also in the case $P(x) \equiv 1$, that is, $y_0 = y_1 = \cdots = y_n = 1$ (since every polynomial of degree $\leq n$ is faithfully reproduced). Thus,

$$1 = \left\{ \prod_{i=0}^{n} (x - x_i) \right\} \left\{ \sum_{k=0}^{n} \frac{w_k}{x - x_k} \right\},$$

and dividing (8) by this identity, there follows

$$P(x) = \frac{\sum\limits_{k=0}^{n} \dfrac{w_k}{x - x_k} y_k}{\sum\limits_{k=0}^{n} \dfrac{w_k}{x - x_k}} \tag{9}$$

[that is, $P(x)$ is a weighted average of the y_k with, to be sure, partly negative weights]. This is the *second (proper) form of the barycentric formula.*

Once the w_k have been computed (which requires an expense proportional to n^2), only $O(n)$ operations are necessary for the evaluation of $P(x)$, *as long as the nodes are not changed.*

Programming. The w_k are given as **array** $w[0:n]$, the x_k, y_k as **array** $x,y[0:n]$, and x as **real** xx. Then (9) can be programmed as follows:

```
real procedure baryz2(n, x, y, w, xx);
    value n, xx;
    integer n; real xx; array x, y, w;
    begin
        real den, num, s, t;
        integer k;
        den := num := 0;
        for k := 0 step 1 until n do
        begin
            s := xx − x[k];
            if s = 0 then s := 10−30;(¹)
            t = w[k]/s;
            den := den + t;
            num := num + t × y[k]
        end;
        baryz2 := num/den
    end baryz2;
```

§6.3. Divided differences

The interpolation problem, already solved in principle by the Lagrange formula, is intimately related to the divided differences of a function $f(x)$. Given fixed nodes $x_0, x_1, \ldots, x_n$, one first forms the first difference quotients

$$f(x_0, x_1) = \frac{f(x_0) - f(x_1)}{x_0 - x_1}, \quad f(x_1, x_2) = \frac{f(x_1) - f(x_2)}{x_1 - x_2}, \text{ etc. },$$

---

[1] Avoiding the division by 0 and maintaining the correct result, without the use of a jump command. (Editors' remark)

and then the higher difference quotients

$$f(x_0,x_1,x_2) = \frac{f(x_0,x_1) - f(x_1,x_2)}{x_0 - x_2} \quad , \quad \text{etc.}$$

These values are arranged in a scheme:

$$
\begin{array}{ll}
x_0\ f(x_0) \\
\quad\quad f(x_0,x_1) \\
x_1\ f(x_1) \quad\quad f(x_0,x_1,x_2) \\
\quad\quad f(x_1,x_2) \quad\quad\quad f(x_0,x_1,x_2,x_3) \\
x_2\ f(x_2) \quad\quad f(x_1,x_2,x_3) \\
\quad\quad f(x_2,x_3) \\
\quad\quad\quad\quad\quad\quad\quad\quad\quad\quad\quad\quad\quad\quad\quad\quad f(x_0,x_1,\ldots,x_n),\quad (10) \\
x_3\ f(x_3) \\
\quad\quad\quad\quad\quad\quad\quad f(x_{n-3},x_{n-2},x_{n-1},x_n) \\
\vdots\quad\vdots\quad\quad f(x_{n-2},x_{n-1},x_n) \\
\quad\quad f(x_{n-1},x_n) \\
x_n\ f(x_n)
\end{array}
$$

the general rule of formation being as follows:

$$f(x_i,x_{i+1},\ldots,x_j) = \frac{f(x_i,x_{i+1},\ldots,x_{j-1}) - f(x_{i+1},x_{i+2},\ldots,x_j)}{x_i - x_j} \quad (11)$$

$$(i,j = 0,\ldots,n, \quad i < j).$$

The formula (11) means that a new element E of the scheme (10) is formed by computing the difference of the two elements E_1 and E_2 above and below immediately to the left of E; the denominator is the difference of those arguments x_i and x_j which one finds on the left margin by proceeding from E diagonally to the upper and lower left, respectively:

$$
\begin{array}{l}
x_i\ f(x_i) \\
\quad\quad\quad\quad E_1 \\
\quad\quad\quad\quad\quad\quad E = \dfrac{E_1 - E_2}{x_i - x_j}\ . \\
\quad\quad\quad\quad E_2 \\
x_j\ f(x_j)
\end{array}
$$

Definition. *One calls* (10) *the scheme of divided difference of the function* $f(x)$ *for the nodes* $x_0, x_1, \ldots, x_n$; *specifically,* $f(x_i, \ldots, x_j)$ *is called a divided difference of order* $j - i$.

Example. Suppose we are given the function $f(x) = e^x$ at the nodes $x_0 = -.4$, $x_1 = -.2$, $x_2 = 0$, $x_3 = .2$, $x_4 = .4$ (i.e., $n = 4$):

x_i	$f(x_i)$				
−.4	.670320				
		.742055			
−.2	.818731		.410725		
		.906345		.151583	
0	1.000000		.501675		.041900
		1.107015		.185103	
.2	1.221403		.612737		
		1.352110			
.4	1.491825				

Here, for example, $f(x_1, x_2, x_3, x_4)$ is computed as

$$\frac{.501675 - .612737}{-.2 - .4} = .185103.$$

The x_i, however, need by no means be equally spaced, in fact not even ordered; for example, we can form also the following scheme:

x_i	$f(x_i)$					
0	1.000000					
		.906345				
−.2	.818731		.501675			
		1.006680		.151588		
.2	1.221403		.441040		.041885	(12)
		.918472		.168342		
−.4	.670320		.542045			
		1.026881				
.4	1.491825					

One notices here that the underlined values occur in both schemes, although with minor differences caused by rounding errors. This is no accident but follows from the

Theorem 6.1 (Symmetry property). *The divided difference* $f(x_i, x_{i+1}, \ldots, x_j)$ *is a symmetric function of its arguments.*

In the above example, e.g., one has

$$f(0, -.2, .2) = f(-.2, 0, .2) = .501675.$$

A first difference can be written in the form

$$f(x_0, x_1) = \frac{f(x_1) - f(x_0)}{x_1 - x_0} = \frac{f(x_0)}{x_0 - x_1} + \frac{f(x_1)}{x_1 - x_0},$$

a second analogously as

$$f(x_0, x_1, x_2) = \frac{f(x_0)}{(x_0 - x_1)(x_0 - x_2)} + \frac{f(x_1)}{(x_1 - x_0)(x_1 - x_2)} + \frac{f(x_2)}{(x_2 - x_0)(x_2 - x_1)},$$

from which the symmetry is evident. In general, one has:

$$f(x_i, x_{i+1}, \ldots, x_j) = \sum_{\mu=i}^{j} \frac{f(x_\mu)}{\prod\limits_{\substack{\nu=i \\ \nu \neq \mu}}^{j} (x_\mu - x_\nu)}. \tag{13}$$

Proof by mathematical induction: The induction basis has already been established. Now (13) is assumed for $f(x_i, x_{i+1}, \ldots, x_{j-1})$ and $f(x_{i+1}, \ldots, x_j)$; then (11) is applied:

$$\frac{\displaystyle\sum_{\mu=i}^{j-1} \frac{f(x_\mu)}{\prod\limits_{\substack{\nu=i \\ \nu \neq \mu}}^{j-1} (x_\mu - x_\nu)} - \sum_{\mu=i+1}^{j} \frac{f(x_\mu)}{\prod\limits_{\substack{\nu=i+1 \\ \nu \neq \mu}}^{j} (x_\mu - x_\nu)}}{x_i - x_j}$$

$$= \frac{1}{x_i - x_j} \left\{ \frac{f(x_i)}{\prod_{\substack{v=i \\ v \neq i}}^{j-1} (x_i - x_v)} - \frac{f(x_j)}{\prod_{\substack{v=i+1 \\ v \neq j}}^{j} (x_j - x_v)} \right.$$

$$\left. + \sum_{\mu=i+1}^{j-1} f(x_\mu) \left[\frac{1}{\prod_{\substack{v=i \\ v \neq \mu}}^{j-1} (x_\mu - x_v)} - \frac{1}{\prod_{\substack{v=i+1 \\ v \neq \mu}}^{j} (x_\mu - x_v)} \right] \right\} .$$

The expression in brackets, however, is equal to

$$\frac{(x_\mu - x_j) - (x_\mu - x_i)}{\prod_{\substack{v=i \\ v \neq \mu}}^{j} (x_\mu - x_v)} = \frac{(x_i - x_j)}{\prod_{\substack{v=i \\ v \neq \mu}}^{j} (x_\mu - x_v)} ;$$

one therefore obtains

$$\frac{f(x_i)}{\prod_{\substack{v=i \\ v \neq i}}^{j} (x_i - x_v)} + \frac{f(x_j)}{\prod_{\substack{v=i \\ v \neq j}}^{j} (x_j - x_v)} + \sum_{\mu=i+1}^{j-1} \frac{f(x_\mu)}{\prod_{\substack{v=i \\ v \neq \mu}}^{j} (x_\mu - x_v)} ,$$

and this is, up to notations, exactly the asserted formula (13). From its validity for $j - i$ arguments thus follows the correctness for $j - i + 1$ arguments. The symmetry property can now be read off immediately from (13).

§6.4. Newton's interpolation formula

In order to represent $f(x)$, one can add yet an $(n+2)$nd node x (for the present, x is a fixed, but arbitrary, node different from x_0, $x_1, \ldots, x_n$), by means of which the scheme (10) is enlarged to

$$
\begin{array}{ll}
x \quad f(x) & \\
\quad\quad f(x,x_0) & \\
x_0 \; f(x_0) \quad\quad f(x,x_0,x_1) & \\
\quad\quad f(x_0,x_1) \quad\quad f(x,x_0,x_1,x_2) & \\
x_1 \; f(x_1) \quad\quad f(x_0,x_1,x_2) & \\
\quad\quad f(x_1,x_2) & \\
x_2 \; f(x_2) & \\
\end{array}
$$

$$f(x,x_0,\ldots,x_n).$$

$$f(x_{n-3},x_{n-2},x_{n-1},x_n)$$

$$f(x_{n-2},x_{n-1},x_n)$$

$$f(x_{n-1},x_n)$$

$$x_n \; f(x_n)$$

We then have

$$
f(x,x_0) = \frac{f(x) - f(x_0)}{x - x_0} \; , \text{ i.e., } \; f(x) = f(x_0) + (x - x_0)f(x,x_0) ,
$$

$$
f(x,x_0,x_1) = \frac{f(x,x_0) - f(x_0,x_1)}{x - x_1} \; , \text{ i.e., } \; f(x,x_0) = f(x_0,x_1) + (x - x_1)f(x,x_0,x_1),
$$

$$
\begin{array}{c}
. \\
. \\
.
\end{array}
$$

$$
f(x,x_0,x_1,\ldots,x_n) = \frac{f(x,x_0,x_1,\ldots,x_{n-1}) - f(x_0,x_1,\ldots,x_n)}{x - x_n} \; ,
$$

i.e., $f(x,x_0,\ldots,x_{n-1}) = f(x_0,x_1,\ldots,x_n) + (x - x_n)f(x,x_0,\ldots,x_n) .$

Putting this together yields

$$
f(x) = f(x_0) + (x - x_0)[f(x_0,x_1) + (x - x_1)[f(x_0,x_1,x_2) + \cdots
$$

$$
+ (x - x_{n-1})[f(x_0,x_1,\ldots,x_n) + (x - x_n)f(x,x_0,x_1,x_2,\ldots,x_n)] \cdots]]
$$

$$
= \sum_{k=0}^{n} f(x_0,\ldots,x_k) \left[\prod_{\ell=0}^{k-1} (x - x_\ell) \right] + f(x,x_0,\ldots,x_n) \prod_{\ell=0}^{n} (x - x_\ell) ,
$$

$$
f(x) = P(x) + f(x,x_0,\ldots,x_n) \prod_{\ell=0}^{n} (x - x_\ell) . \tag{14}
$$

Here, $P(x)$ is a polynomial which at the nodes $x_0, x_1, \ldots, x_n$ agrees with $f(x)$, since the second term evidently vanishes([1]) at these points, i.e., $P(x)$ is the uniquely determined interpolation polynomial for the ordinates $y_k = f(x_k)$; it is given by *Newton's interpolation formula*

$$P(x) = \sum_{k=0}^{n} f(x_0, x_1, \ldots, x_k) \prod_{\ell=0}^{k-1} (x - x_\ell) . \tag{15}$$

The coefficients $f(x_0, \ldots, x_k)$ in this formula lie precisely on the top (descending) diagonal of the scheme (10) of divided differences. Now when $f(x)$ is replaced by $P(x)$, one commits an error which is exactly determined by the *remainder term*

$$f(x) - P(x) = f(x, x_0, x_1, \ldots, x_n) \prod_{\ell=0}^{n} (x - x_\ell) . \tag{16}$$

Observing that $f(x) - P(x)$ has the $n+1$ zeros $x_0, x_1, \ldots, x_n$, it follows from the theorem of Rolle that the nth derivative $f^{(n)}(x) - P^{(n)}(x)$, too, must have at least one zero ξ between the x_i (i.e., between $\min\{x_i\}$ and $\max\{x_i\}$). One then has $f^{(n)}(\xi) = P^{(n)}(\xi)$, but in differentiating (15) n times, only the term with $k = n$ gives a contribution,

$$P^{(n)}(x) = \left[\frac{d}{dx}\right]^n f(x_0, x_1, \ldots, x_n) \prod_{\ell=0}^{n-1} (x - x_\ell) = n! f(x_0, x_1, \ldots, x_n) \tag{17}$$

(which of course is independent of x); therefore,

$$f(x_0, x_1, \ldots, x_n) = \frac{f^{(n)}(\xi)}{n!} ,$$

where ξ is a certain point between the $x_0, \ldots, x_n$. Now of course, one has likewise

$$f(x, x_0, x_1, \ldots, x_n) = \frac{f^{(n+1)}(\eta)}{(n+1)!} , \tag{18}$$

---

[1] It is easily shown by induction that a divided difference tends to a finite limit as two nodes merge into one, provided f has a derivative at the point of merger. (Translator's remark)

where η is a certain point between the arguments $x, x_0, \ldots, x_n$, from which, together with (16), there follows the error estimate

$$|f(x) - P(x)| \le \left| \prod_{\ell=0}^{n} (x - x_\ell) \right| \max_{\eta} \left| \frac{f^{(n+1)}(\eta)}{(n+1)!} \right|. \qquad (19)$$

Here the maximum is to be taken over all possible η, thus over the interval between the smallest and largest of the values $x, x_0, x_1, \ldots, x_n$.

Example. Let e^x be interpolated at the nodes $-.4, -.2, 0, .2, .4$. On the basis of the scheme (12) of divided differences, one obtains

$$e^x \approx P(x) = 1 + .906345x + .501675x(x + .2)$$
$$+ .151588x(x + .2)(x - .2)$$
$$+ .041885x(x + .2)(x - .2)(x + .4).$$

The maximum error $|e^x - P(x)|$ can be estimated by (19):

$$|e^x - P(x)| < \frac{e^{.4}}{120} \left| x(x^2 - .04)(x^2 - .16) \right|.$$

In the interval $|x| < .4$ this error remains below $.00005$.

Programming. Denoting the divided difference $f(x_i, x_{i+1}, \ldots, x_{i+p})$ by $f[i,p]$, hence the given function values $f(x_k)$ accordingly by $f[k, 0]$, one has in

for $p := 1$ **step** 1 **until** n **do**
    for $i := 0$ **step** 1 **until** $n - p$ **do** $\qquad$ (20)
        $f[i,p] := (f[i+1, p-1] - f[i,p-1])/(x[i+p] - x[i]);$

a possible algorithm for the construction of the scheme of divided differences. Since, however, for Newton's interpolation formula one does not need the whole scheme, but only its top diagonal, one can get by with fewer storage locations: in fact, $f[i+1, p-1]$ is no longer needed, once the quantities

$$f[q, p-1] \quad (q = 0, 1, \ldots, i),$$
$$f[q, p] \quad (q = i, \ldots, n-p) \qquad (21)$$

have been computed. At this point in time, therefore, one can overwrite

$f[i + 1, p - 1]$ by $f[i, p]$, which is realized by storing the quantities $f[0, k]$, $f[1, k - 1]$, ..., $f[k, 0]$ all as $g[k]$. The index i, however, must be run backwards in order to conform with the assumption (21):

> **for** $p := 1$ **step** 1 **until** n **do**
>> **for** $i := n - p$ **step** -1 **until** 0 **do**   (22)
>>> $g[i + p] := (g[i + p] - g[i + p - 1])/(x[i + p] - x[i])$;

or, with $j := i + p$,

> **for** $p := 1$ **step** 1 **until** n **do**
>> **for** $j := n$ **step** -1 **until** p **do**   (23)
>>> $g[j] := (g[j] - g[j - 1])/(x[j] - x[j - p])$;

Note: At the beginning, the $g[k]$ must be the function values $f(x_k)$; at the end, they are the desired coefficients c_k, by means of which the value of the polynomial $P(z)$, here denoted by fw, can be computed as follows:

$$fw := g[n];$$
$$\text{for } k := n - 1 \text{ step } -1 \text{ until } 0 \text{ do} \qquad (24)$$
$$fw := g[k] + fw \times (z - x[k]);$$

It is true, though, that this algorithm is not optimal with respect to rounding errors.

§6.5. Specialization to equidistant x_i

If $x_k = x_0 + kh$, it is convenient to introduce a new variable $t = (x - x_0)/h$; to the point $x = x_k$ then corresponds the value $t = k$. Using the abbreviation $f_k = f(k)$, the divided differences now are

$$f(k, k + 1) = f_{k+1} - f_k = \Delta f_k,$$

$$f(k, k + 1, k + 2) = \frac{\Delta f_{k+1} - \Delta f_k}{2} = \frac{\Delta^2 f_k}{2},$$

and in general,

$$f(k, k + 1, \ldots, k + p) = \frac{\Delta^p f_k}{p!}, \qquad (25)$$

having introduced in the usual way the *ordinary difference scheme*

$$
\begin{array}{l}
f_0 \\
\quad \Delta f_0 \\
f_1 \qquad \Delta^2 f_0 \\
\quad \Delta f_1 \qquad \Delta^3 f_0 \\
f_2 \qquad \Delta^2 f_1 \\
\quad \Delta f_2 \qquad\qquad\qquad \Delta^n f_0 \\
f_3 \\
\vdots \qquad\qquad \Delta^3 f_{n-3} \\
\qquad \Delta^2 f_{n-2} \\
\quad \Delta f_{n-1} \\
f_n
\end{array}
\tag{26}
$$

(Differences here are always understood as *lower value minus upper value*.)

The interpolation polynomial therefore becomes

$$
P(x_0 + th) = f_0 + t\Delta f_0 + t(t-1)\frac{\Delta^2 f_0}{2} + \cdots
$$
$$
+ t(t-1)(t-2) \cdots (t-n+1)\frac{\Delta^n f_0}{n!} = \sum_{p=0}^{n} \begin{bmatrix} t \\ p \end{bmatrix} \Delta^p f_0.
\tag{27}
$$

This is the *interpolation formula of Newton and Gregory*. There is still a multitude of specializations of the classical Newton formula, which are named after Bessel, Stirling, Everett, Gauss, but they all, in the end, still only produce one and the same interpolation polynomial.

§6.6. The problematic nature of Newton interpolation

What we still want to discuss here is interpolation after Newton-Gregory in an extensive mathematical table[1]. Suppose, for example, we are given the rounded function values

---

[1] The problems exhibited here are independent of the chosen representation of the interpolation polynomial, but not of the choice of nodes. (Editor's remark)

$$
\begin{aligned}
\log 1 \quad &= 0 \\
\log 1.01 &= .0043214 \\
\log 1.02 &= .0086002 \\
\log 1.03 &= .0128372
\end{aligned}
$$

.

.

.

$$\log 10 \quad = 1.$$

It would be foolish, with these 901 values, to set up the interpolation polynomial of degree 900, since it behaves completely pathologically, fluctuating between nodes by amounts of up to $\pm\, 10^{100}$. Rather, interpolation polynomials of high degree must be avoided as a matter of principle; one should proceed instead as follows:

Having determined that x lies in the interval $x_k \le x < x_{k+1}$, one extracts from the table the nodes and ordinates

$$
\begin{array}{cc}
x_{k-m+1} & y_{k-m+1} \\
x_{k-m+2} & y_{k-m+2} \\
. & . \\
. & . \\
. & . \\
x_{k-1} & y_{k-1} \\
x_k & y_k \\
x_{k+1} & y_{k+1} \\
. & . \\
. & . \\
. & . \\
x_{k+m} & y_{k+m}
\end{array}
$$

(thus m of them on each side of the point x); with these, one then constructs the difference scheme and evaluates the Newton-Gregory formula:

$$P_k(x) = \sum_{j=0}^{2m-1} \binom{t}{j} \Delta^j y_{k-m+1} \left[\text{with } t = \frac{x - x_{k-m+1}}{h} \right] . \tag{28}$$

Here, m should not be too large; perhaps 2 to 5, depending on the accuracy of computation.

Regardless, however, of how one effects this interpolation, in each of the intervals (x_k, x_{k+1}) one uses a different interpolation polynomial $P_k(x)$, so that the *global interpolation function* $F(x)$ is piecewise composed of polynomials of degree $2m - 1$; more precisely, given the ordinates $f(x_0), \ldots, f(x_N)$, one has

$$F(x) = \begin{cases} P_{m-1}(x), & \text{if } x < x_{m-1}, \\ P_k(x), & \text{if } x_k \le x < x_{k+1} \ (k = m-1, \ldots, N-m), \\ P_{N-m}(x), & \text{if } x \ge x_{N-m+1} . \end{cases} \tag{29}$$

But now, the following holds:

Theorem 6.2. *If the $2m + 1$ ordinates $f(x_j)$, $j = k - m$, $k - m + 1$, $\ldots$, $k + m - 1$, $k + m$, do not belong to a polynomial of degree $2m - 1$, then $F'(x)$ is discontinuous at the node x_k.*

Proof. $F(x)$ in the interval (x_{k-1}, x_{k+1}) is composed of $P_{k-1}(x)$ and $P_k(x)$ (both of degree $2m - 1$), which adjoin at the node x_k. For their difference $d(x)$, on account of $P_k(x_j) = P_{k-1}(x_j)$ $(j = k - m + 1, \ldots, k + m - 1)$, one clearly has

$$d(x) = 0 \text{ for } x = x_{k-m+1}, x_{k-m+2}, \ldots, x_{k+m-1} .$$

Since $d(x)$ is of degree $2m - 1$, and these are already $2m - 1$ zeros, either $d(x) \equiv 0$ or $d'(x_k) \ne 0$, q.e.d.

The global interpolation function $F(x)$, therefore, is not everywhere continuously differentiable if the ordinates $f(x_j)$ $(j = 0, \ldots, N)$ do not already belong to a polynomial of degree $2m - 1$. Thus $F(x)$, for example, can have the appearance shown in Figure 6.3.

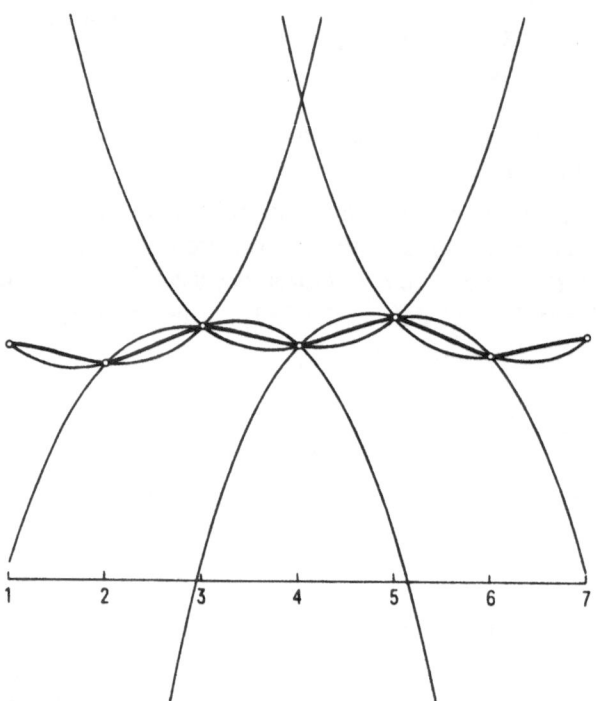

Figure 6.3. *Interpolation by polynomial pieces of degree* 3

If one interpolates a smooth function, the kinks are very small for sufficiently small h, but if one has to interpolate empirical data, large kinks are unavoidable.

§6.7. Hermite interpolation

For a function $f(x)$, which we assume to be $(p-1)$-times continuously differentiable, let the following data be given:

$$f(x_j),\ f'(x_j),\ f''(x_j)\ ,\ldots,\ f^{(p-1)}(x_j)\quad (j = 0, 1, \ldots, N),\quad (30)$$

thus p pieces of data at each node. If we consider only two nodes x_k, x_{k+1}, then for both together we have $2p$ data elements, from which a polynomial $P_k(x)$ (as interpolation polynomial for the interval $x_k \leq x \leq x_{k+1}$) of degree $2p - 1$ can be constructed. Since this polynomial agrees at the points x_k, x_{k+1} with $f(x)$ and all its $p - 1$ first derivatives, the polynomial constructed similarly for the neighboring interval has the same $p - 1$ derivatives at the transition point. The global interpolation function $F(x)$ obtained by combining the $P_k(x)$ for the intervals $[x_k, x_{k+1}]$ $(k = 0, 1, \ldots, N - 1)$ is then everywhere $(p - 1)$-times continuously differentiable. In this way, the disadvantage mentioned in §6.6 of Newton interpolation can be avoided.

For the actual *construction of the Hermite interpolation* polynomial corresponding to the data $f(x_k), f'(x_k), \ldots, f^{(p-1)}(x_k), f(x_{k+1}), f'(x_{k+1}),$ $\ldots, f^{(p-1)}(x_{k+1})$, a new variable

$$t = (x - x_k)/h \quad (h = x_{k+1} - x_k) \tag{31}$$

is introduced, so that the nodes x_k and x_{k+1} are transformed to 0 and 1. The problem is then reduced to the task of constructing the polynomial $Q(t)$ of degree $2p - 1$ subject to the "boundary conditions"

$$\left. \begin{array}{l} Q^{(j)}(0) = h^j f^{(j)}(x_k) \\ Q^{(j)}(1) = h^j f^{(j)}(x_{k+1}) \end{array} \right\} \quad j = 0, 1, \ldots, p - 1 . \tag{32}$$

Then, approximately,

$$f(x_k + th) = Q(t) \quad (0 \le t \le 1).$$

As a preliminary exercise we first determine the interpolation polynomial $Q(t)$ of degree $2p - 1$ for the ordinates $Q(t_j)$ $(j = 0, 1, \ldots, 2p - 1)$, and we do this by means of the difference scheme

The numbers in the top diagonal we denote by $c_0, c_1, \ldots, c_{2p-1}$, i.e., we let (in view of the symmetry property)

$$c_{2\ell} = Q(t_{p-\ell-1}, t_{p-\ell}, \ldots, t_{p+\ell-1}),$$
$$c_{2\ell+1} = Q(t_{p-\ell-1}, t_{p-\ell}, \ldots, t_{p+\ell}) \tag{33}$$

$(\ell = 0, \ldots, p - 1)$. Then, by (15),

$$Q(t) = c_0 + c_1(t - t_{p-1}) + c_2(t - t_{p-1})(t - t_p) + \cdots$$
$$+ c_{2p-1}(t - t_{p-1})(t - t_p) \cdots (t - t_0).$$

If one constructs the usual scheme of divided differences

$$
\begin{array}{ll}
t_0 & Q(t_0) \\
 & \qquad Q(t_0, t_1) \\
t_1 & Q(t_1) \\
\vdots & \vdots \\
t_{p-1} & Q(t_{p-1}) \\
 & \qquad Q(t_{p-1}, t_2) \qquad\qquad Q(t_0, \ldots, t_{2p-1}), \\
t_p & Q(t_p) \\
\vdots & \vdots \\
t_{2p-1} & Q(t_{2p-1})
\end{array}
$$

the coefficients are found in it again, namely the c_{2j} in the same row as t_{p-1}, the c_{2j+1} half a row lower:

$$
\begin{array}{ccccccc}
t_{p-1} & c_0 & & c_2 & & c_4 & \cdots & c_{2p-2} \\
 & & c_1 & & c_3 & & & & c_{2p-1} \cdot \\
t_p & * & & * & & * & \cdots & *
\end{array} \tag{34}
$$

Now, finally, we carry out the limit processes

$$t_0, t_1, t_2, \ldots, t_{p-1} \to 0$$
$$t_p, t_{p+1}, \ldots, t_{2p-1} \to 1, \tag{35}$$

for which, by (17),

$$
Q(t_i, t_{i+1}, \ldots, t_{j-1}, t_j) \to
\begin{cases}
\dfrac{Q^{(j-i)}(0)}{(j-i)!} & \text{if } j \le p - 1, \\[3mm]
\dfrac{Q^{(j-i)}(1)}{(j-i)!} & \text{if } i \ge p,
\end{cases}
\tag{36}
$$

while for $i \le p - 1$, $j \ge p$, the difference relationships remain preserved also in the limit, since in the denominator we always have the difference of an element of the first group (which tend to 0) and an element of the second group (which tend to 1), so that the difference relationship in the limit takes on the form

$$
Q(t_i, t_{i+1}, \ldots, t_{j-1}, t_j) = Q(t_{i+1}, \ldots, t_j) - Q(t_i, \ldots, t_{j-1}). \tag{37}
$$

The quantities $c_0, c_1, \ldots, c_{2p-1}$, therefore, can be computed as follows: the values

$$
\begin{aligned}
a_j &= \frac{Q^{(j)}(0)}{j!} = \frac{h^j}{j!}\, f^{(j)}(x_k) \\[2mm]
b_j &= \frac{Q^{(j)}(1)}{j!} = \frac{h^j}{j!}\, f^{(j)}(x_{k+1})
\end{aligned}
\qquad (j = 0, 1, \ldots, p - 1)
\tag{38}
$$

are arranged as shown below, and the lozenge-shaped area between them filled in by ordinary differencing:

$$
\tag{39}
$$

Then the $c_{2\iota}$ are the quantities at the level of a_0 in this *Hermite scheme,* while the $c_{2\iota + 1}$ are located half a row below.

Finally, also the interpolation polynomial

$$Q(t) = c_0 + c_1(t - t_{p-1}) + c_2(t - t_{p-1})(t - t_p) + \cdots$$

tends in the limit to

$$Q(t) = c_0 + c_1 t + c_2 t(t - 1) + c_3 t^2(t - 1) + \cdots + c_{2p-1} t^p (t - 1)^{p-1},$$

or

$$Q(t) = \sum_{\iota = 0}^{p-1} (c_{2\iota} + c_{2\iota + 1} t)[t(t - 1)]^{\iota}, \qquad (40)$$

still with $t = (x - x_k)/h$.

The *remainder term* also follows from the general Newton formula; one obtains with the same considerations as in §6.4 (with a certain ξ between x_k and x_{k+1}):

$$f(x_k + th) - Q(t) = \frac{f^{(2p)}(\xi)}{2p!} h^{2p}[t(t - 1)]^p. \qquad (41)$$

Programming.   In   the   following   we   assume   the   abscissae $x_k = x_0 + kh$ equidistant (which up to now was not really necessary), and we assume that the derivatives

$$\frac{h^{\iota} f^{(\iota)}(x_k)}{\iota!}$$

for all $k = 0, 1, \ldots, N; \iota = 0, 1, \ldots, p$ are stored as $f[k, \iota]$.[1] Then the following piece of program yields the computation of $Q(t)$ for a given x:

```
begin
    t := (x − x0)/h;
    k := entier (t);
    t := t − k;
    for ℓ := 0 step 1 until p do
    begin
```

---

[1] The program which follows produces the Hermite interpolation polynomial of degree $2p + 1$, not of degree $2p - 1$, as previously discussed. (Translator's remark)

$$a[\ell] := f[k,\ell];$$
$$b[\ell] := f[k+1,\ell]$$
end *for* ℓ;
for $\ell := 0$ **step** 1 **until** p **do**
begin
    if $\ell = 0$ **then goto** $\ell 1$;
    $a[\ell] := b[\ell - 1] - a[\ell]$;
    for $j := \ell + 1$ **step** 1 **until** p **do**
       $a[j] := a[j - 1] - a[j]$;
$\ell 1$: $b[\ell] := b[\ell] - a[\ell]$
    for $j := \ell + 1$ **step** 1 **until** p **do**
       $b[j] := b[j] - b[j - 1]$;
end *for* ℓ;
$x := t \times (t - 1)$;
$s := 0$;
for $\ell := p$ **step** $- 1$ **until** 0 **do**
    $s := s \times x + a[\ell] + b[\ell] \times t$;
end;   (42)

It is perhaps worth noting how (in the second ℓ-loop) the Hermite scheme is built up from the a_ℓ, b_ℓ. It is indeed possible to get by with $2p$ storage locations through systematically overwriting quantities that are no longer needed, as is shown below for $p = 4$:

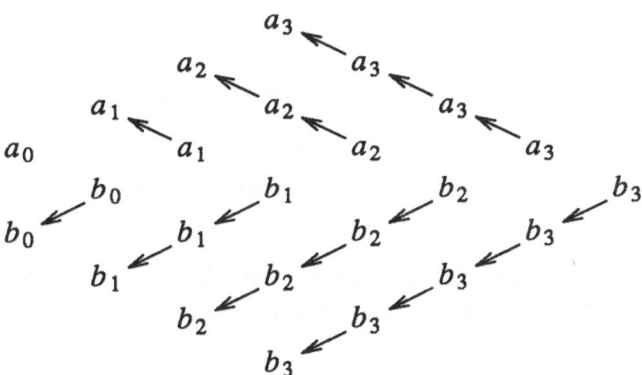

At the end one has $c_{2\ell} = a_\ell$, $c_{2\ell + 1} = b_\ell$.

Numerical example. Suppose the function $f(x) = \ln(x)$ together with the quantities $f^{(\ell)}(x)/\ell!$ for $\ell = 1,2,3$ is tabulated for $x = 1,2,3, \ldots$:

x	f	f'	$f''/2!$	$f'''/3!$
1	.000000	1.000000	−.500000	.333333
2	.693147	.500000	−.125000	.041667
3	1.098612	.333333	−.055556	.012346
4	1.386294	.250000	−.031250	.005208
5	1.609438	.200000	−.020000	.002667

Here the Hermite scheme for the interval (2,3) reads as follows (every-thing rounded and in units of the 6th position after the decimal point):

```
                                 41667
                  −125000                  −11202
          500000                  30465                  3140
    693147          − 94536                − 8062                −904
          405465                  22403                  2235                226
  1098612          − 72132                − 5827                −639
          333333                  16576                  1596
                  − 55556                − 4231
                                 12346
```

Therefore,

$$Q(t) = .693147 + .405465t$$
$$+ (−.094535 + .022403t)t(t − 1)$$
$$+ (−.008062 + .002235t)[t(t − 1)]^2$$
$$+ (−.000904 + .000266t)[t(t − 1)]^3.$$

Evaluation (with 14-digit coefficients) at the point $t = .5$ then gives .91629110, while the 8-digit value of ln (2.5) is .91629073.

§6.8. Spline interpolation

Hermite interpolation can only be applied for functions whose derivatives up to a certain order are known. To let also empirical func-tions partake in the favorable properties of Hermite interpolation, the given function values $f(x_i)$ must be supplemented in a suitable way by derivatives:

$$
\begin{array}{llll}
x_0 & f(x_0) & f'(x_0) \; f''(x_0) & \cdots & f^{(p-1)}(x_0) \\
x_1 & f(x_1) & f'(x_1) \; f''(x_1) & \cdots & f^{(p-1)}(x_1) \\
x_2 & f(x_2) & f'(x_2) \; f''(x_2) & \cdots & f^{(p-1)}(x_2) \\
\cdot & \cdot & \cdot \qquad \cdot & & \cdot \\
\cdot & \cdot & \cdot \qquad \cdot & & \cdot \\
\cdot & \cdot & \cdot \qquad \cdot & & \cdot \\
x_N & f(x_N) & f'(x_N) \; f''(x_N) & \cdots & f^{(p-1)}(x_N)
\end{array}
$$

$$
\underbrace{}_{\text{given}} \quad \underbrace{}_{\text{to be supplemented}}
$$

Of course, these additions should not be made arbitrarily, but according to certain principles. A very elegant method of supplying the missing derivatives is *spline interpolation* which derives from the following type of problem:

We are given function values $f(x_k)$ at $N+1$ nodes x_k $(k = 0, 1, \ldots, N)$. Desired is a function $g(x)$ with the following properties (where $p \le N + 1$):

a) $g(x_k) = f(x_k)$ for $k = 0, 1, \ldots, N$;

b) except at the nodes $x_0, x_1, \ldots, x_N$, $g(x)$ is $2p$-times continuously differentiable, but even at the nodes, $g(x)$ is still $(p-1)$-times continuously differentiable and $g^{(p)}(x)$ is bounded;

c) $E = \frac{1}{2} \int_{x_0}^{x_N} |g^{(p)}(x)|^2 dx$ is minimal.

One thus seeks the smoothest function $g(x)$ through the prescribed points in the sense that the mean square of the pth derivative is minimal.

Now integrating p-times by parts the variation of E, one obtains

$$\delta E = g^{(p)} \delta g^{(p-1)} \bigg|_{x_0}^{x_N} - \sum_{k=1}^{N-1} [g^{(p)}(x_k + 0) - g^{(p)}(x_k - 0)] \delta g^{(p-1)}(x_k)$$

$$- g^{(p+1)} \delta g^{(p-2)} \bigg|_{x_0}^{x_N} + \sum_{k=1}^{N-1} [g^{(p+1)}(x_k + 0) - g^{(p+1)}(x_k - 0)] \delta g^{(p-2)}(x_k)$$

$$\vdots$$

$$(-1)^p g^{(2p-1)} \delta g \bigg|_{x_0}^{x_N} + (-1)^p \sum_{k=1}^{N-1} [g^{(2p-1)}(x_k + 0) - g^{(2p-1)}(x_k - 0)] \delta g(x_k)$$

$$+ (-1)^p \int_{x_0}^{x_N} g^{(2p)}(x) \delta g(x) dx.$$

Since, with the exception of $\delta g(x_k)$, all variations $\delta g^{(j)}(x_k)$ are free, there follows from the minimality requirement:

1) $g^{(p)}(x) = g^{(p+1)}(x) = \cdots = g^{(2p-2)}(x) = 0$ for $x = x_0$, $x = x_N$.
2) $g^{(j)}(x_k + 0) - g^{(j)}(x_k - 0) = 0$ for $j = p, p + 1, \ldots, 2p - 2$, $k = 1, 2, \ldots, N - 1$; i.e., $g(x)$ is actually $(2p - 2)$-times continuously differentiable in the whole interval (x_0, x_N).
3) But $g^{(2p-1)}(x)$ (since the values $g(x_k)$ cannot be varied) may be discontinuous at the nodes $x_1, x_2, \ldots, x_{N-1}$, and its boundary values at x_0 and x_N need not vanish.
4) $g^{(2p)}(x) = 0$ for $x \neq x_0, x_1, \ldots, x_N$.

Therefore, $g(x)$ is piecewise made up of polynomials of degree $2p - 1$ in such a way that at the joints $x_1, x_2, \ldots, x_{N-1}$ only the $(2p - 1)$st derivatives has jumps.

As to the determination of the component polynomials of $g(x)$, note that:

a) yields $N + 1$ conditions,
1) yields $2p - 2$ conditions,
3) involves $N - 1$ degrees of freedom, namely the jumps of the $(2p - 1)$st derivative at the nodes $x_1, \ldots, x_{N-1}$,

4) involves $2p$ degrees of freedom (differential equation of order $2p$).

Altogether one thus has $2p + N - 1$ conditions and equally many degrees of freedom, so that one can hope for uniqueness of the solution. Now, since $g(x)$ in each of the intervals (x_i, x_{i+1}) is a polynomial $P_i(x)$ of degree $2p - 1$, one can express this polynomial uniquely in terms of g_i, g'_i, g''_i, ..., $g_i^{(p-1)}$, $g_{i+1}, g'_{i+1}, \ldots, g_{i+1}^{(p-1)}$, for example by means of the Hermite interpolation formula. Obviously, $P_i(x)$, and hence also $P_i^{(p)}(x)$, is linear in the $g_i^{(k)}$, $g_{i+1}^{(k)}$. This means, however, that $\int |P_i^{(p)}|^2 dx$ is a quadratic form in the same quantities. By still summing over all N subintervals, one obtains the following fact: E *is a quadratic form in all* $g_i^{(k)} = g^{(k)}(x_i)$ $(i = 0, 1, \ldots, N; k = 0, 1, \ldots, p - 1)$, *thus*

$$E = \frac{1}{2} \sum_{i,j=0}^{N} \sum_{k,\ell=0}^{p-1} a_{ikj\ell} g_i^{(k)} g_j^{(\ell)}. \tag{43}$$

Since the $N + 1$ quantities $g_i = g_i^{(0)} = g(x_i)$ are prescribed, only a reduced quadratic function,

$$E^* = \frac{1}{2} \sum_{i,j=0}^{N} \sum_{k,\ell=1}^{p-1} a_{ikj\ell} g_i^{(k)} g_j^{(\ell)} + \sum_{i,j=0}^{N} \sum_{k=1}^{p-1} a_{ikj0} g_i^{(k)} g_j. \tag{44}$$

has in fact to be minimized with respect to the $g_i^{(k)}$ $(k \neq 0)$, which leads to the linear system of equations

$$\sum_{j=0}^{N} \sum_{\ell=1}^{p-1} a_{ikj\ell} g_j^{(\ell)} + \sum_{j=0}^{N} a_{ikj0} = 0 \quad (i = 0, 1, \ldots, N; k = 1, 2, \ldots, p-1) \tag{45}$$

with positive definite symmetric coefficient matrix

$$\mathbf{A} = \{a_{ikj\ell}, \quad i,j = 0, \ldots, N; \quad k, \ell = 1, \ldots, p - 1\}$$

(cf. §3.1). While symmetry is obvious, positive definiteness is obtained as follows: The quadratic form for $\mathbf{A}$ is nothing but $2E$, if one puts there all $g_i^{(0)} = 0$, thus

$$Q = \int_{x_0}^{x_N} |g^{(p)}(x)|^2 dx \quad \text{for} \quad g(x_0) = g(x_1) = \cdots = g(x_N) = 0.$$

Q attains a minimum 0 for $g \equiv 0$; this minimum, however, cannot be attained for any other $(p-1)$-times continuously differentiable function g, since $Q = 0$ would mean that g is a polynomial of degree $p-1$ which then could not be 0 at $N + 1 > p - 1$ points, q.e.d.

The system of equations (45) therefore admits a unique solution; as a result, one has in each of the subintervals $x_i \leq x \leq x_{i+1}$ the necessary $2p$ data elements for Hermite interpolation according to §6.7.

Practical implementation for $p = 2$ and equidistant nodes.

Here, $g(x)$ in each subinterval is a polynomial of degree 3. As one easily verifies on the basis of (38), (39), (40), this polynomial, in (x_1, x_{i+1}), is given by

$$P_i(x) = g_i(1-t)^2(1+2t) + hg_i'(1-t)^2 t$$
$$+ g_{i+1}(3t^2 - 2t^3) + hg_{i+1}'t^2(t-1), \tag{46}$$

where $t = (x - x_i)/h$. Therefore,

$$h^2 P_i''(x) = (12t-6)g_i + (6t-4)hg_i' + (6-12t)g_{i+1} + (6t-2)hg_{i+1}',$$

and thus finally

$$\frac{1}{2} \int_{x_i}^{x_{i+1}} |P_i''(x)|^2 dx = Q(g_i, hg_i', g_{i+1}, hg_{i+1}'),$$

where the form Q, up to a constant factor, has the following coefficient matrix:

$$
\mathbf{M}_i =
\begin{bmatrix}
6 & 3 & -6 & 3 \\
3 & 2 & -3 & 1 \\
-6 & -3 & 6 & -3 \\
3 & 1 & -3 & 2
\end{bmatrix}
\begin{matrix}
g_i \\
hg_i' \\
g_{i+1} \\
hg_{i+1}'
\end{matrix}
$$
$$\quad\quad g_i \quad hg_i' \quad g_{i+1} \quad hg_{i+1}'$$

The matrix of the quadratic form E is obtained through summation over i, whereby the matrices $\mathbf{M}_i$ add up with overlaps; e.g., for $N = 3$:

$$g_0 \quad hg_0' \quad g_1 \quad hg_1' \quad g_2 \quad hg_2' \quad g_3 \quad gh_3'$$

6	3	−6	3				
3	2	−3	1				
−6	−3	12	0	−6	3		
3	1	0	4	−3	1		
		−6	−3	12	0	−6	3
		3	1	0	4	−3	1
				−6	−3	6	−3
				3	1	−3	2

In the matrix of the reduced quadratic form E^* the rows corresponding to the g_i drop out [cf. (44), (45)]; therefore, the following system of equations results ([1]):

$$hg_0' \quad hg_1' \quad hg_2' \quad \cdots \quad hg_{N-1}' \quad hg_N'$$

2	1					$= 3g_1 - 3g_0$
1	4	1				$= 3g_2 - 3g_0$
	1	4	1			$= 3g_3 - 3g_1$
		⋱	⋱	⋱		⋮
		1	4	1		$= 3g_{N-1} - 3g_{N-3}$
			1	4	1	$= 3g_N - 3g_{N-2}$
				1	2	$= 3g_N - 3g_{N-1}$

$$(47)$$

The solution of this system, even for large N, does not cause any difficulties, since the matrix is banded and very well-conditioned. (Cf. §10.6; the eigenvalues lie between 1 and 6.)

Numerical example. We are given 7-digit logarithms to the base 10 (see 2nd column of Table 6.1).

---

[1] A simpler derivation of (47) can be had by starting with (46) and imposing on $g''(x)$ the following conditions, earlier recognized as necessary: continuity in $x_1, \ldots, x_{N-1}$, vanishing at x_0, x_N. (Editors' remark)

Table 6.1. *Spline interpolation of logarithms*

x_k	f_k	hg'_k	hf'_k
2	.3010300	.0214011	.0217147
2.1	.3222193	.0207657	.0206807
2.2	.3424227	.0197143	.0197407
2.3	.3617278	.0189028	.0188824
2.4	.3802112	.0180402	.0180956
2.5	.3979400	.0175731	.0173718

The system of equations for the hg'_k reads here as follows:

hg'_0	hg'_1	hg'_2	hg'_3	hg'_4	hg'_5		
2	1					=	.0635679
1	4	1				=	.1241781
	1	4	1			=	.1185255
		1	4	1		=	.1133655
			1	4	1	=	.1086366
				1	2	=	.0531864

and has the hg'_k-values given in the third column of Table 6.1 as solution. (For comparison, the corresponding exact derivatives are quoted in the last column.)

In the interval $2.1 \le x \le 2.2$ one obtains as Hermite difference scheme to g_1, hg'_1, g_2, hg'_2 (in units of the 7th position after the decimal point):

$$
\begin{array}{ccccc}
 & & 207657 & & \\
3222193 & & & -5623 & \\
\underline{} & & 202034 & & 732, \\
 & & \underline{} & & \\
3424227 & & & -4891 & \\
 & & 197143 & & \\
\end{array}
$$

and thus,

$$g(2.1 + .1t) = .3222193 + .0202034t + (-.0005623 + .0000732t)[t(t-1)],$$

in particular, for example, $g(2.15) = .3324524$, while $\log(2.15) = .3324385$.

As a further example, we show in Figure 6.4 the interpolating spline $g(x)$, together with its derivatives $g'(x)$ and $g''(x)$, which belongs to the support points already used in Figure 6.3. $g''(x)$ is piecewise linear, $g'''(x)$ therefore would be piecewise constant.

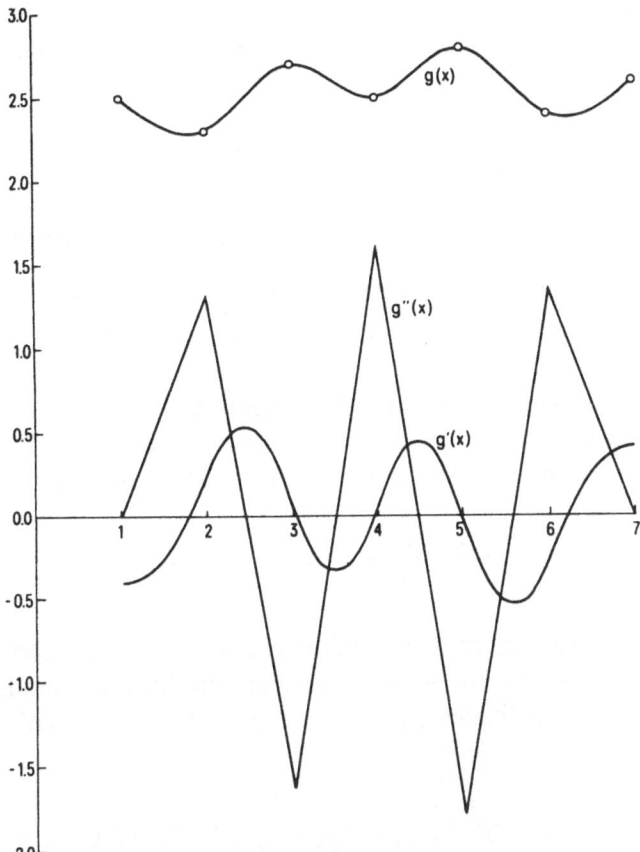

Figure 6.4. *Spline interpolation: the interpolating function g and its derivatives g' and g''*

Physical interpretation. Spline interpolation in the case $p = 2$ admits the following simple interpretation: given $N + 1$ points (x_i, y_i), the deflection of a thin beam ("spline") placed through these points is characterized by the following conditions, provided the linear theory of elasticity is applicable:

a) $y(x_i) = y_i$,

b) y, y', y'' everywhere continuous,

c) $y^{(4)}(x) = 0$ for $x \neq x_i$.

These, however, are precisely the conditions imposed on the function $g(x)$ in spline interpolation with $p = 2$. One can therefore accomplish spline interpolation also with a spline (and actually does so).

§6.9. Smoothing

If the ordinates $f(x_0), f(x_1), \ldots, f(x_N)$ to be interpolated are inaccurate measurements and exhibit an irregular behavior, one can try, through small changes in the $f(x_i)$, to enforce a somewhat smoother behavior of the interpolated function, in short: to first smooth the values $f(x_i)$.

The interpolation function $g(x)$ therefore – regardless of how one wants to interpolate – will not be forced through the prescribed ordinates $f(x_j) = f_j$, but deviations are tolerated, though such that [if $g_j = g(x_j)$]

$$\alpha) \quad \sum_{k=0}^{N} |g_k - f_k|^2 \text{ remains small .}$$

On the other hand, in case of equidistant nodes, the second differences $g_{k+1} - 2g_k + g_{k-1}$ presumably will be a measure for the smoothness of the new sequence of ordinates $g_0, g_1, \ldots, g_N$. One thus will have to be careful that

$$\beta) \quad \sum_{k=1}^{N-1} |g_{k+1} - 2g_k + g_{k+1}|^2 \text{ remains small .}$$

A compromise between the conflicting requirements $\alpha)$, $\beta)$ can be achieved by determining the adjusted ordinates g_i such that

$$\sum_{k=0}^{N} (g_k - f_k)^2 + \gamma \sum_{k=1}^{N-1} (g_{k+1} - 2g_k + g_{k-1})^2 \tag{48}$$

becomes minimal. Here, γ is the so-called *smoothing coefficient*, about which we have to say more later. The above minimum requirements immediately leads to a linear system of equations

$$(I + \gamma A)g = f,$$

where f, g denote the vectors $[f_0, f_1, \ldots, f_N]^T$ and $[g_0, g_1, \ldots, g_N]^T$ formed with the old and new ordinates, respectively, and

$$
A = \begin{bmatrix}
1 & -2 & 1 & & & & & & \\
-2 & 5 & -4 & 1 & & & & & \\
1 & -4 & 6 & -4 & 1 & & & & \\
 & 1 & -4 & 6 & -4 & 1 & & & \\
 & & \cdot & \cdot & \cdot & \cdot & \cdot & & \\
 & & & \cdot & \cdot & \cdot & \cdot & \cdot & \\
 & & & & \cdot & \cdot & \cdot & \cdot & \cdot \\
 & & & & 1 & -4 & 6 & -4 & 1 \\
 & & & & & 1 & -4 & 5 & -2 \\
 & & & & & & 1 & -2 & 1
\end{bmatrix}. \tag{49}
$$

The matrix $I + \gamma A$ is positive definite; the eigenvalues lie in the interval[1] $1 \le \lambda \le 1 + 16\gamma$. Having determined the g_k, one can interpolate with them by whatever method, for example by means of spline interpolation.

The *choice of the smoothing coefficient* is completely free, to begin with, but the following should be observed: $\gamma < .01$ means weak smoothing; the new values g_k follow the given values $f_0, \ldots, f_N$ essentially still point for point. $\gamma > 10$ produces a strong smoothing; the shape of the curve determined by the g_k follows the given values f_k only globally. For still larger γ also the condition of the matrix $I + \gamma A$ gradually worsens[2].

---

[1] As matrix of the quadratic form of the second sum in (48), A is positive semidefinite; on the other hand, the upper limit of the interval follows from the row sum norm, cf. §10.7. (Editors' remark)

[2] The stepsize h of the nodes, too, has an influence on the choice of γ : starting with the problem of minimizing

$$\int_{x_0}^{x_N} (g(x) - f(x))^2 dx + \tilde{\gamma} \int_{x_0}^{x_N} (g''(x))^2 dx,$$

discretization indeed yields (48), but with $\gamma = \tilde{\gamma}/h^4$. Small h therefore requires very strong smoothing γ. (Editors' remark)

Numerical example. We consider the 5-digit values of the common logarithm of the numbers 1.001, 1.002 ,..., 2.000. This function defined on 1000 nodes is of course slightly irregular, if one considers the rounded values as exact. We therefore want to subject them to the smoothing process described. The result is summarized in Table 6.2 for various γ.

Table 6.2. *Smoothing of rounded logarithms*

nodes x_i	given function values $f_i \times 10^5$	smoothed function values $g_i \times 10^5$					7-digit logarithms $\times 10^5$
		$\gamma = .01$	$\gamma = .1$	$\gamma = 1$	$\gamma = 10$	$\gamma = 100$	
1.001	43	43.01	43.07	43.17	43.31	43.65	43.41
1.002	87	86.98	86.87	86.69	86.69	86.88	86.77
1.003	130	130.00	130.01	130.03	130.03	130.12	130.09
1.004	173	173.03	173.16	173.35	173.35	173.34	173.37
1.005	217	216.97	216.86	216.73	216.63	216.54	216.61
1.006	260	260.01	260.02	259.95	259.84	259.72	259.80
1.007	303	303.00	303.01	303.02	302.97	302.86	302.95
1.008	346	346.00	346.00	346.03	346.05	345.97	346.05
1.009	389	389.00	389.00	389.02	389.08	389.03	389.12
1.010	432	432.00	432.00	432.01	432.09	432.07	432.14
⋮							
1.041	1745	1745.00	1745.00	1745.02	1745.07	1745.11	1745.07
1.042	1787	1786.97	1786.84	1786.71	1786.75	1786.80	1786.77
1.043	1828	1828.03	1828.16	1828.31	1828.40	1828.46	1828.43
1.044	1870	1870.00	1870.00	1870.02	1870.05	1870.09	1870.05
1.045	1912	1911.97	1911.85	1911.71	1911.67	1911.68	1911.63
⋮							
1.501	17638	17638.00	17638.00	17638.02	17638.03	17638.01	17638.07
1.502	17667	17667.00	17667.00	17667.03	17666.99	17666.94	17666.99
1.503	17696	17696.00	17696.01	17696.02	17695.93	17695.86	17695.90
1.504	17725	17725.01	17725.02	17724.93	17724.81	17724.74	17724.78
1.505	17754	17753.97	17753.85	17753.70	17753.64	17753.61	17753.65
⋮							
1.995	29994	29994.03	29994.15	29994.34	29994.38	29994.33	29994.29
1.996	30016	30015.99	30015.99	30016.10	30016.14	30016.10	30016.05
1.997	30038	30038.01	30038.01	30037.93	30037.89	30037.87	30037.81
1.998	30060	30059.97	30059.86	30059.69	30059.62	30059.63	30059.55
1.999	30081	30081.03	30081.15	30081.28	30081.31	30081.38	30081.28
2.000	30103	30102.99	30102.95	30102.94	30103.01	30103.14	30103.00

§6.10. **Approximate quadrature**

For the determination of $\int f(x)dx$ one uses different methods, depending on whether the integral is a definite or an indefinite one. In the first case, a number is produced, in the second a numerical table (the integral as function of the upper limit).

If the *definite integral* $\int_a^b f(x)dx$ is to be computed, one could begin by subdividing the interval $[a,b]$ into n subintervals $[x_{i-1}, x_i]$ ($i = 1, \ldots, n$) of equal length h, then from the $n + 1$ ordinates $f(x_i)$ construct the interpolation polynomial $P(x)$ of degree n, and finally integrate the latter from a to b:

$$\int_a^b f(x)dx \approx \int_a^b P(x)dx = \int_a^b \sum_{k=0}^n \binom{t}{k} \Delta^k f_0 dx = h \sum_{k=0}^n \Delta^k f_0 \int_0^n \binom{t}{k} dt \ . (50)$$

Examples. For $n = 1$, because of

$$\int_0^1 \binom{t}{0} dt = 1, \quad \int_0^1 \binom{t}{1} dt = \frac{1}{2},$$

one obtains the formula

$$\int_{x_0}^{x_1} f(x)dx \approx hf_0 + \frac{h}{2} \Delta f_0 = \frac{h}{2}(f_0 + f_1). \tag{51}$$

Now, of course, one can joint together m such individual intervals, which leads to

$$\int_{x_0}^{x_m} f(x)dx \approx \frac{h}{2}\{f_0 + 2f_1 + 2f_2 + \cdots + 2f_{m-1} + f_m\}. \tag{52}$$

This is the so-called *trapezoidal rule*.

Similarly, in the case $n = 2$, from

$$\int_0^2 \binom{t}{0} dt = 2, \quad \int_0^2 \binom{t}{1} dt = 2, \quad \int_0^2 \binom{t}{2} dt = \frac{1}{3}$$

there first follows the formula

$$\int_{x_0}^{x_2} f(x)dx \approx 2hf_0 + 2h\Delta f_0 + \frac{h}{3}\Delta^2 f_0$$

$$= 2hf_0 + 2h(f_1 - f_0) + \frac{h}{3}(f_2 - 2f_1 + f_0)$$

$$= \frac{h}{3}(f_0 + 4f_1 + f_2).$$

Now by joining together m such interval pairs (cf. Fig. 6.5) one obtains the *Simpson rule*

$$\int_{x_0}^{x_{2m}} f(x)dx \approx \frac{h}{3}\{f_0 + 4f_1 + 2f_2 + 4f_3 + 2f_4 + \cdots + 4f_{2m-1} + f_{2m}\}. \quad (53)$$

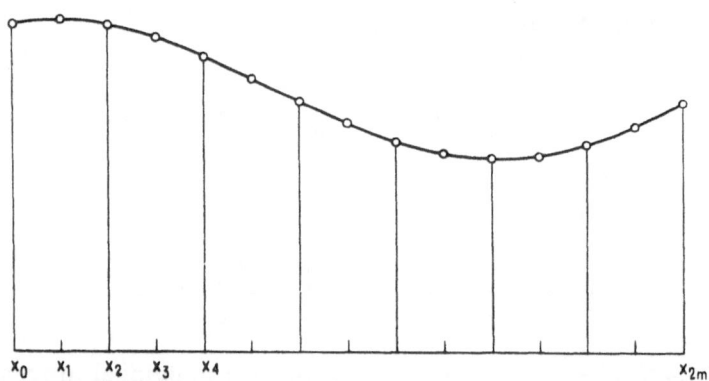

Figure 6.5. *Quadrature according to Simpson*

One can also pass through five consecutive ordinates a polynomial of degree four and integrate it; in this way one obtains the *Newton-Cotes formula*

$$\int_{x_0}^{x_4} f(x)dx \approx \frac{2h}{45}\{7f_0 + 32f_1 + 12f_2 + 32f_3 + 7f_4\}. \quad (54)$$

But all these efforts of increasing the accuracy are by far outdone by the following consideration:

Dividing the interval $[a,b]$, as above, in n subintervals of length $h = (b - a)/n$ and applying the trapezoidal rule, then — provided that the function f is $2m$-times continuously differentiable — the error will satisfy an asymptotic expansion for $h \to 0$ $(n \to \infty)$ of the following form([1]):

$$\int_a^b f(x)dx - T(h) = c_1 h^2 + c_2 h^4 + \cdots + c_m h^{2m} + o(h^{2m}), \quad (55)$$

where of course

$$T(h) = \frac{h}{2} \{f_0 + 2f_1 + 2f_2 + \cdots + 2f_{n-1} + f_n\} \quad (56)$$

denotes the approximate value of the integral computed by the trapezoidal rule.

Now this holds for all h; hence

$$\int_a^b f(x)dx - T\left[\frac{h}{2}\right] = c_1 \frac{h^2}{4} + c_2 \frac{h^2}{16} + \cdots + c_m \frac{h^{2m}}{4^m} + o(h^{2m}).$$

Multiplying this by $\frac{4}{3}$ and subtracting a third of (55), one gets

$$\int_a^b f(x)dx - \left[\frac{4}{3} T\left[\frac{h}{2}\right] - \frac{1}{3} T(h)\right] = c_1 \left[\frac{4}{3} \frac{h^2}{4} - \frac{1}{3} h^2\right] +$$

$$c_2 \left[\frac{4}{3} \frac{h^4}{16} - \frac{1}{3} h^4\right] + c_3 \left[\frac{4}{3} \frac{h^6}{64} - \frac{1}{3} h^6\right] + \cdots + o(h^{2m}),$$

and one sees that the first term on the right drops out. With the definition

$$T_1(h) = \frac{4 T\left[\frac{h}{2}\right] - T(h)}{3} \quad (57)$$

---

[1] This can be derived from the Euler-Maclaurin summation formula; see, e.g., Stoer J., Bulirsch R.: *Introduction to Numerical Analysis*, Springer-Verlag, New York 1980, Sections 3.3 and 3.4. (Translator's remark)

one thus obtains

$$\int_a^b f(x)dx - T_1(h) = c_2' h^4 + c_3' h^6 + \cdots + c_m' h^{2m} + o(h^{2m}), \quad (58)$$

and likewise, if $T\left[\dfrac{h}{4}\right]$ is the result of the trapezoidal rule applied with step $h/4$, with

$$T_1\left[\frac{h}{2}\right] = \frac{4\,T\left[\dfrac{h}{4}\right] - T\left[\dfrac{h}{2}\right]}{3},$$

$$\qquad\qquad (59)$$

$$\int_a^b f(x)dx - T_1\left[\frac{h}{2}\right] = c_2' \frac{h^4}{16} + c_3' \frac{h^6}{64} + \cdots + c_m' \frac{h^{2m}}{4^m} + o(h^{2m}).$$

Now again, one combines linearly: the relation (58) is to be multiplied by $-\dfrac{1}{15}$, the last one by $\dfrac{16}{15}$:

$$\int_a^b f(x)dx - \left[\frac{16}{15} T_1\left[\frac{h}{2}\right] - \frac{1}{15} T_1(h)\right] = c_2' \left[\frac{16}{15} \frac{h^4}{16} - \frac{1}{15} h^4\right] +$$

$$c_3' \left[\frac{16}{15} \frac{h^6}{64} - \frac{1}{15} h^6\right] + \cdots + o(h^{2m});$$

thus,

$$\int_a^b f(x)dx - T_2(h) = c_3'' h^6 + c_4'' h^8 + \cdots + c_m'' h^{2m} + o(h^{2m}), \quad (60)$$

if one introduces

$$T_2(h) = \frac{16\, T_1\left[\dfrac{h}{2}\right] - T_1(h)}{15}. \qquad (61)$$

The next step would employ

$$T_3(h) = \frac{64\, T_2\left[\dfrac{h}{2}\right] - T_2(h)}{63}. \qquad (62)$$

In this way one obtains approximations to $\int_a^b f(x)dx$ which converge very rapidly to the exact value. It is convenient to arrange these approximations in the following *Romberg scheme*:

$$
\begin{array}{cccc}
T(h) & & & \\
 & T_1(h) & & \\
T\left[\dfrac{h}{2}\right] & & T_2(h) & \\
 & T_1\left[\dfrac{h}{2}\right] & & T_3(h) \\
T\left[\dfrac{h}{2}\right] & & T_1\left[\dfrac{h}{4}\right] & \quad \cdot \quad \cdot \\
 & T_1\left[\dfrac{h}{4}\right] & \quad \cdot & \quad \cdot \\
T\left[\dfrac{h}{8}\right] & & \quad \cdot & \\
 & \quad \cdot & & \\
\quad \cdot & \quad \cdot & & \\
\quad \cdot & & &
\end{array}
\tag{63}
$$

Example. Let us compute $\int_1^2 \dfrac{dx}{x}$. One first determines

$$T(1) = \tfrac{1}{2} \{f(1) + f(2)\} = .75,$$

$$T\left[\tfrac{1}{2}\right] = \tfrac{1}{4} \{f(1) + 2\,f(1.5) + f(2)\} = .708333,$$

$$T\left[\tfrac{1}{4}\right] = .697024,$$

$$T\left[\tfrac{1}{8}\right] = .694122.$$

These numbers are put into the first column of the Romberg scheme, which can now be built up by utilizing (57), (61) and (62):

```
.750000
            .694444
.708333                 .693175
            .693254                 .693147
.697024                 .693148
            .693155
.694122
```

The value $T_3(h)$ at the tip is exact to 6 digits. (With an additional row, the result would be obtained to 9 correct digits.)

Of course, one also has to pay for this: for each new value that is added at the bottom in the first column of the Romberg scheme (in order to be able to add a new value also in every other column), one must apply the trapezoidal rule with twice as many terms as before. For

$$\int_0^{10} \frac{dx}{1 + x^2} = \tan^{-1}(10) = 1.4711276743037,$$

for example, the expense, as can be seen from Table 6.3, is rather substantial. It must be remarked, though, that we are dealing here with a pathological case[2].

In contrast, when dealing with a well-behaved case, the accuracy of the value at the tip, with each halving, can be enormously enhanced. For

---

[2] For the improper integral

$$I = \int_0^\infty \frac{dx}{1 + x^2} = \frac{\pi}{2}$$

there indeed holds

$$T(h) - I = \frac{\pi}{2} \coth \frac{\pi}{2} - \frac{\pi}{2} \approx \pi \exp\left[-\frac{2\pi}{h}\right],$$

so that

$$\lim_{h \to 0} \frac{T\left[\frac{h}{2}\right] - I}{(T(h) - I)^2} = \frac{1}{\pi}.$$

If the infinite series for the $T(h)$ were evaluated exactly, the first column of the associated Romberg scheme would thus converge quadratically. [In contrast, (55) implies linear convergence of this column in the general case.] Romberg extrapolation, therefore, does not make sense. The finite interval considered, behaves similarly in practice. Compare Section 8 in Bauer F.L., Rutishauser H., Stiefel E.: New aspects in numerical quadrature, *Proc. Symp. Appl. Math.* **15**, 199–218 (1963), Amer. Math. Soc., Providence, R.I. (Editors' remark)

example, in the computation of $\int_1^2 \frac{dx}{x}$, these values have the following errors:

$$\text{error of } T_1(h): \quad .0013 \ldots$$
$$\text{error of } T_2(h): \quad .000027 \ldots$$
$$\text{error of } T_3(h): \quad .00000030 \ldots$$
$$\text{error of } T_4(h): \quad .0000000014 \ldots$$

Table 6.3. *Romberg scheme for* $\int_0^{10} \frac{dx}{1+x^2}$

(The triangle at the bottom must be thought of as the tip, to be joined on the right to the trapezoid above it.)

5.0495049504951					
2.7170601675552	1.9395785732420				
1.7470257922553	1.4236810004886	1.3892878289717			
1.4916226095488	1.4064882153133	1.4053420296350	1.4055968582170		
1.4711991596046	1.4643913429565	1.4682515514661	1.4692501153047	1.4694997359207	
1.4711117278096	1.4710825838780	1.4715286666061	1.4715806843067	1.4715898237930	1.4715918668896
1.4711236856579	1.4711276716073	1.4711306774559	1.4711243601678	1.4711225706614	1.4711221139135
1.4711266771069	1.4711276742565	1.4711276744332	1.4711276267661	1.4711276395763	1.4711276445313
1.4711274250023	1.4711276743008	1.4711276743037	1.4711276743017	1.4711276744881	1.4711276745222
1.4711276119782	1.4711276743036	1.4711276743037	1.4711276743037	1.4711276743037	1.4711276743036
1.4711276587223	1.4711276743037	1.4711276743037	1.4711276743037		1.4711276743037

1.4711219991997				
1.4711276458818	1.4711276462265			
1.4711276745295	1.4711276745313	1.4711276745317		
1.4711276743035	1.4711276743035	1.4711276743035	1.4711276743035	
1.4711276743037	1.4711276743037	1.4711276743037	1.4711276743037	1.4711276743037

Notes to Chapter 6

§6.1 The basic Lagrange polynomials l_k of Eq. (2) not only enter in Lagrange's interpolation formula (4), but also play an important role in metric properties of the Lagrange interpolation operator. Assuming $a \leq x_0 < x_1 < x_2 < \cdots < x_n \leq b$, the interpolation process may be viewed as a projector $P_n\colon C[a,b] \to \mathbf{P}_n$ from the space of continuous functions on $[a,b]$ to polynomials of degree $\leq n$. The norm of this projector, $||P_n|| = \sup(||P_n f|| / ||f||)$, where $||f|| = \max_{a \leq x \leq b} |f(x)|$, can be expressed in terms of the "Lebesgue function" $\lambda_n(x) = \sum_{k=0}^{n} |l_k(x)|$ as $||P_n|| = ||\lambda_n||$. A study of this norm is of both theoretical and practical interest. It yields, for example, information about the interpolation error, by virtue of $||f - P_n f|| \leq (1 + ||P_n||) \operatorname{dist}(f, \mathbf{P}_n)$, where $\operatorname{dist}(f, \mathbf{P}_n)$ is the distance (in the norm $||\cdot||$) of f to $\mathbf{P}_n$, i.e., the error of best uniform approximation of f by polynomials of degree $\leq n$ (see §7.6). Therefore, if this error, as $n \to \infty$, goes to zero faster than $||P_n||$ tends to ∞, the interpolation process converges uniformly. This is so, e.g., if f has a continuous first derivative on $[a,b]$ and x_i are the zeros of the Chebyshev polynomial T_{n+1} (defined in §7.2), adjusted to the interval $[a,b]$. Points x_i that are uniformly spaced on $[a,b]$, on the other hand, may yield divergence, even for functions analytic on $[a,b]$, as is shown by a famous example of Runge; see, e.g., Todd [1962, p. 148], Epperson [1987]. It is known indeed, for equidistant points x_i, that $||P_n|| \sim 2^{n+1}/(e\,n \log n)$ as $n \to \infty$ (cf. Trefethen & Weideman [to appear]). By its very definition, the norm $||P_n||$ also measures the sensitivity of the interpolation polynomial to perturbations in f, since $||P_n f^* - P_n f|| \leq ||P_n|| \cdot ||f^* - f||$.

In the light of these remarks, the following problem is of interest, and in fact, has had a long history: Determine nodes x_i such that $||P_n|| = ||\lambda_n||$ is as small as possible. The problem has recently been solved by de Boor & Pinkus [1978], following work of Kilgore [1978]. To state their principal result, which confirms long outstanding conjectures of Bernstein and Erdös, one should first note that $\lambda_n(x) \geq 1$ for all x, and that $\lambda_n(x)$, for $n \geq 2$, on each interval $[x_{i-1}, x_i]$, has a unique local maximum, say at $x = \xi_i$, $i = 1, 2, \ldots, n$. Then the optimal nodes x_i are characterized, and uniquely determined, by the "equioscillation property" $\lambda_n(\xi_1) = \lambda_n(\xi_2) = \cdots = \lambda_n(\xi_n)$. The computation of these optimal nodes, of course, is not quite easy; numerical values for $n \leq 15$, however, have already been obtained by Hayes & Powell [1969], who took the validity of Bernstein's conjecture for granted. On the other hand, very good approximations (yielding the minimum of $||\lambda_n||$ within a margin of .201, and even, very likely, of .02; see Brutman [1978], Günttner [1980]) are given by the Chebyshev nodes adjusted to the interval $[a,b]$ and expanded such that the smallest and the largest node coincides with the respective endpoint of $[a,b]$.

Other interpolation processes are sometimes more appropriate. For periodic functions, one often employs *trigonometric interpolation*. This consists in passing a trigonometric polynomial of degree n (that is, a linear combination of 1, cos x, sin x, . . . , cos nx, sin nx, if the period is 2π) through given ordinates y_k at $2n + 1$ distinct points x_k in $[0, 2\pi)$. The analogue of Lagrange's interpolation formula, due to Gauss, is now
$$T(x) = \sum_{k=0}^{2n} y_k t_k(x), \quad \text{where} \quad t_k(x) = \prod_{\substack{j=0 \\ j \neq k}}^{2n} \frac{\sin \tfrac{1}{2}(x - x_j)}{\sin \tfrac{1}{2}(x_k - x_j)}.$$ The corresponding projector has minimum norm precisely if the points x_k are equally spaced on $[0, 2\pi)$; see de Boor &

Pinkus [1978]. In this case, there exists also an extensive convergence theory (Zygmund [1968, Ch. 10]). The interpolation process converges, for example, if the function f to be interpolated has an absolutely convergent Fourier series; see also Theorem 7.3. The error, then, can be estimated by $||T - f|| \le 2 \sum_{|k| > n} |c_k(f)|$, where $c_k(f)$ are the complex Fourier coefficients of f (Zygmund [1968, Thm. 5.16]).

For *interpolation by rational functions* there are known continued fraction representations for the interpolant, if numerator and denominator have the same degree, and algorithmic procedures in the general case; see, e.g., Bulirsch & Rutishauser [1968]. The interpolation problem, however, does not always admit solution, and even if it does, the interpolant may have undesirable poles between the nodes. For rational interpolation with prescribed poles in the complex plane, see Walsh [1969, Ch. 8].

Lagrange's interpolation formula (4) remains valid without change for complex-valued functions and arbitrary distinct nodes in the complex plane. *Interpolation in several* (real) *variables*, on the other hand, is more complicated. The problem then involves $n_d = \binom{n + d}{d}$ points in d-dimensional space $\mathbf{R}^d$, there being exactly n_d monomials of degree $\le n$ in d variables. A unique solution exists whenever the given points do not lie on an algebraic hypersurface of degree n. A Lagrange-type formula, involving determinants, can then be constructed (Thacher [1960], Thacher & Milne [1960], Mysovskih [1981, §3.1]). In practice, however, it will be simpler to numerically solve the linear system of equations which express the interpolation property in a convenient basis of polynomials. Nevertheless, to require the interpolant to have total degree $\le n$ places severe restrictions both on the kind of data that can be interpolated and on the number of data points. It seems more natural to associate with any given set of points a suitable space of polynomials from which interpolation is possible, and indeed uniquely so. For a theory developing such spaces, see de Boor & Ron [to appear].

For linear systems, like (6), whose coefficient matrix is a Vandermonde matrix, the triangular decomposition (see §2.2), and more generally, block-triangular decompositions, can be carried out explicitly. For recent work on this subject, see Tang and Golub [1981]. The accuracy attainable, however, is often limited on account of ill-conditioning; see Gautschi [1990] for a survey on the condition of Vandermonde matrices.

§§6.2, 6.4 Computations for the barycentric formula can be arranged so that it becomes easy, just as in Newton's formula, to add one interpolation point at a time; see Werner [1984]. The number of arithmetic operations required is indeed the same for both formulae.

Barycentric formulae are also known for trigonometric interpolation (Salzer [1948], Berrut [1984]); they assume a particularly simple form in the case of equally-spaced nodes (Henrici [1979]).

The remainder term of Lagrange interpolation in the form (16) is valid for arbitrary functions f, but in conjunction with (18) requires continuity of the $(n+1)$st derivative of f. For functions f with low smoothness properties one can apply the theory of *Peano kernels* to obtain alternative representations and estimates of the remainder; see, e.g., Hämmerlin & Hoffmann [1989, Ch. 5, §2.4]. If, on the other hand, f can be extended to a function holomorphic in a domain of the complex plane that includes all interpolation points, then derivative-free estimates of the remainder are available based on contour integrals; see

Walsh [1969, Ch. III, §3.1].

§6.8 The spline function derived in this paragraph is sometimes referred to as the "natural" spline interpolant, or the spline interpolant with "free end conditions" [the conditions 1) on p. 154]. If, at the endpoints, one interpolates not only to the function values, but in addition to the first $p - 1$ derivative values (assumed known), and then again solves the extremal problem a) – c), one is led to the "complete" spline interpolant. Its approximation properties near the end zones of the interval are superior to those of the natural spline.

Spline functions are widely used today, not only for interpolation, but also in connection with other approximation processes (least squares approximation, smoothing of data, harmonic analysis, collocation methods in differential equations, to name a few). In many of these applications, it is important to have good basis functions for representing splines (or more general piecewise polynomial functions). A very elegant, and computationally effective, basis is provided by the *normalized B-splines*. For these, and for a thorough discussion of computational and approximation-theoretic aspects of splines, see de Boor [1978]. This source also contains many useful Fortran subroutines for computation with splines.

§6.9 The idea of smoothing according to (48) goes back to Whittaker [1922/23], who uses the sum of the squares of the third differences as a measure of smoothness. Minimizing (48), in which the second differences are replaced by the second derivative, and summation by integration, allows one to solve the variational problem analytically, and yields the *cubic smoothing spline* of Schoenberg and Reinsch; see de Boor [1978, Ch. 14].

§6.10 The Romberg scheme, of course, is pointless if in the expansion (55) of the error all coefficients c_i are zero. This turns out to be the case if the integrand f is a periodic function with period $b - a$. The trapezoidal rule then integrates exactly trigonometric polynomials of the largest possible degree, and therefore can hardly be improved upon when a smooth periodic function is to be integrated over the full period. This is exploited in Fourier analysis; see §7.4, Eq. (15).

The expansion (55), as stated in the text, holds only for sufficiently smooth functions f. Alternative expansions, and modified Romberg schemes, are known if f exhibits certain types of singularities at one or both endpoints of the interval; see Fox [1967].

Singularities, to be sure, can sometimes be removed by an appropriate change of variables, or can be attenuated by "subtracting them out". Alternatively, one may account for the singularity by introducing an appropriate weight function and by using *weighted quadrature rules*, particularly those of Gaussian type. The latter are based on unequally spaced nodes (the zeros of appropriate orthogonal polynomials) and achieve the highest algebraic degree of exactness; see Stroud & Secrest [1966], Gautschi [1981]. A detailed treatment of the subject of numerical integration, also in higher dimensions, can be found in Davis & Rabinowitz [1984]. Specialized texts on the numerical evaluation of *multiple integrals* are Stroud [1971], Sobolev [1974], Mysovskih [1981].

References

Berrut, J.-P. [1984]: Baryzentrische Formeln zur trigonometrischen Interpolation I, Z. *Angew. Math. Phys.* **35**, 91–105.

de Boor, C. [1978]: *A Practical Guide to Splines*, Springer, New York.

de Boor, C. and Pinkus, A. [1978]: Proof of the conjectures of Bernstein and Erdös concerning the optimal nodes for polynomial interpolation, *J. Approx. Theory* **24**, 289–303.

de Boor, C. and Ron, A. [to appear]: On multivariate polynomial interpolation, *Constructive Approx.*

Brutman, L. [1978]: On the Lebesgue function for polynomial interpolation, *SIAM J. Numer. Anal.* **15**, 694–704.

Bulirsch, R. and Rutishauser, H. [1968]: Interpolation und genäherte Quadratur, in *Mathematische Hilfsmittel des Ingenieurs* (R. Sauer and I. Szabó, eds.), Teil III, pp. 232–319. Springer, Berlin.

Davis, P.J. and Rabinowitz, P. [1984]: *Methods of Numerical Integration*, 2nd ed., Academic Press, New York.

Epperson, J.F [1987]: On the Runge example, *Amer. Math. Monthly* **94**, 329–341.

Fox, L. [1967]: Romberg integration for a class of singular integrands, *Comput. J.* **10**, 87–93.

Gautschi, W. [1981]: A survey of Gauss-Christoffel quadrature formulae, in *E.B. Christoffel – The Influence of his Work on Mathematics and the Physical Sciences* (P.L. Butzer and F. Fehér, eds.), pp. 72–147. Birkhäuser, Basel.

Gautschi, W. [1990]: How (un)stable are Vandermonde systems? in *Asymptotic and Computational Analysis* (R. Wong, ed.), pp. 193–210. Marcel Dekker.

Günttner, R. [1980]: Evaluation of Lebesgue constants, *SIAM J. Numer. Anal.* **17**, 512–520.

Hämmerlin, G. and Hoffmann, K.-H. [1989]: *Numerische Mathematik*, Grundwissen Mathematik 7, Springer, Berlin.

Hayes, W.E. and Powell, M.J.D. [1969]: Unpublished data, Atomic Energy Research Establishment, Harwell, England.

Henrici, P. [1979]: Barycentric formulas for interpolating trigonometric polynomials and their conjugates, *Numer. Math.* **33**, 225–234.

Kilgore, T.A. [1978]: A characterization of the Lagrange interpolating projection with minimal Tchebycheff norm, *J. Approx. Theory* **24**, 273–288.

Mysovskih, I.P. [1981]: *Interpolatory Cubature Formulae* (Russian), Izdat. "Nauka", Moscow.

Salzer, H.E. [1948]: Coefficients for facilitating trigonometric interpolation, *J. Math. and Phys.* **27**, 274–278.

Sobolev, S.L. [1974]: *Introduction to the Theory of Cubature Formulae* (Russian), Izdat. "Nauka", Moscow.

Stroud, A.H. [1971]: *Approximate Calculation of Multiple Integrals,* Prentice-Hall, Englewood Cliffs, N.J.

Stroud, A.H. and Secrest, D. [1966]: *Gaussian Quadrature Formulas,* Prentice-Hall, Englewood Cliffs, N.J.

Tang, W.P. and Golub, G.H. [1981]: The block decomposition of a Vandermonde matrix and its applications, *BIT* **21**, 505–517.

Thacher, H.C., Jr. [1960]: Derivation of interpolation formulas in several independent variables, *Ann. New York Acad. Sci.* **86**, 758–775.

Thacher, H.C., Jr. and Milne, W.E. [1960]: Interpolation in several variables, *J. Soc. Indust. Appl. Math.* **8**, 33–22.

Todd, J. [1962]: The constructive theory of functions, in *Survey of Numerical Analysis* (J. Todd, ed.), pp. 119–159. McGraw-Hill, New York.

Trefethen, L.N. and Weideman, J.A.C. [to appear]: Two results on polynomial interpolation in equally spaced points, *J. Approx. Theory.*

Walsh, J.L. [1969]: *Interpolation and Approximation by Rational Functions in the Complex Plane,* AMS Colloquium Publications **20**, 5th. ed., American Mathematical Society, Providence, R.I.

Werner, W. [1984]: Polynomial interpolation: Lagrange versus Newton, *Math. Comp.* **43**, 205–217.

Whittaker, E.T. [1922/23]: On a new method of graduation, *Proc. Edinburgh Math. Soc.* **41**, 63–75.

Zygmund, A. [1968]: *Trigonometric Series,* 2nd ed., Cambridge University Press, Cambridge.

CHAPTER 7

Approximation

While interpolation attempts to approximate a function *piecewise* by polynomials which pass exactly through prescribed support points, we shall now try to approximate a given function $f(x)$ on a (relatively large) interval I by *one* polynomial. Such an approximation polynomial, naturally, must be of a higher degree than in the case where $f(x)$ is approximated by polynomial pieces.

We limit ourselves here to polynomial approximation and do not consider approximation by rational and still more general functions.

§7.1. Critique of polynomial representation

A general fact about the approximation by polynomials is furnished by the following

Theorem 7.1([1]) (Weierstrass approximation theorem). *If $f(x)$ is a continuous function on the interval $a \le x \le b$, then for each $\varepsilon > 0$ there exists (at least) one polynomial $P(x)$ such that*

$$|f(x) - P(x)| < \varepsilon \ \text{for} \ a \le x \le b.$$

Every continuous function thus can be approximated arbitrarily closely by polynomials.

---

[1] Proof, e.g., in Achieser N.I.: *Theory of Approximation*, F. Ungar Publ. Co., New York 1956, §20.

Example. For $f(x) = e^{-x}$ on the interval $[0,a]$, the polynomial

$$P(x) = \sum_{k=0}^{N} \frac{(-x)^k}{k!}$$

achieves the desired approximation, provided N is chosen so large that $a^N/N! < \varepsilon \, (\leq 1)$.

But the approximating polynomial now is to be used in a numerical calculation as substitute for $f(x)$; this requires that $P(x)$ be represented in a form suitable for computation. If one simply writes $P(x)$ in the form

$$P(x) = \sum_{k=0}^{n} c_k x^k \tag{1}$$

(with given c_k), this requirement is not necessarily met, as is shown by the following example:

$$P(x) = .9869 - 11.8245x + 86.4317x^2 - 352.9509x^3$$
$$+ 807.1695x^4 - 1025.4367x^5 + 674.8324x^6 - 179.1590x^7.$$

This looks like an awful polynomial, but is nothing but a polynomial which approximates $f(x) = 1/(1 + 15x)$ on the interval $[0,1]$ with a maximum deviation of .0132.

The polynomial is thus quite harmless; it has gotten such large coefficients only through the unfortunate choice of powers $1, x, x^2, \ldots$ as polynomial basis. In this way, indeed, the numerical evaluation of the polynomial becomes inaccurate; as a matter of fact, the function values in this example, in 5-digit computation, will exhibit errors up to 5% in the vicinity of $x = 1$.

This situation can be significantly improved upon by means of other representations of polynomials, that is, through a choice of other bases for spanning the polynomial space. Indeed, $\sum c_k x^k$ is only one of many possible ways of representing a polynomial, and by no means the best when it comes to approximating a function $f(x)$ on an interval $a \leq x \leq b$ by a polynomial. Better for this purpose are always the Chebyshev polynomials. Further possibilities are: Legendre polynomials, Newton's interpolation formula. In the latter case, the polynomial is determined by the divided differences.

§7.2. Definition and basic properties of Chebyshev polynomials

The *Chebyshev polynomials* (*T*-polynomials) arise from the fact that $\cos(n\phi)$ can be expressed as a polynomial in $\cos\phi$; we have indeed, for example,

$$\cos(2\phi) = 2\cos^2\phi - 1,$$

$$\cos(3\phi) = 4\cos^3\phi - 3\cos\phi,$$

$$\cos(4\phi) = 8\cos^4\phi - 8\cos^2\phi + 1, \quad \text{etc.}$$

In general, $\cos(k\phi)$ is a polynomial of degree k in $\cos\phi$, which we denote by $T_k(\cos\phi)$. After the substitution $x = \cos\phi$ one has

$$T_0(x) = 1,$$

$$T_1(x) = x,$$

$$T_2(x) = 2x^2 - 1, \tag{2}$$

$$T_3(x) = 4x^3 - 3x,$$

$$T_4(x) = 8x^4 - 8x^2 + 1, \quad \text{etc.}$$

Because of $T_k(\cos\phi) = \cos(k\phi)$, many properties of trigonometric functions can be carried over to T-polynomials, namely:

1) $|T_k(x)| \leq 1$ for $|x| \leq 1$.

2) $T_k(x)$ has (relative) extreme values ± 1 at $x = x_j = \cos\left[\dfrac{\pi}{k}j\right]$
$(j = 1,2,\ldots,k-1)$.

3) $T_k(x)$ has zeros at $x = z_j = \cos\left[\dfrac{\pi}{k}j - \dfrac{\pi}{2k}\right]$ $(j = 1,2,\ldots,k)$.

It follows from this that all zeros of these polynomials are simple and real, and lie in the interval $|x| < 1$. One obtains, for example, the

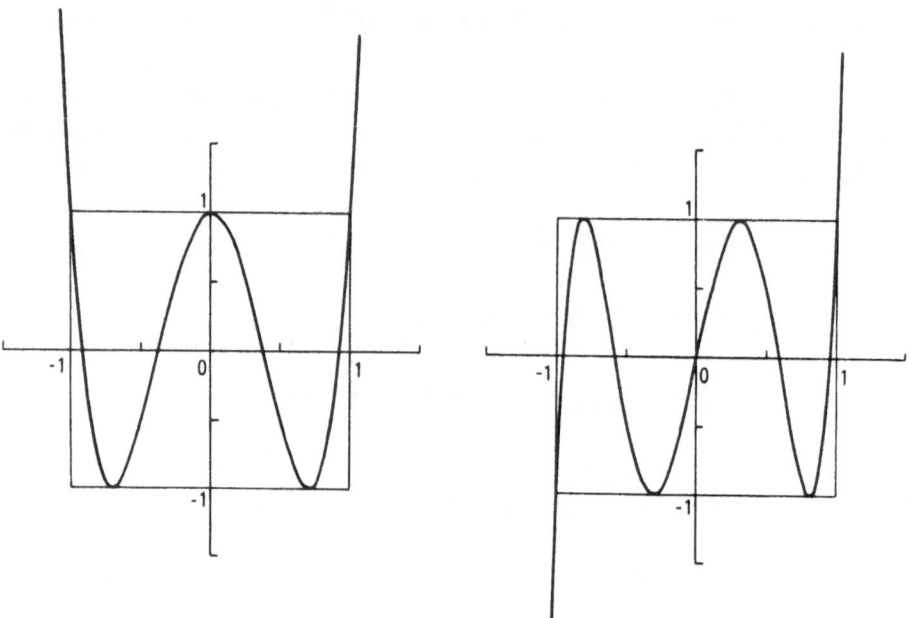

Figure 7.1. *The Chebyshev polynomial $T_4(x)$ and $T_5(x)$*

curves depicted in Figure 7.1. (The parts contained in the square are special Lissajous figures([1]).)

4) From the trigonometric identity

$$\cos (k + 1)\phi + \cos (k - 1)\phi = 2 \cos \phi \cos (k\phi)$$

there follows immediately the identity

$$T_{k+1}(x) = 2xT_k(x) - T_{k-1}(x), \qquad (3)$$

which can be used for the recursive computation of the T_k. For example, with $k = 4$, one obtains [cf. (2)]:

$$
\begin{array}{rccccc}
2xT_4 &=& 16x^5 &-& 16x^3 &+& 2x \\
-T_3 &=& & & -4x^3 &+& 3x \\
\hline
T_5 &=& 16x^5 &-& 20x^3 &+& 5x
\end{array}
$$

---

[1] Cf., e.g., French A.P.: *Vibrations and Waves*, W.W. Norton and Co., New York 1971, pp. 34f. (Translator's remark)

furthermore:

$$
\begin{array}{rl}
2xT_5 = & 32x^6 \;-\; 40x^4 \;+\; 10x^2 \\
-\,T_4 = & \;-\; 8x^4 \;+\; 8x^2 \;-1 \\
\hline
T_6 = & 32x^6 \;-\; 48x^4 \;+\; 18x^2 \;-1, \quad \text{etc.}
\end{array}
$$

5) From the recurrence formula one also notes at once the general fact that $T_k(x)$ is a polynomial of degree k with leading coefficient 2^{k-1},

$$T_k(x) = 2^{k-1}x^k - \cdots + \cdots$$

[exception: $T_0(x) \equiv 1$], and that $T_k(x)$ for even (odd) k is an even (odd) function.

Now the idea of the recurrence formula, however, is not to express the polynomials $T_k(x)$ in terms of powers of x. One would then, in fact, create the very calamity that one has tried to avoid by means of the T-polynomials. For example,

$$T_{20}(x) = 524288x^{20} - 2621440x^{18} + 5570560x^{16} - \cdots + \cdots ,$$

which is a polynomial with large coefficients, but small function values.

The T_k, therefore, should not be expressed in powers of x, but should be considered as irreducible basic elements through which one expresses other polynomials, as for example in

$$x^5 = .0625T_5 + .3115T_3 + .625T_1 ;$$

the recurrence formula, on the other hand, should be used to evaluate *numerically* the $T_k(x)$ for given x.

Example. Computation of $T_6(.7)$: for $x = .7$ the recurrence formula (3) reads: $T_{k+1} = 1.4T_k - T_{k-1}$. This is applied for $k = 1,2,3,4,5$, setting initially $T_0 = 1$, $T_1 = .7$. One obtains $T_2 = -.02$, $T_3 = -.728$, $T_4 = -.9992$, $T_5 = -.67088$, $T_6 = .059968$.

It is possible to also accelerate the computation by means of

$$T_{k+\ell}(x) = 2T_k(x)T_\ell(x) - T_{k-\ell}(x) \quad (k \geq \ell) \tag{4}$$

(for example, above, with $T_6 = 2T_3^2 - T_0 = 2 \times .728^2 - 1$).

   The Chebyshev polynomials are also defined for $|x| > 1$; the relation $T_k (\cos \phi) = \cos (k\phi)$ is continued outside of $|x| \leq 1$ as follows[2]:

$$T_k (\cosh \psi) = \cosh (k\psi), \quad \text{for} \quad x = \cosh \psi \geq 1,$$

$$T_k (-\cosh \psi) = (-1)^k \cosh (k\psi), \quad \text{for} \quad x = -\cosh \psi \leq -1. \tag{5}$$

The recurrence formula holds there unchanged. For $x = 1.1$, e.g., it reads: $T_{k+1} = 2.2\ T_k - T_{k-1}$, and thus yields the following values ($T_0 = 1$, $T_1 = 1.1$): $T_2 = 1.42$, $T_3 = 2.024$, $T_4 = 3.0328$, $T_5 = 4.64816$, $T_6 = 7.19315$, $T_7 = 11.1768$, etc. One can see from this that the $T_k(x)$ of high degree grow very rapidly outside of $|x| \leq 1$ [$T_7(1) = 1$, $T_7(1.1) = 11.1768$], just as inside of $|x| \leq 1$ they strongly oscillate.

   We note in passing that the T-polynomials can also be defined in terms of the expression

$$(z + \sqrt{z^2 - 1})^k.$$

One has indeed, as can be seen immediately by multiplying out,

$$(z + \sqrt{z^2 - 1})^k = T_k(z) + \sqrt{z^2 - 1}\ U_{k-1}(z). \tag{6}$$

   The polynomials $U_{k-1}(z)$ (*T-polynomials of the second kind*) are of degree $k - 1$ and satisfy the identity

$$U_{k-1} (\cos \phi) = \frac{\sin (k\phi)}{\sin \phi}. \tag{7}$$

Example. For $k = 4$ one has

$$(z + \sqrt{z^2 - 1})^4 = z^4 + 4z^3\sqrt{z^2 - 1} + 6z^2(z^2 - 1) + 4z(z^2 - 1)\sqrt{z^2 - 1}$$

$$+ (z^2 - 1)^2 = (8z^4 - 8z^2 + 1) + \sqrt{z^2 - 1}\,(8z^3 - 4z),$$

thus, $T_4 = 8z^4 - 8z^2 + 1$, $U_3 = 8z^3 - 4z$.

---

[2] For the T-polynomials one has (cf. (12) in §7.3):

$$T_k(z) = \frac{1}{2}(w^k + w^{-k}), \quad \text{if} \quad z = \frac{w + w^{-1}}{2}.$$

For, with $w = e^{i\phi}$, there follows $T_k (\cos \phi) = \cos (k\phi)$; with $w = \pm e^\psi$ one obtains precisely (5). (Editors' remark)

§7.3. Expansion in T-polynomials

By *T-expansion* of a function $f(x)$ on the interval[1] $[-1,1]$ we mean a representation

$$f(x) = \frac{c_0}{2} + \sum_{k=1}^{\infty} c_k T_k(x). \tag{8}$$

The convergence properties of this series on the interval $|x| \le 1$ can be read off at once from the coefficients (*T-coefficients*):

Theorem 7.2. *If*

$$\sum_{k=1}^{\infty} |c_k| \tag{9}$$

converges, then the series in (8) *converges uniformly and absolutely for all x with* $|x| \le 1$.

On the other hand, convergence of (9) is not a necessary condition for convergence of the series in (8). For example,

$$T_1(x) - \frac{1}{3} T_3(x) + \frac{1}{5} T_5(x) - \frac{1}{7} T_7(x) + \cdots - \cdots$$

is convergent for all x with $|x| \le 1$ [to $f(x) = \frac{\pi}{4}$ sign (x)], although not uniformly. Nevertheless, only those T-expansions for which (9) converges are useful in practice.

The practical importance of a T-expansion, in fact, derives precisely from the convergence of this series (9): for any $\varepsilon > 0$, we can then indeed find a $N(\varepsilon)$ such that

$$\sum_{k=N+1}^{\infty} |c_k| < \varepsilon,$$

hence also

$$\left| \sum_{k=N+1}^{\infty} c_k T_k(x) \right| < \varepsilon \quad \text{for } |x| \le 1.$$

---

[1] Every other finite interval can be transformed to $[-1,1]$ in a trivial way.

Now

$$P(x) = \frac{c_0}{2} + \sum_{k=1}^{N} c_k T_k(x),$$

as sum of polynomials of degrees at most N, is itself a polynomial in x of degree $\leq N$, and, if (8) holds, one has

$$f(x) - P(x) = \sum_{k=N+1}^{\infty} c_k T_k(x),$$

thus

$$|f(x) - P(x)| < \varepsilon \quad \text{for } |x| \leq 1.$$

By truncating the T-expansion of $f(x)$ one thus obtains arbitrarily accurate approximations to $f(x)$ over the whole interval $|x| \leq 1$. It is true, however, that the truncated T-expansion, as a rule, is *not* the best approximation in the sense of Chebyshev (cf. §7.6).

Example. For $|x| \leq 1$ one has (derivation later)

$$f(x) = \frac{4}{17 + 15x} = \frac{1}{2} + \sum_{k=1}^{\infty} \left(-\frac{3}{5}\right)^k T_k(x).$$

Truncating the series after the T_6-term yields

$$P(x) = .5 - .6T_1(x) + .36T_2(x) - .216T_3(x)$$
$$+ .1296T_4(x) - .07776T_5(x) + .046656T_6(x),$$

a polynomial which, since $\sum_{k=7}^{\infty} |c_k| = .069984$, deviates from $f(x)$ on the interval $|x| \leq 1$ by at most .07.

Our efforts, in the following, will be directed towards obtaining T-expansions in as simple a way as possible. To begin with, we remark that from *Fourier series* one obtains T-expansions in a trivial way. Through the substitution $x = \cos \phi$, indeed, (8) transforms into

$$f(\cos \phi) = \frac{c_0}{2} + \sum_{k=1}^{\infty} c_k \cos(k\phi). \tag{10}$$

On the left we have a periodic even function of ϕ, which is developed into a pure cosine-series. As is evident from the right-hand side, the Fourier coefficients of this function are the desired T-coefficients.

Example. T-expansion of $\sin x$ in $|x| \leq \dfrac{\pi}{2}$. The interval $|x| \leq \dfrac{\pi}{2}$ must first be transformed to $|x| \leq 1$, which is done by seeking a representation

$$\sin\left[\frac{\pi}{2} x\right] = \frac{c_0}{2} + \sum_{k=1}^{\infty} c_k T_k(x) \quad \text{for} \quad |x| \leq 1.$$

One then has indeed

$$\sin x = \frac{c_0}{2} + \sum_{k=1}^{\infty} c_k T_k\left[\frac{2x}{\pi}\right] \quad \text{for} \quad |x| \leq \frac{\pi}{2}.$$

Now, with $x = \cos \phi$, we have

$$\sin\left[\frac{\pi}{2} \cos \phi\right] = \frac{c_0}{2} + \sum_{k=1}^{\infty} c_k \cos(k\phi),$$

with

$$c_k = \frac{2}{\pi} \int_0^{\pi} \sin\left[\frac{\pi}{2}\cos \phi\right] \cos(k\phi)d\phi = \begin{cases} 0 \text{ if } k \text{ is even,} \\ 2(-1)^n J_k\left[\dfrac{\pi}{2}\right] \text{ if } k = 2n+1 \text{ is odd,} \end{cases}$$

where J_k is the Bessel function of order $k^{(2)}$. One thus obtains

$$\sin\left[\frac{\pi}{2} x\right] = 1.13364818T_1(x) - .13807178T_3(x) + .00449071T_5(x)$$

$$- .00006770T_7(x) + .00000059T_9(x) - \cdots + \cdots$$

Because of the rapid convergence of this series one has, already with the terms given here, an approximation of $\sin\left[\dfrac{\pi}{2} x\right]$ accurate to about 8

---

[2] Cf., e.g., Watson G.N.: *A Treatise on the Theory of Bessel Functions*, 2nd ed., University Press, Cambridge, 1948, Section 2.2. (Editors' remark)

digits.

T-expansions can also be obtained easily with the help of *Laurent series*. Let indeed $f(z)$ be real for real z and analytic in an ellipse with foci at -1 and $+1$. By means of

$$z = \frac{1}{2}\left[w + \frac{1}{w}\right]$$

this ellipse, whose larger half axis shall be a, is mapped into the annulus

$$a - \sqrt{a^2 - 1} < |w| < a + \sqrt{a^2 - 1},$$

whereby to each z there correspond two points (w and w^{-1}). Then

$$g(w) = f\left[\frac{w + w^{-1}}{2}\right] \tag{11}$$

is analytic in this annulus and has the additional property $g(w) = g(w^{-1})$, so that the Laurent series

$$g(w) = \frac{1}{2} \sum_{k=-\infty}^{\infty} c_k w^k$$

has real coefficients with $c_k = c_{-k}$. Consequently,

$$g(w) = \frac{1}{2} c_0 + \sum_{k=1}^{\infty} c_k \frac{w^k + w^{-k}}{2} .$$

Now with $w = e^{i\phi}$ one obtains $z = \cos \phi$, hence

$$\frac{w^k + w^{-k}}{2} = \frac{e^{ik\phi} + e^{-ik\phi}}{2} = \cos(k\phi) = T_k(z),$$

$$f(z) = \frac{c_0}{2} + \sum_{k=1}^{\infty} c_k T_k(z) . \tag{12}$$

Example. Let us derive the T-expansion of

$$f(z) = \frac{4}{17 + 15z} .$$

Here we get

$$g(w) = f\left[\frac{w+w^{-1}}{2}\right] = \frac{4}{17+\frac{15}{2}\left[w+w^{-1}\right]} = \frac{8w}{15w^2+34w+15} = \frac{\frac{5}{2}}{3w+5} - \frac{\frac{3}{2}}{5w+3}$$

$$= \frac{1}{2}\left[1 - \frac{3}{5}w + \left(\frac{3}{5}\right)^2 w^2 - \cdots\right] - \frac{1}{2}\left[\frac{3}{5}\frac{1}{w} - \left(\frac{3}{5}\right)^2 \frac{1}{w^2} + \cdots\right]$$

$$= \frac{1}{2} - \frac{3}{5}\frac{w+w^{-1}}{2} + \left(\frac{3}{5}\right)^2 \frac{w^2+w^{-2}}{2} - \left(\frac{3}{5}\right)^3 \frac{w^3+w^{-3}}{2} + \cdots ,$$

thus

$$f(z) = \frac{1}{2} + \sum_{k=1}^{\infty}\left(-\frac{3}{5}\right)^k T_k(z).\tag{13}$$

§7.4. Numerical computation of the T-coefficients

Since the Fourier coefficients of the periodic function $f(\cos\phi)$ are the desired T-coefficients of $f(z)$, one finds the latter by

$$c_k = \frac{2}{\pi}\int_0^\pi f(\cos\phi)\cos(k\phi)\,d\phi.\tag{14}$$

[We used here the fact that (10) is a pure cosine-series.] The integrals can be evaluated with the trapezoidal rule: by introducing the nodes $\phi_j = j\frac{\pi}{N}$ $(j = 0, 1, \ldots, N)$ one gets

$$c_k \approx \frac{2}{N}\sum_{j=0}^{N}{}''f\left[\cos\frac{\pi j}{N}\right]\cos\left[k\frac{\pi j}{N}\right],\tag{15}$$

where $''$ means that the terms for $j = 0$ and $j = N$ are to be added with only half their values.

If the expression on the right of (15) is denoted by $c_{N,k}$, one has, if $f(\cos \phi)$ is Riemann integrable:

$$\lim_{N \to \infty} c_{N,k} = c_k \quad (k = 0,1, \ldots). \tag{16}$$

Based on this relation (16), there is a primitive method for the calculation of the T-coefficients:

If the first $m + 1$ coefficients $c_0, c_1, \ldots, c_m$ of the T-expansion of $f(x)$ are desired, one computes for an increasing sequence of N-values (e.g. $N = 4, 8, 16, 32, \ldots$) the $c_{N,k}$ ($k = 0,1, \ldots, m$) and one continues to do so until the $c_{N,k}$ practically no longer change.

More precise information about the quality of the convergence $c_{N,k} \to c_k$ can be obtained from the following representation, in which the $c_{N,k}$ are expressed in terms of the exact T-coefficients. (This relation always holds when $\sum |c_k| < \infty$, which is the case, e.g., if $f(x)$ is twice continuously differentiable([1]).)

$$c_{N,k} = c_k + c_{2N-k} + c_{2N+k} + c_{4N-k} + c_{4N+k} + c_{6N-k} + \cdots . \tag{17}$$

One sees from this that, in general, $c_{N,k}$ can be a good approximation to c_k only if $|c_{2N-k}| << |c_k|$. If the c_k have a tendency to decrease, this means that in any case k must be $< N$; in other words: with the formula (15) one can at best approximate the coefficients $c_0, c_1, \ldots, c_{N-1}$.

Connection with interpolation. The function

$$P_N(x) = \frac{c_{N,0}}{2} + \sum_{k=1}^{N-1} c_{N,k} T_k(x) + \frac{c_{N,N}}{2} T_N(x) \tag{18}$$

is a polynomial of degree N, for which (17) yields the following representation:

---

[1] Derivation, e.g., in Fox L., Parker I.B.: *Chebyshev Polynomials in Numerical Analysis*, Oxford University Press, London 1968, Section 4.3. (Editors' remark)

$$P_N(x) = \frac{1}{2} c_0 + c_{2N} + c_{4N} + c_{6N} + \cdots$$

$$+ \sum_{k=1}^{N-1} (c_k + c_{2N-k} + c_{2N+k} + c_{4N-k} + \cdots)T_k(x)$$

$$+ (c_N + c_{3N} + c_{5N} + c_{7N} + \cdots)T_N(x).$$

In this expression every coefficient c_k occurs exactly once, only c_k is not multiplied by $T_k(x)$, but by $T_\ell(x)$, where $\ell = \ell(k)$ is the function depicted in Fig. 7.2.

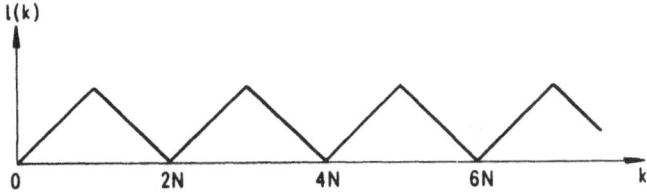

Figure 7.2. *The index function $\ell(k)$*

Therefore,

$$P_N(x) = \frac{c_0}{2} + \sum_{k=1}^{\infty} c_k T_{\ell(k)}(x),$$

and

$$P_N(x) - f(x) = \sum_{k=N+1}^{\infty} c_k[T_{\ell(k)}(x) - T_k(x)]$$

(note that $\ell(k) = k$ for $k \leq N$).

From this, there follow two facts:

1) If $\sum_{k=1}^{\infty} |c_k| < \infty$, then $\lim_{N \to \infty} P_N(x) = f(x)$, uniformly for $|x| \leq 1$.

2) For $x = x_j = \cos\left[\dfrac{\pi j}{N}\right]$ one has $P_N(x_j) = f(x_j)$.

Proof of 2): One has

$$T_{\ell(k)}(x) - T_k(x) = \cos(\ell(k)\phi) - \cos(k\phi)$$

$$= -2\sin\left[\frac{\ell(k)-k}{2}\phi\right]\sin\left[\frac{\ell(k)+k}{2}\phi\right],$$

or, for $x = x_j$,

$$T_{\ell(k)}(x_j) - T_k(x_j) = -2\sin\left[\frac{\ell(k)-k}{2N}j\pi\right]\sin\left[\frac{\ell(k)+k}{2N}j\pi\right].$$

However, since always either $\ell(k) - k$ or $\ell(k) + k$ is a multiple of $2N$, one has for all k

$$T_{\ell(k)}(x_j) - T_k(x_j) = 0 \quad (j = 0,1,\ldots,N), \quad \text{q.e.d.}$$

The statement 2) means that the polynomial $P_N(x)$ is nothing but the interpolation polynomial of the function $f(x)$ for the nodes $x_j = \cos\left[\dfrac{\pi j}{N}\right]$ $(j = 0,\ldots,N)$, the so-called *Chebyshev abscissas*. One could therefore compute P_N also with one of the known interpolation procedures. The route via the coefficients $c_{N,k}$ is called *Chebyshev interpolation*.

Together with statement 1) one obtains

Theorem 7.3. *The interpolation polynomial $P_N(x)$ of degree N for the nodes $x_0, \ldots, x_N$, where $x_j = \cos\dfrac{\pi j}{N}$, tends to $f(x)$ as $N \to \infty$ (uniformly in x for $|x| \leq 1$), provided f has the expansion (8) and the series (9) converges.*

We are witnessing here the remarkable fact that the interpolation polynomial for a *suitable* distribution of the nodes converges towards the function, and in fact uniformly on a certain interval (here $|x| \leq 1$). This theorem would not be true for an arbitrary distribution of nodes, especially not for

$$x_j = -1 + \frac{2}{N}j, \quad j = 0,\ldots,N \quad (N \to \infty).$$

Numerical illustration. The T-expansion for the function $f(z) = 4/(17 + 15z)$ was already computed analytically and given in (13); one has $c_k = (-.6)^k$. For comparison we give in Table 7.1 approximations $c_{N,k}$ for these coefficients, computed numerically according to formula (15), whereby for various N the $c_{N,0}, \ldots, c_{N,N}$ were determined each time from $N + 1$ ordinates. At the bottom of the table one finds the (rounded) exact c_k ($k = 0, \ldots, 20$).

Table 7.1. *Numerical computation of the T-coefficients*

N	$c_{N,0}, \ldots, c_{N,N}$					
1	2.1250000	−.9375000				
2	1.2977941	−.9375000	.4136029			
3	1.0978786	−.7109291	.5135607	−.2265709		
4	1.0341662	−.6387217	.4136029	−.2987783	.1318140	
5	1.0121668	−.6137890	.3790884	−.2454779	.1773282	−.0782330
6	1.0043631	−.6049448	.3668452	−.2265709	.1467155	−.1059843
	.0467578					
7	1.0015685	−.6017776	.3624608	−.2198002	.1357530	−.0879006
	.0635019	−.0280156				
8	1.0005644	−.6006396	.3608855	−.2173674	.1318140	−.0814109
	.0527175	−.0380820	.0168009			
9	1.0002031	−.6002302	.3603187	−.2164922	.1303969	−.0790741
	.0488377	−.0316248	.0228451	−.0100787		
10	1.0000731	−.6000829	.3601147	−.2161772	.1298869	−.0782330
	.0474414	−.0293007	.0189736	−.0137062	.0060468	
20	1.0000000	−.6000000	.3600000	−.2160000	.1296000	−.0777600
	.0466560	−.0279936	.0167962	−.0100778	.0060468	−.0036283
	.0021774	−.0013071	.0007853	−.0004730	.0002868	−.0001772
	.0001147	−.0000829	.0000366			
∞	1.0000000	−.6000000	.3600000	−.2160000	.1296000	−.0777600
	.0466560	−.0279936	.0167962	−.0100777	.0060466	−.0036280
	.0021768	−.0013061	.0007836	−.0004702	.0002821	−.0001693
	.0001016	−.0000609	.0000366	...		

§7.5. The use of T-expansions

If the T-coefficients decrease rapidly enough, the polynomial $P(x)$ obtained by truncating the T-expansion can be reexpanded in powers of x. Consider, for example, the T-expansion for $\tan^{-1}x$ in the interval $|x| \leq 1$:

$$\tan^{-1}x = 2\sum_{k=0}^{\infty} (-1)^k \frac{(\sqrt{2} - 1)^{2k+1}}{2k + 1} T_{2k+1}(x). \tag{19}$$

Truncating this series after the T_9-term and rounding the coefficients to 4 places after the decimal point, one obtains

$$P(x) = .8284T_1 - .0474T_3 + .0049T_5 - .0006T_7 + .0001T_9.$$

The deviation $P(x) - f(x)$ comes from two sources:

a) Omission of the terms with $T_{11}, T_{13}, T_{15}, \ldots$. The sum of the moduli of the omitted coefficients equals $138_{10}-7$.

b) Rounding of the remaining coefficients c_1, c_3, c_5, c_7, c_9 to 4 places after the decimal point. The sum of the moduli of the changes equals $930_{10}-7$.

The total change amounts to $1068_{10}-7$, hence

$$|P(x) - f(x)| < .00011 \quad \text{for} \quad |x| \leq 1.$$

For the transformation into a power series one considers the rounded coefficients as exact and also makes use of the exact integer coefficients of the T_k; then the transformation becomes exact.

$$
\begin{array}{rcl}
8284T_1 &=& 8284x \\
-474T_3 &=& 1422x - 1896x^3 \\
49T_5 &=& 245x - 980x^3 + 784x^5 \\
-6T_7 &=& 42x - 336x^3 + 672x^5 - 384x^7 \\
T_9 &=& 9x - 120x^3 + 432x^5 - 576x^7 + 256x^9 \\
\hline
P(x) &=& 1.0002x - .3332x^3 + .1888x^5 - .0960x^7 + .0256x^9
\end{array}
$$

This is now a polynomial of degree 9 which for $|x| \leq 1$ deviates from $\tan^{-1}x$ by at most .00011. Note that it does not agree with the beginning

of the $\tan^{-1}$ series; the small deviations, indeed, are of enormous importance, because for

$$Q(x) = x - .3333x^3 + .2x^5 - .1429x^7 + .1111x^9$$

the maximum error at $x = 1$ is $Q(1) - \pi/4 = .8349 - .7854 = .0495$,
which is about as poor as the first term alone of the T-series, i.e.,
$P_1(x) = .8284x$.

Reexpansion in powers of x, however, becomes disadvantageous as
soon as the T-coefficients decrease more slowly than $(\sqrt{2} - 1)^k$; more precisely: as soon as

$$\Gamma = \frac{\displaystyle\sum_{k=0}^{m} (2.414)^k |c_k|}{\displaystyle\sum_{k=0}^{m} |c_k|} \tag{20}$$

becomes much larger than 1. In the case of

$$f(x) = \frac{1}{\sqrt{3}} \tan^{-1}\left[\sqrt{48}\,x\right] = \sum_{k=0}^{\infty} (-1)^k \frac{.75^k}{2k+1} T_{2k+1}(x),$$

for example, one obtains a rough approximation by

$$P(x) = T_1 - .25T_3 + .11T_5 - .06T_7 + .03T_9 - .02T_{11} + .01T_{13},$$

from which by reexpansion as above there results

$$P(x) = 3.34x - 18.20x^3 + 75.20x^5 - 177.28x^7 + 230.40x^9$$
$$- 153.60x^{11} + 40.96x^{13}.$$

Since here $|P(x)| < 1$ (for $|x| \le 1$), upon evaluation of this polynomial
one obtains the value as a difference of large numbers, hence with cancellation. This danger is already signaled by the fact that

$$\Gamma = \frac{\displaystyle\sum_{k=1}^{13} (2.414)^k |c_k|}{\displaystyle\sum_{k=1}^{13} |c_k|} = 943.85 .$$

Computation with T-expansions. In cases where Γ is large, reexpansion in powers of x is ill-advised, numerically; in such cases one should rather compute directly with the polynomial given in the form

$$P(x) = \frac{c_0}{2} + \sum_{k=1}^{N} c_k T_k(x).$$

We mention here only two operations:

a) *Computation of a function value.* Introducing the auxiliary functions

$$p_\ell(x) = \frac{c_\ell}{2} + \sum_{k=1}^{\infty} c_{\ell+k} T_k(x) \quad (\ell = 0, 1, \ldots), \tag{21}$$

where $c_{N+1} = c_{N+2} = \cdots = 0$, one obtains [by using (3)]

$$p_{\ell+1}(x) = \frac{c_{\ell+1}}{2} + \sum_{k=1}^{\infty} c_{\ell+k+1} T_k(x) = \sum_{k=1}^{\infty} c_{\ell+k} T_{k-1}(x) - \frac{c_{\ell+1}}{2},$$

$$p_{\ell-1}(x) = \frac{c_{\ell-1}}{2} + c_\ell T_1(x) + \sum_{k=1}^{\infty} c_{\ell+k} T_{k+1}(x)$$

$$= \frac{c_{\ell-1}}{2} + c_\ell x + \sum_{k=1}^{\infty} c_{\ell+k} (2x T_k(x) - T_{k-1}(x))$$

$$= \frac{c_{\ell-1}}{2} + c_\ell x + 2x \left[p_\ell(x) - \frac{c_\ell}{2} \right] - \left[p_{\ell+1}(x) + \frac{c_{\ell+1}}{2} \right],$$

or

$$p_{\ell-1}(x) = \frac{c_{\ell-1} - c_{\ell+1}}{2} + 2x p_\ell(x) - p_{\ell+1}(x). \tag{22}$$

Noting that $p_{N+1} = p_{N+2} = 0$, and applying this *formula of Clenshaw*([1]) (with a fixed numerical value of x) for $\ell = N + 1, N, \ldots, 1$,

---

[1] Clenshaw C.W.: A note on the summation of Chebyshev series. *Math. Tables Aids Comput.* **9**, 118–120 (1955).

one obtains directly the desired function value $P(x) = p_0(x)$. The algorithm corresponds to the Horner scheme([2]). The derivative $P'(x) = p_0'(x)$ is computed analogously by the recursion

$$p_{\ell-1}'(x) = 2p_\ell(x) + 2xp_\ell'(x) - p_{\ell+1}'(x), \tag{23}$$

which is obtained by differentiation of (22).

b) *Multiplication* of two T-expansions:

$$f(x)g(x) = \left[\frac{c_0}{2} + \sum_{k=1}^{\infty} c_k T_k(x)\right] \left[\frac{d_0}{2} + \sum_{\ell=1}^{\infty} d_\ell T_\ell(x)\right].$$

When multiplying out this product, there occur terms of the kind $T_k(x)T_\ell(x)$; according to (4), however,

$$T_k T_\ell = \frac{1}{2}\left[T_{k+\ell} + T_{k-\ell}\right],$$

where one has to define $T_{-j}(x) = T_j(x)$. In this way one finds

$$f(x)g(x) = \frac{e_0}{2} + \sum_{k=1}^{\infty} e_k T_k(x) \text{ with } e_k = \frac{1}{2}\sum_{\ell=-\infty}^{+\infty} d_{|k-\ell|}c_{|\ell|}. \tag{24}$$

c) The remaining operations can be carried out in the same way; only the expansion of a quotient

$$\frac{\dfrac{c_0}{2} + \sum_{k=1}^{\infty} c_k T_k(x)}{\dfrac{d_0}{2} + \sum_{k=1}^{\infty} d_k T_k(x)},$$

---

[2] With the notation analogous to (21),

$$p_\ell(x) = \sum_{k=1}^{-} c_{\ell+k}x^k, \text{ where } c_{N+1} = c_{N+2} = \cdots = 0,$$

the formation rule of the Horner scheme (cf. footnote ([1]) in §4.4) reads: $p_{\ell-1}(x) = c_{\ell-1} + xp_\ell(x)$. (Editors' remark)

again in a *T*-series, gives some trouble. *Division with remainder* of two polynomials, on the other hand, does not present any particular difficulties.

§7.6. Best approximation in the sense of Chebyshev (T-approximation)

In contrast to the *T*-expansion, which aims at a representation

$$f(x) = \frac{c_0}{2} + \sum_{k=1}^{\infty} c_k T_k(x)$$

of the given function, and from which by truncation one obtains a rough approximation $P(x)$ of $f(x)$, we now propose to solve the following problem (*T-approximation*):

Let $f(x)$ be continuous on the interval $I = [a,b]$. From among all polynomials $P(x)$ of degree $\leq n$ determine the one for which $||\varepsilon|| = ||p - f||$ becomes minimum. Here, $||\varepsilon||$ is defined by

$$||\varepsilon|| = \max_{x \in I} |\varepsilon(x)| ,$$

i.e., $||\cdot||$ denotes the *maximum norm*([1]).

This problem, as we shall see, has exactly one solution ([2]), which as a rule, however, does not coincide with the *T*-expansion truncated after the T_n-term.

Theorem 7.4 (Alternation theorem). *Let $P(x)$ be the nth degree polynomial of best approximation for the (continuous) function $f(x)$ on the interval I. Then there exist in I (at least) $n + 2$ points $x_0 > x_1 > x_2 > \cdots > x_{n+1}$ in which the error function $\varepsilon(x) = P(x) - f(x)$ alternately attains its extreme values $\pm ||\varepsilon||$. That is,*

$$\varepsilon(x_k) = (-1)^k h, \quad k = 0, \ldots, n + 1, \quad with \quad h = \pm ||\varepsilon|| . \qquad (25)$$

---

[1] Here and in analogous cases the author wrote $||\varepsilon(x)||$ in place of $||\varepsilon||$.

[2] The *existence* of the polynomial of best approximation, not actually proved below, follows from a standard compactness argument in functional analysis; see, e.g., Todd J. (ed.): *Survey of Numerical Analysis*, McGraw-Hill, New York, 1962, pp. 129f. (Translator's remark)

These $n + 2$ points $x_0, x_1, \ldots, x_{n+1}$ are called an *alternation* of f. Even though $P(x)$ is uniquely determined, there can be several alternations.

Example. Let $I = [-1,1]$, $f(x) = T_{n+1}(x)$. The nth degree polynomial of best approximation is here $P(x) \equiv 0$. The alternation (here the only one) consists of the points

$$x_k = \cos\left[\frac{k\pi}{n+1}\right] \quad (k = 0, 1, \ldots, n + 1),$$

in which

$$\varepsilon(x_k) = T_{n+1}(x_k) = (-1)^k;$$

one thus has $||\varepsilon|| = 1$.

Proof of the alternation theorem. It is always possible to subdivide I, beginning from the upper end, into subintervals $I_0, I_1, \ldots, I_\ell$, such that $\varepsilon(x)$ in I_k assumes only extrema $(-1)^k h$, whereby for all k either $h = ||\varepsilon||$ or $h = -||\varepsilon||$ (cf. Figure 7.3). Now either there exists the asserted alternation with $n + 2$ points, or such a subdivision is possible with $\ell \leq n$, thus with at most $n + 1$ subintervals. If $\varepsilon(x)$ has the form shown in Figure 7.3, then 3 intervals suffice, since the 3 consecutive extrema of equal sign can be collected in one interval.

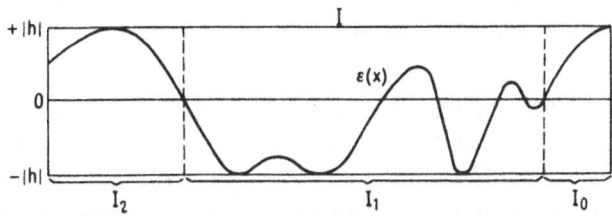

Figure 7.3. *To the proof of the alternation theorem: subdivision of I into subintervals*

If now a subdivision with $\ell \leq n$ is possible, there is a polynomial $R(x)$ of degree $\ell \leq n$ with the property

$$\text{sign}\,(R(x)) = (-1)^k \,\text{sign}\,(h) \quad \text{for } x \in I_k, \quad k = 0, \ldots, \ell,$$

(sign (h) is the sign of the extrema in I_0, thus $+1$ in the above figure) and therefore one has for each $\gamma > 0$:

$\varepsilon(x) - \gamma R(x) < \varepsilon(x)$ in the intervals with positive extrema;

$\varepsilon(x) - \gamma R(x) > \varepsilon(x)$ in the intervals with negative extrema.

Consequently, for sufficiently small $\gamma > 0$,

$$|\varepsilon(x) - \gamma R(x)| < ||\varepsilon|| \quad \text{for all} \quad x \in I, \text{ i.e.,}$$
$$||P - \gamma R - f|| < ||P - f||,$$

so that the polynomial $P(x) - \gamma R(x)$, contrary to the assumption, is a better approximation.

Theorem 7.5. (Uniqueness theorem). *If a polynomial of degree n has the property that in $n + 2$ points $x_0 > x_1 > x_2 > \cdots > x_{n+1}$ of the interval I*

$$P(x_k) - f(x_k) = (-1)^k h \quad (k = 0, \ldots, n + 1),$$

and in addition

$$|P(x) - f(x)| \leq |h| \quad \text{for all} \quad x \in I,$$

then $P(x)$ is the uniquely determined nth degree polynomial of best approximation for the (continuous) function $f(x)$ on the interval I.

Proof. The curve $y = \varepsilon(x) = P(x) - f(x)$ traverses the strip $I \times (-|h| \leq y \leq |h|)$ at least $(n + 1)$-times, if it has $n + 2$ extrema $\pm h$ (cf. Fig. 7.4). The function $\varepsilon_1(x) = P_1(x) - f(x)$ formed with another polynomial must intersect each of these $n + 1$ branches at least once, if one wants $||\varepsilon_1|| \leq ||\varepsilon||$; we therefore have, at $n + 1$ points,

$$\varepsilon_1(x) = \varepsilon(x), \quad P_1(x) - f(x) = P(x) - f(x),$$

thus $P_1(x) = P(x)$. Since both polynomials are of degree n, it follows, necessarily, that $P_1(x) \equiv P(x)$.

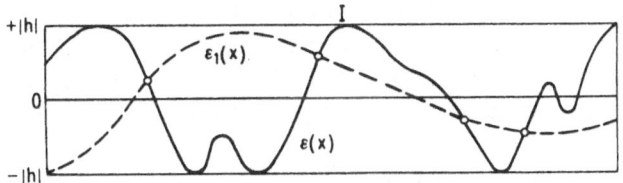

Figure 7.4. *To the proof of the uniqueness theorem: points of intersection of* $\varepsilon(x)$ *and* $\varepsilon_1(x)$

There remains the possibility that $\varepsilon_1(x)$ intersects two branches at the same time (in an extremum), so that these two branches then contribute only one point of intersection S (cf. Figure 7.5). Near S, however, one has $\varepsilon_1(x) \geq \varepsilon(x)$, hence also $P_1(x) \geq P(x)$, so that S is a double point of intersection. With this, the uniqueness is proved.

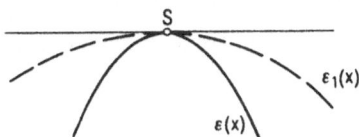

Figure 7.5. *To the proof of the uniqueness theorem: double point of intersection of* $\varepsilon(x)$ *and* $\varepsilon_1(x)$

As to the *construction* of the polynomial of best approximation, we first need some terminology: $n + 2$ points $x_0 > x_1 > \cdots > x_{n+1}$ in I are called a *reference*, and a polynomial $R(x)$ of degree n which at these points alternately has the same (in absolute value) deviations $\pm h$ from $f(x)$, hence the property

$$R(x_k) - f(x_k) = (-1)^k h, \tag{26}$$

is the associated *reference polynomial*. (h can be positive or negative.) $|h|$ is called the *reference deviation* for the reference $x_0, x_1, \ldots, x_{n+1}$.

Theorem 7.6. *The reference polynomial $R(x)$ (and with it also h) is uniquely determined by $f(x)$ and the reference $x_0, x_1, \ldots, x_{n+1}$.*

Proof. The ordinates $S(x_k) = f(x_k)$ $(k = 0, 1, \ldots, n + 1)$ determine uniquely a polynomial S of degree $n + 1$; let α be its leading coefficient. Furthermore, the $n + 2$ conditions $T(x_k) = (-1)^k$ also determine uniquely a polynomial T of degree $n + 1$; let β be its leading coefficient. We have $\beta \neq 0$, since $T(x)$ already has $n + 1$ sign changes between x_0 and x_{n+1} and therefore cannot degenerate to a polynomial of degree n. But then

$$R(x) = S(x) - \frac{\alpha}{\beta} T(x)$$

is a polynomial of degree n which at the points x_k assumes the following values:

$$R(x_k) = f(x_k) - \frac{\alpha}{\beta} (-1)^k .$$

We thus have

$$R(x_k) - f(x_k) = - (-1)^k \frac{\alpha}{\beta} ,$$

i.e., $R(x)$ is a reference polynomial with $h = - \frac{\alpha}{\beta}$. A second reference polynomial $R^*(x)$ cannot exist, since otherwise $R - R^*$ would be a polynomial of degree n with $n + 1$ zeros; q.e.d.

It is to be noted, however, that the extrema of the error function $\varepsilon(x) = R(x) - f(x)$ in general exceed $\pm h$, as for example in Fig. 7.6.

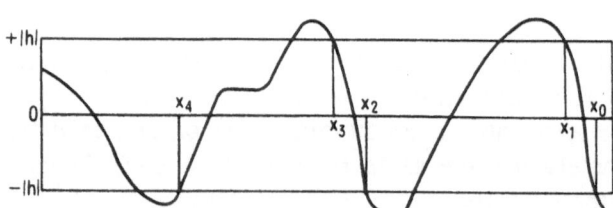

Figure 7.6. *Reference and error function*

Indeed, if the reference deviation $|h|$ is at the same time the maximum deviation $||R - f||$, then by the uniqueness theorem, $R(x)$ must be the polynomial of best approximation. More precisely, one has:

Theorem 7.7. *If the reference polynomial $R(x)$ has the reference deviation $|h|$ and the maximum deviation $||R - f|| > |h|$, then the* *polynomial P (x) of best approximation satisfies*

$$|h| < ||P - f|| \le ||R - f||.\tag{27}$$

Proof. In the same way as in the proof of the uniqueness theorem 7.5 one can refute the existence of a polynomial Q of degree n with $||Q - f|| \le |h|$; such a polynomial, namely, would have at least $n + 1$ points of intersection with $R(x)$ and therefore would have to be identical to R, which however is not possible. Therefore, $P(x)$ has a maximum deviation larger than $|h|$, but of course at most equal to $||R - f||$.

Corollary: *Among all polynomials of degree n the polynomial of best approximation is the one that has the smallest maximum deviation, but the largest reference deviation.*

§7.7. The Remez algorithm

Since the polynomial of best approximation has the largest reference deviation, one proceeds as follows to construct it: *Choose an arbitrary reference* $x_0 > x_1 > \cdots > x_{n+1}$ ($x_k \in I$) *and determine the reference polynomial R and its reference deviation* $|h|$. *Then vary the reference in such a way that* $|h|$ *increases.* This yields an iterative process which generates a sequence of reference polynomials R_ℓ with monotonically increasing reference deviations $|h_\ell|$. One hopes that

$$\lim_{\ell \to \infty} R_\ell(x) = P(x),$$

where P is the polynomial of best approximation.

Note that the reference deviation is $|h| = |1 - \alpha/\beta|$, where α and β are the leading coefficients of the polynomials S and T (introduced in the proof of Theorem 7.6). One can compute α and β by means of the Lagrange interpolation formula (see §6.1):

$$\alpha = \sum_{k=0}^{n+1} w_k f(x_k), \quad \beta = \sum_{k=0}^{n+1} w_k (-1)^k, \quad \text{where } w_k = \frac{1}{\prod_{j \ne k} (x_k - x_j)}.$$

But since $w_k = (-1)^k |w_k|$, hence

$$h = -\frac{\alpha}{\beta} = -\frac{\sum\limits_{k=0}^{n+1} (-1)^k f(x_k) |w_k|}{\sum\limits_{k=0}^{n+1} |w_k|} , \tag{28}$$

h is a weighted mean of the quantities $-(-1)^k f(x_k)$.

Denoting by $\bar{x}_0 > \bar{x}_1 > \cdots > \bar{x}_{n+1}$ $(\bar{x}_k \in I)$ a second, new, reference, and by $|\bar{h}|$ the corresponding reference deviation, one has of course likewise([1]):

$$\bar{h} = -\frac{\sum\limits_{k=0}^{n+1} (-1)^k f(\bar{x}_k) |\bar{w}_k|}{\sum\limits_{k=0}^{n+1} |\bar{w}_k|} , \quad \text{where} \quad \bar{w}_k = \frac{1}{\prod\limits_{j \neq k} (\bar{x}_k - \bar{x}_j)} . \tag{29}$$

Since the polynomial R belonging to the first reference has only degree n, the leading coefficient of the $(n + 1)$st-degree interpolation polynomial for the ordinates $R(\bar{x}_k)$, $k = 0, \ldots, n + 1$, vanishes, that is,

$$\sum_{k=0}^{n+1} \bar{w}_k R(\bar{x}_k) = 0. \tag{30}$$

With $\varepsilon(x) = R(x) - f(x)$, one thus has

$$\bar{h} = \frac{\sum\limits_{k=0}^{n+1} (-1)^k \varepsilon(\bar{x}_k) |\bar{w}_k|}{\sum\limits_{k=0}^{n+1} |\bar{w}_k|} . \tag{31}$$

Therefore, $\bar{h}$ may be viewed not only as weighted mean of the quantities $-(-1)^k f(\bar{x}_k)$ (cf. (29)), but also as weighted mean of the quantities $(-1)^k \varepsilon(\bar{x}_k)$, where $\varepsilon(x)$ is the error function belonging to the old reference polynomial R.

Because of $h = (-1)^k \varepsilon(x_k)$, $k = 0, \ldots, n + 1$, we have in addition, trivially,

---
[1] The editors here had to slightly deviate from the original.

$$h = \frac{\sum\limits_{k=0}^{n+1} (-1)^k \, \varepsilon \, (x_k) \, |\overline{w}_k|}{\sum\limits_{k=0}^{n+1} |\overline{w}_k|}. \tag{32}$$

Comparison with (31) now shows immediately how the x_k are to be changed in order to have $|\overline{h}| > |h|$ when passing to the new reference $\overline{x}_0, \ldots, \overline{x}_{n+1}$. For this it suffices, e.g., to replace *one* of the x_k, say x_j, by an $\overline{x}$ for which

$$(-1)^j \, \varepsilon(\overline{x}) > (-1)^j \, \varepsilon(x_j) > 0, \text{ resp. } (-1)^j \, \varepsilon(\overline{x}) < (-1)^j \, \varepsilon(x_j) < 0.$$

Then, indeed, the new reference, consisting of $x_0, x_1, \ldots, x_{j-1}$, $x_{j+1}, \ldots, x_{n+1}$ and $\overline{x}$ (in place of x_j), is such that the new reference deviation $|\overline{h}|$ is the weighted mean of quantities which, with the only exception of one, occur also in the weighted mean (32) for $|h|$; this latter, missing, quantity has been replaced by a larger one, causing the mean to increase in absolute value, so that indeed $|\overline{h}| > |h|$. It is to be noted, however, that $\varepsilon(\overline{x})$ and $\varepsilon(x_j)$ must have the same sign.

In this way we obtain a particularly transparent variant of the so-called *Remez algorithm*[2]; it consists of the following steps:

1) Take any reference $x_0 > x_1 > \cdots > x_{n+1}$ and determine h according to formula (28); thereupon, the reference polynomial $R(x)$ can be evaluated through interpolation at $n + 1$ of the $n + 2$ reference points, in which one knows, after all, that $R(x_k) = f(x_k) + (-1)^k h$.

2) Determine the maximum $||\varepsilon||$ of the modulus of the error function $\varepsilon(x) = R(x) - f(x)$ on the interval I. *Either* this maximum is equal to $|h|$; then it is attained at $n + 2$ points with alternating signs of $\varepsilon(x)$, so that R is the polynomial of best approximation. *Or* one has $||\varepsilon|| > |h|$; then this maximum of $|\varepsilon(x)|$ is attained at a point $\overline{x}$ ($\neq x_j$, $j = 0, \ldots, n + 1$).

---

[2] In the most common variant of the Remez algorithm one chooses all reference points afresh when passing to a new reference, and in fact alternately equal to the maximum and minimum abscissas of the error function (or first approximations thereof). See, e.g., Murnaghan F.D., Wrench J.W., Jr.: The determination of the Chebyshev approximation polynomial for a differential function, *Math. Tables Aids Comput.* **13**, 185–193 (1959). (Editors' remark)

3) Construct a new reference $\bar{x}_0, \bar{x}_1, \ldots, \bar{x}_{n+1}$ consisting of $\bar{x}$ and $n + 1$ of the current reference points, i.e., a reference point x_j is replaced by $\bar{x}$. The choice of x_j is dictated by the conditions

$$\bar{x}_0 > \bar{x}_1 > \cdots > \bar{x}_{n+1} ,$$

$(-1)^k \, \varepsilon \, (\bar{x}_k), \quad k = 0, \ldots, n + 1, \ \text{have the same sign} \ .$

We distinguish three cases:

a) $x_\ell > \bar{x} > x_{\ell +1}$.

In the sequence $\varepsilon(x_\ell)$, $\varepsilon(\bar{x})$, $\varepsilon(x_{\ell +1})$ we then have two equal signs in succession; in order to reestablish the alternating sign sequence, that point of the pair x_ℓ, $x_{\ell +1}$ must be dropped for which $\varepsilon(x)$ has the same sign as $\varepsilon(\bar{x})$, thus:

$$\bar{x}_\ell \qquad = \bar{x} \quad \text{if sign} \, (\varepsilon(x_\ell)) = \text{sign} \, (\varepsilon(\bar{x})),$$

$$\bar{x}_{\ell +1} \quad = \bar{x} \quad \text{if sign} \, (\varepsilon(x_{\ell +1})) = \text{sign} \, (\varepsilon(\bar{x})).$$

b) $\bar{x} > x_0$, sign $(\varepsilon(x_0)) \neq$ sign $(\varepsilon(\bar{x}))$.

Since the sequence $\varepsilon(\bar{x})$, $\varepsilon(x_0)$, $\varepsilon(x_1)$, $\ldots$ already has alternating signs, only x_{n+1} can be dropped in order to satisfy the sign condition. Then

$$\bar{x}_0 = \bar{x}, \quad \bar{x}_k = x_{k-1} \quad (k = 1, \ldots, n + 1) .$$

In this case, by the way, one has sign $(\bar{h}) = -$ sign (h).

c) $\bar{x} > x_0$, sign $(\varepsilon(x_0)) =$ sign $(\varepsilon(\bar{x}))$.

Since here the sequence $\varepsilon(\bar{x})$, $\varepsilon(x_0)$, $\varepsilon(x_1)$, $\ldots$ has two equal signs at the beginning, the sign condition is fulfilled by dropping x_0. Thus:

$$\bar{x}_0 = \bar{x}, \quad \bar{x}_k = x_k \quad (k = 1, \ldots, n + 1) .$$

If $\bar{x} < x_{n+1}$, one proceeds similarly as in the cases b) and c).

The new reference, of course, then again undergoes the same process, etc.; in this way one obtains a sequence of references with the property that the corresponding reference deviations $|h_\ell|$ form a monotonically increasing sequence, which therefore converges ($|\varepsilon(\bar{x})|$ is an upper bound for all $|h_\ell|$). It can be proved that $\lim |h_\ell|$ is equal to the reference deviation of the polynomial of best approximation, which, as we know, coincides with the maximum error of this polynomial. A more detailed analysis even shows that the reference polynomials R_ℓ converge uniformly to the polynomial of best approximation[3].

Numerical example. Approximate the function

$$f(x) = \frac{1}{1+x}$$

on the interval $[0,1]$ by a polynomial of degree 2. As initial reference we choose

$$x_0 = 1, \quad x_1 = .75, \quad x_2 = .25, \quad x_3 = 0.$$

The quantities occurring during the computation of the reference polynomial are summarized in the following schema:

x_k	$f(x_k)$	w_k	$R(x_k)$
1	.5	5.33333	.50714
.75	.57143	−10.66667	.56429
.25	.8	10.66667	.80714
0	1	− 5.33333	.99286

$$\alpha = -.22857 \quad \beta = 32.00000 \quad => h = .00714$$

One gets $R(x) = .99286 - .82858x + .34285x^2$. The error function $\varepsilon(x) = R(x) - f(x)$ has its maximum, in absolute value, approximately at $\bar{x} = .2$ with the value $\varepsilon(\bar{x}) = .00752$ (cf. Fig. 7.7).

---

[3] See, e.g., Meinardus G.: *Approximation of Functions: Theory and Numerical Methods*, Springer-Verlag, New York 1967, Theorem 83. (Translator's remark)

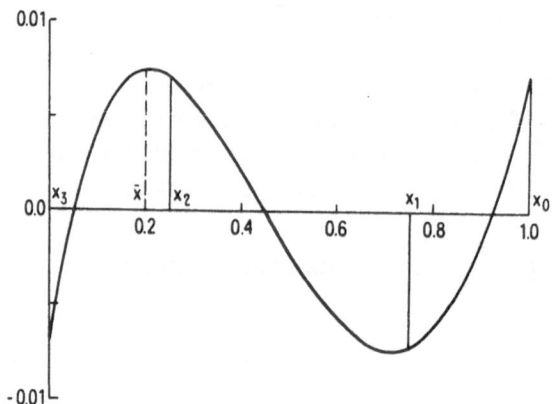

Figure 7.7. *The error function $\varepsilon(x)$ for the initial reference*

Since $\varepsilon\,(.25) = .00714$, one must replace $x_2 = .25$ by $\bar{x} = .2$:

x_k	$f\,(x_k)$	w_k	$R\,(x_k)$
1	.5	5	.50728
.75	.57143	−9.69697	.56415
.2	.83333	11.36364	.84061
0	1	−6.66667	.99272

$$\alpha = -.23814 \quad \beta = 32.72728 \quad \Rightarrow h = .00728$$

The extremum of $|\varepsilon(x)|$ lies approximately at $\bar{x} = .7$, where $\varepsilon(\bar{x}) = -.00755$. One must replace $x_1 = .75$ by $\bar{x} = .7$:

x_k	$f\,(x_k)$	w_k	$R\,(x_k)$
1	.5	4.166667	.50735
.7	.58824	−9.52381	.58088
.2	.83333	12.50000	.84068
0	1	−7.14286	.99265

$$\alpha = -.24510 \quad \beta = 33.33333 \quad \Rightarrow h = .00735$$

At this point, the error function $\varepsilon(x)$ is practically leveled. The resulting polynomial is

$$R\,(x) = .99265 - .82849x + .34321x^2 \; .$$

Notes to Chapter 7

§7.1 While Weierstrass's theorem is of considerable theoretical interest, it is of little use in practice, since it gives no indication of how large a degree the polynomial $P(x)$ may have to have in order to achieve a given accuracy, let alone how one might go about constructing it. The questions alluded to near the end of this section are related to the condition of polynomial bases; for this, see, e.g., Gautschi [1984].

§7.2 The classical source on Chebyshev polynomials and their applications is Lanczos's introduction in National Bureau of Standards [1952]. More recent accounts can be found in the books of Fox & Parker [1968], Rivlin [1974] and Paszkowski [1975], the last containing the most extensive treatment of computational methods related to Chebyshev polynomials and Chebyshev series.

§7.3 For many special functions in current use, the T-coefficients have been tabulated extensively; see Clenshaw [1962], Clenshaw & Picken [1966], Luke [1969, Vol. II, Ch. 17]. Gautschi [1975, §1.2.3] has references to more recent tables. An important technique of computing T-coefficients is based on the fact that these coefficients often satisfy a linear difference equation of some given order and, in fact, constitute a solution of minimum growth. They can therefore be computed very effectively by backward recurrence algorithms; see, e.g., Paszkowski [1975, §15].

§7.4 The approximation in Eq. (15) is a special instance of the discrete Fourier transform. For large and highly composite integers N (for example, powers of 2), the discrete Fourier transform can be evaluated very efficiently by algorithms which have come to be known as *Fast Fourier Transforms* (Brigham [1974], Nussbaumer [1981]). Rather than the N^2 operations that one would expect, they require only of the order of $N \log_2 N$ operations, and therefore have found important applications in many problems of applied analysis and engineering; see, e.g., Henrici [1979]. In numerical weather prediction it is not uncommon to compute as many as 15 million real Fourier transforms with $N = 192$, just to arrive at a 10-day forecast (Temperton [1983]).

The polynomial (18) interpolates to f at the *extreme values* on $[-1,1]$ of the Chebyshev polynomial T_N. The polynomial interpolating at the *zeros* of T_{N+1} can be similarly expressed; see Fox & Parker [1968, p. 32].

§7.6 There is an analogous theory of *best approximation by rational functions* (Achieser [1956, Ch. 2]). One again has uniqueness of the best rational approximant, for arbitrary prescribed numerator and denominator degrees. It can be characterized by an alternation property analogous to the one in Theorem 7.4, but slightly more complicated because of the possibility of common factors in numerator and denominator. Also Theorem 7.7 has its analogue in rational approximation and so does, therefore, the Remez algorithm; see, e.g., Ralston [1967].

§7.7 In practice, best rational approximations are usually preferred over best polynomial approximations, because they yield better approximations for the same degree of freedom. A collection of best (or nearly best) rational approximations to some of the common special functions can be found in Hart et al. [1968], and additional references in Gautschi [1975, §1.1.2]. A number of computer programs for generating best rational approximations are available and are referenced in Gautschi [1975, §1.1.3].

While the construction of best approximations (by polynomials or rationals) is cost-effective for functions that are to be evaluated many times, good approximations such as those described in the earlier sections of this chapter usually suffice for occasional use.

References

Achieser, N.I. [1956]: *Theory of Approximation* (Translated from the Russian by C.J. Hyman), Frederick Ungar Publ. Co., New York.

Brigham, E.O. [1974]: *The Fast Fourier Transform*, Prentice-Hall, Englewood Cliffs, N.J.

Clenshaw, C.W. [1962]: *Chebyshev Series for Mathematical Functions*, National Physical Laboratory Mathematical Tables **5**, Her Majesty's Stationery Office, London.

Clenshaw, C.W. and Picken, S.M. [1966]: *Chebyshev Series for Bessel Functions of Fractional Order*, National Physical Laboratory Mathematical Tables **8**, Her Majesty's Stationery Office, London.

Fox, L. and Parker, I.B. [1968]: *Chebyshev Polynomials in Numerical Analysis*, Oxford University Press, London.

Gautschi, W. [1975]: Computational methods in special functions – a survey, in *Theory and Application of Special Functions* (R.A. Askey, ed.), pp. 1–98, Academic Press, New York.

Gautschi, W. [1984]: Questions of numerical condition related to polynomials, in *Studies in Numerical Analysis* (G.H. Golub, ed.), pp. 140–177. Studies in Mathematics **24**, The Mathematical Association of America.

Hart, J.F., Cheney, E.W., Lawson, C.L., Maehly, H.J., Mesztenyi, C.K., Rice, J.R., Thacher, H.C., Jr. and Witzgall, C. [1968]: *Computer Approximations*, Wiley, New York.

Henrici, P. [1979]: Fast Fourier methods in computational complex analysis, *SIAM Rev.* **21**, 481–527.

Luke, Y.L. [1969]: *The Special Functions and Their Approximations*, Vols. I, II, Academic Press, New York.

National Bureau of Standards [1952]: *Tables of Chebyshev Polynomials $S_n(x)$ and $C_n(x)$*, Appl. Math. Ser. **9**, U.S. Government Printing Office, Washington, D.C.

Nussbaumer, H.J. [1981]: *Fast Fourier Transform and Convolution Algorithms*, Springer, Berlin.

Paszkowski, S. [1975]: *Numerical Applications of Chebyshev Polynomials and Series* (Polish), Państwowe Wydawnictwo Naukowe, Warsaw. [Russian translation in: "Nauka", Fiz.-Mat. Lit., Moscow, 1983].

Ralston, A. [1967]: Rational Chebyshev approximation, in *Mathematical Methods for Digital Computers*, Vol. 2 (A. Ralston and H.S. Wilf, eds.), pp. 264–284. Wiley, New York.

Rivlin, T.J. [1974]: *The Chebyshev Polynomials*, Wiley, New York.

Temperton, C. [1983]: Self-sorting mixed-radix fast Fourier transforms, *J. Comput. Phys.* **52**, 1–23.

CHAPTER 8

Initial Value Problems For Ordinary Differential Equations

It is a well-known fact that differential equations occurring in science and engineering can generally not be solved exactly, that is, by means of analytical methods. Even when this is possible, it may not necessarily be useful. For example, the second-order differential equation with two initial conditions,

$$y'' + 5y' + 4y = 1 - e^x, \quad y(0) = y'(0) = 0, \tag{1}$$

has the *exact* solution

$$y = \frac{1}{4} - \frac{1}{3} xe^{-x} - \frac{2}{9} e^{-x} - \frac{1}{36} e^{-4x}, \tag{2}$$

but when this formula is evaluated, say at the point $x = .01$, one obtains with 8-digit computation

$$y = .25 - .00330017 - .22001107 - .02668860 = .00000016 ,$$

which is no longer very accurate.

In such cases, and in others where an "exact" solution, i.e., a solution in closed form, does not exist, one must resort to numerical methods which admittedly yield the solution only *approximately*, but then right in finished tabular form. With such "inaccurate" methods one indeed succeeds in obtaining a much more accurate approximation to $y(.01) = .000000164138 \dots$.

§8.1. Statement of the problem

As a basic model we consider a *differential equation of the first order with one initial condition,*

$$y' = f(x,y), \quad y(x_0) = y_0. \tag{3}$$

Given here are the value y_0 and the function $f(x,y)$, which, depending on the context, must be required to have certain continuity properties (e.g., continuity and Lipschitz condition in the case of Euler's method).

However, we treat also *systems of differential equations,*

$$\frac{dy_\ell}{dx} = f_\ell(x, y_1(x), y_2(x), \dots, y_n(x)) \quad (\ell = 1, \dots, n), \tag{4}$$

with initial conditions $y_\ell(x_0) = y_{0\ell}$ $(\ell = 1, \dots, n)$, where n unknown functions $y_1(x), y_2(x), \dots, y_n(x)$ are to be determined. We are given here the n initial values $y_{0\ell}$ and the n functions $f_\ell(x, y_1, y_2, \dots, y_n)$. Such a system (4) can also be written in vector form as

$$\mathbf{y}' = \mathbf{f}(x, \mathbf{y}), \quad \mathbf{y}(0) = \mathbf{y}_0. \tag{5}$$

The *higher-order differential equation*

$$y^{(n)} = f(x, y, y', \dots, y^{(n-1)}) \tag{6}$$

with initial conditions for $y(x_0), y'(x_0), \dots, y^{(n-1)}(x_0)$ can be reduced to the case (4) by introducing new variables: one puts

$$y_1 = y, \quad y_2 = y', \quad y_3 = y'', \dots, y_n = y^{(n-1)};$$

then $y_\ell' = (y^{(\ell-1)})' = y^{(\ell)} = y_{\ell+1}$ $(\ell = 1, \dots, n-1)$ and $y_n' = (y^{(n-1)})' = y^{(n)} = f(x, y, y', \dots, y^{(n-1)})$, that is, one obtains

$$f_1(x, y_1, \ldots, y_n) \equiv y_2$$
$$f_2(x, y_1, \ldots, y_n) \equiv y_3$$

.

.   (7)

.

$$f_{n-1}(x, y_1, \ldots, y_n) \equiv y_n$$
$$f_n(x, y_1, \ldots, y_n) \equiv f(x, y_1, \ldots, y_n).$$

Example. From the second-order differential equation with initial conditions,

$$y'' + xy = 0, \quad y(0) = 0, \quad y'(0) = 1,$$

one obtains in this way the system

$$y_1' = y_2, \qquad\qquad y_1(0) = 0,$$
$$y_2' = -xy_1, \qquad\qquad y_2(0) = 1.$$

§8.2. The method of Euler

Now in order to integrate $y' = f(x, y)$, $y(x_0) = y_0$ numerically, the x-axis is discretized, that is, partitioned regularly or also irregularly, beginning with x_0 (see Fig. 8.1).

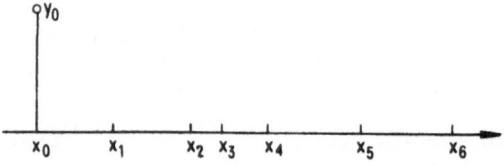

Figure 8.1. *Discretization of the x-axis*

The subdivision points x_k are called *support points*, while the support values $y_1, y_2, \ldots$ are now precisely the desired quantities. Often, the support points are chosen equally spaced; one then has $x_k = x_0 + kh$.

In general, a numerical method for the integration of (3) consists of a computational rule for determining the function value y_{k+1} (at the point

x_{k+1}) from the values $y_k, y_{k-1}, \ldots, y_1, y_0$ which are assumed already computed.

In the method of Euler one determines y_{k+1} by extending the tangent of the slope field at the point (x_k, y_k) until it intersects the ordinate at x_{k+1} (see Fig. 8.2).

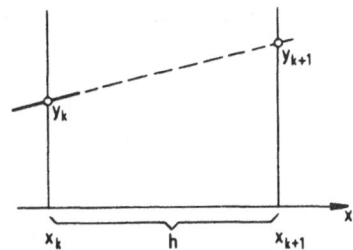

Figure 8.2. *The method for Euler*

In formulae, this means that

$$y_{k+1} = y_k + hf_k \quad \text{(where } f_k = f(x_k, y_k)), \tag{8}$$

where h denotes the length of the subinterval (x_k, x_{k+1}).

Example. For $y' = e^{-y}$, $y(0) = 0$, Euler's method with constant stepsize $h = .1$ yields the function table

x	y	y'
0	0	1
.1	.1	.90484
.2	.19048	.82656
.3	.27314	.
.	.	.
.	.	.
.	.	.

The exact solution here would be $y(x) = \ln(1 + x)$ with $\ln(1.3) = .26236$. The method is thus, unfortunately, rather inaccurate.

The inaccuracy of Euler's method is especially transparent in the example $y' = y$. With the initial condition $y(0) = 1$ and the constant stepsize h, thus $x_k = kh$, one obtains

$$y_{k+1} = y_k + hf_k = y_k(1+h),$$

hence in general,

$$y_n = (1+h)^n.$$

At a fixed point x, therefore, one finds upon integration with n steps of equal length $h = x/n$,

$$y(x) \approx y_n = \left(1 + \frac{x}{n}\right)^n.$$

As $n \to \infty$, $h \to 0$, this indeed converges toward the exact value $y(x) = e^x$, but convergence is slow. In first approximation,

$$y_n = \exp\left[n \ln\left(1 + \frac{x}{n}\right)\right] = \exp\left[x - \frac{x^2}{2n} + \frac{x^3}{3n^2} - \cdots\right]$$

$$\approx e^x e^{-x^2/2n} \approx e^x\left(1 - \frac{x^2}{2n}\right) = e^x\left(1 - h\,\frac{x}{2}\right).$$

This shows: the numerical integration yields the solution with a relative error of $hx/2$; thus, in order to obtain $y(1)$ with an error of 1%, one must choose $h = .02$ and thus needs 50 steps (i.e., 50 applications of the formula (8)); but in order to bring down the relative error to 10^{-6}, one already needs 500 000 steps. Upon further reduction of h – and thus increase in the number of steps – the rounding errors begin to become more and more noticeable, so that eventually the accuracy again deteriorates.

If we have a system (4) that is to be integrated by Euler, we must first establish some notation, namely the indexing of the integration steps on the one hand, and of the unknown functions $y_\ell(x)$ on the other:

We let $y_{k\ell}$ denote the numerically computed value of the function $y_\ell(x)$ at the support point x_k; in other words: if we denote the solution vector at the point x_k by $\mathbf{y}_k$, then $y_{k\ell}$ is its ℓth component. Similarly, we let $f_{k\ell}$ be the ℓth component of $\mathbf{f}(x_k, \mathbf{y}_k)$, i.e., $f_{k\ell} = f_\ell(x_k, y_{k1}, y_{k2}, \ldots, y_{kn})$.

In place of (8) we then have the formula

$$y_{k+1,\ell} = y_{k\ell} + hf_{k\ell} .$$

(9)

This has to be evaluated in the computer for all ℓ and all k.

As an example, we once again treat the differential equation $y' = e^{-y}$, $y(0) = 0$, which, for the purpose of eliminating the transcendental function e^{-y}, is now transformed into a system of two differential equations. With $y_2 = e^{-y_1}$ we have indeed

$$\frac{dy_2}{dx} = -e^{-y_1} \frac{dy_1}{dx} = -y_2^2 ,$$

so that

$$y_1' = y_2, \qquad y_1(0) = 0,$$
$$y_2' = -y_2^2, \qquad y_2(0) = e^0 = 1.$$

In the first three steps we now get:

x	y_1	y_2	y_1'	y_2'
0	0	1	1	−1
.1	.1	.9	.9	−.81
.2	.19	.819	.819	−.67076
.3	.2719	.75192	.75192	−.56538

Here, an important principle becomes apparent: it is often worthwhile to put up with an inflated system of differential equations, if it is possible, in this way, to eliminate complicated functions. The evaluation of such a function in a computer takes more time than carrying along an additional unknown function. The fact that $y_2 = e^{-y_1}$ is here also integrated inaccurately is of no consequence, since the integration of the function y_1 is inaccurate anyway. Incidentally, $y_1(.3) = .2719$ turns out to be even a bit more accurate than above with direct integration.

Instability. Stability is a concept that in the theory of differential equations has been in use for a long time. A solution of a differential equation is said to be unstable, if there are neighboring solutions which diverge away from it. For example,

$$y'' = 6y^2, \quad y(1) = 1, \quad y'(1) = -2,$$

has the unstable solution $y = 1/x^2$. There are solutions which diverge away from $1/x^2$; eight of them are depicted in Fig. 8.3, where those traced as solid curves belong to initial conditions $y(1) = 1 + \varepsilon$, $y'(1) = -2$, the dotted curves to $y(1) = 1$, $y'(1) = -2(1 + \varepsilon)$, with $\varepsilon = \pm\ .01$, $\pm\ .001$ in each case.

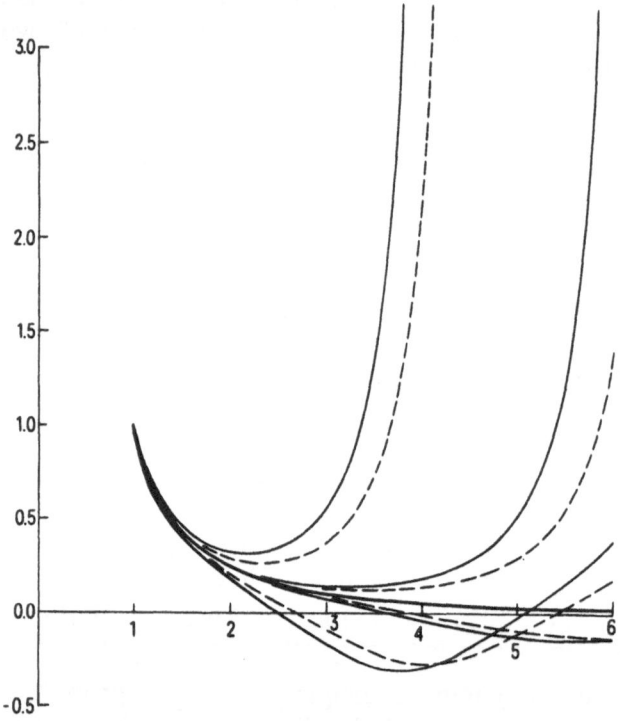

Figure 8.3. *Instability for a differential equation of 2nd order*

Here, however, we are not concerned with this kind of instability; what we have in mind, rather, is the phenomenon whereby a numerically computed solution during the process of integration almost explosively diverges away from the exact solution, without the latter being in any way unusual.

If one integrates, for example,

$$y' + 10y = 0, \quad y(0) = 1,$$

with the constant steplength $h = .2$, one obtains the following completely absurd values (the exact solution is $y(x) = e^{-10x}$):

x	y	y'	e^{-10x}
0	1	−10	1
.2	−1	10	.13534
.4	1	−10	.01832
.6	−1	10	.00248

This is what we call *instability of the numerical integration method.* It occurred here only because the step h was too large. For h sufficiently small, Euler's method is stable; as we are about to show, the numerical solution as $h \to 0$ would indeed converge toward the exact solution, if at the same time the number of decimal digits were continually increased.

Convergence of Euler's method. We want to show now, in the case of a single first-order differential equation (3), that the solution determined by Euler's method converges to the exact solution of the differential equation if one chooses all subintervals equally long and lets their length h tend to zero (and disregards rounding errors).

Let $Y(x)$ be the exact solution of (3), i.e.,

$$Y' = f(x, Y), \qquad Y(x_0) = y_0, \tag{10}$$

and $y(x)$ the numerical solution obtained by the method of Euler with steplength h. For the proof of the assertion $|y(x) - Y(x)| \to 0$ as $h \to 0$, we must first of all make a few assumptions: We assume that numbers K, L, $M > 0$ exist such that for $x_0 \le x \le x_0 + L$, $|y - y_0| \le LM$ the following is true:

$$
\begin{aligned}
&|f(x,y)| \le M, \\
&|f(x,y) - f(x,\eta)| \le K|y - \eta|, \\
&|f(x,y) - f(\xi,y)| \le K|x - \xi|.
\end{aligned}
\tag{11}
$$

Putting $x_k = x_0 + kh$, $y(x_k) = y_k$, $Y(x_k) = Y_k$ and $\varepsilon_k = y_k - Y_k$, one obtains first from (8) and (10):

$$\Delta\varepsilon_k = \varepsilon_{k+1} - \varepsilon_k = (y_{k+1} - y_k) - (Y_{k+1} - Y_k)$$
$$= hf(x_k,y_k) - \int_{x_k}^{x_{k+1}} f(x,Y)dx \; ,$$

from which there follows([1]):

$$|\Delta\varepsilon_k| \le h\,|f(x_k,y_k) - f(x_k,Y_k)| + |hf(x_k,Y_k) - \int_{x_k}^{x_{k+1}} f(x,Y)dx|$$
$$\le hK\,|\varepsilon_k| + \int_{x_k}^{x_{k+1}} |f(x_k,Y_k) - f(x,Y)|\,dx \; . \tag{12}$$

("$hf(x_k,Y_k)$ is smeared over the whole interval".) Now, however,

$$|f(x_k,Y_k) - f(x,Y)| \le |f(x_k,Y_k) - f(x_k,Y)| + |f(x_k,Y) - f(x,Y)|,$$

which, because of (11), reduces to

$$|f(x_k,Y_k) - f(x,Y)| \le K\,|Y_k - Y| + K\,|x_k - x| \; .$$

Furthermore,

$$Y - Y_k = Y(x) - Y(x_k) = \int_{x_k}^{x} f(x,Y)dx \; ,$$

thus $|Y - Y_k| \le hM$, as long as $x_k \le x \le x_{k+1} \le x_0 + L$. Therefore,

$$|f(x_k,Y_k) - f(x,Y)| \le KMh + Kh \; ,$$

hence by (12),

$$|\Delta\varepsilon_k| \le hK\,|\varepsilon_k| + KMh^2 + Kh^2 = hK\,|\varepsilon_k| + Ch^2 \tag{13}$$

---

[1] The first assumption in (11), as is easily seen, implies $|y_k - y_0| \le LM$ for $x_k \le x_0 + L$ and $|Y(x) - y_0| \le LM$ for $x_0 \le x \le x_0 + L$. (Translator's remark)

(with $C = KM + K$). This means that

$$|\varepsilon_{k+1}| \le |\varepsilon_k| + |\Delta\varepsilon_k| \le (1 + hK)|\varepsilon_k| + Ch^2 , \tag{14}$$

or, with $q = 1 + hK$,

$$|\varepsilon_n| \le q|\varepsilon_{n-1}| + Ch^2 ,$$
$$q|\varepsilon_{n-1}| \le q^2|\varepsilon_{n-2}| + Ch^2 q ,$$

$$\cdot$$
$$\cdot$$
$$\cdot$$

$$q^{n-1}|\varepsilon_1| \le q^n|\varepsilon_0| + Ch^2 q^{n-1} .$$

Addition of these inequalities, with $\varepsilon_0 = 0$, yields

$$|\varepsilon_n| \le Ch^2(1 + q + q^2 + \cdots + q^{n-1}) = \frac{Ch^2(q^n - 1)}{q - 1} .$$

But now, $q - 1 = hK$, and, if $x = x_0 + nh$,

$$q^n = (1 + hK)^n = \left[1 + \frac{x - x_0}{n} K\right]^n \le e^{K(x - x_0)} .$$

Therefore,

$$|\varepsilon_n| \le Ch^2 \frac{e^{K(x-x_0)} - 1}{Kh} = h\frac{C}{K}\left[e^{K(x-x_0)} - 1\right] = h(M+1)\left[e^{K(x-x_0)} - 1\right] . \tag{15}$$

Thus, $\varepsilon_n \to 0$ as $h \to 0$ (n and h are related by $x - x_0 = nh$), in fact uniformly for all x in the interval $x_0 \le x \le x_0 + L$. Moreover, the error bound as a function of h goes to 0 proportional to h.

§8.3. The order of a method

The basic formula for Euler's method,

$$y_{k+1} = y_k + hf_k = y_k + hy'_k \, ,$$

simply corresponds to the beginning of the Taylor series

$$y_{k+1} = \sum_{v=0}^{\infty} \frac{h^v}{v!} \, y_k^{(v)} \, ,$$

which, in case of convergence, would yield the exact value of y_{k+1}. One could just as well take more than two terms of this series and, for example, compute y_{k+1} according to the formula (Taylor polynomial)

$$y_{k+1} = y_k + hy'_k + \frac{h^2}{2} \, y''_k + \cdots + \frac{h^N}{N!} \, y_k^{(N)} \, . \tag{16}$$

Admittedly, this requires $y''_k, y'''_k, \ldots, y_k^{(N)}$, which can only be obtained by differentiating the differential equation analytically.

Example. If the differential equation $y' = x^2 + y^2$, $y(0) = -1$, is to be integrated by the formula (16) with $N = 3$, one needs

$$y'' = 2x + 2yy', \quad y''' = 2 + 2yy'' + 2(y')^2 \, .$$

The first three steps (with $h = .1$) then give:

x	y	y'	y''	y'''
0	-1	1	-2	8
.1	$-.9086667$	.8356752	-1.3187004	5.7932244
.2	$-.8307271$	.7301075	$-.8130402$	4.4169430
.3	$-.7610454$			

These y_k are correct to 3–4 digits; the exact values are:

$$Y(.1) = -.90877245..., \ Y(.2) = -.83088131..., \ Y(.3) = -.76121865... \ .$$

Such differentiation, however, is often tedious or even impossible; therefore, this method is not in use, and also not recommended. It will serve us, however, as a model.

We apply it to $y' = y$, $y(0) = 1$, and integrate with stepsize $h = x/n$ from 0 to x. One gets

$$y_{k+1} = y_k + hy_k' + \ \cdots \ + \frac{h^N}{N!} y_k^{(N)} = y_k \left[1 + h + \ \cdots \ + \frac{h^N}{N!} \right] ,$$

thus

$$y_n = \left[1 + h + \frac{h^2}{2} + \ \cdots \ + \frac{h^N}{N!} \right]^n , \qquad (17)$$

$$\ln y_n = n \ln \left[1 + h + \frac{h^2}{2} + \ \cdots \ + \frac{h^N}{N!} \right] = n \ln \left[e^h - \sum_{k=N+1}^{\infty} \frac{h^k}{k!} \right]$$

$$= nh + n \ln \left[1 - e^{-h} \sum_{k=N+1}^{\infty} \frac{h^k}{k!} \right] .$$

For $h \to 0$, one therefore has in first approximation:

$$\ln y_n \approx nh - n \frac{h^{N+1}}{(N+1)!} = x - x \frac{h^N}{(N+1)!} ,$$

$$y_n \approx e^x \exp \left[-x \frac{h^N}{(N+1)!} \right] \approx e^x \left[1 - x \frac{h^N}{(N+1)!} \right] . \qquad (18)$$

The relative error at the point x therefore is $xh^N/(N + 1)!$, that is, proportional to h^N.

Quite generally, one finds that by integrating a given differential equation with different stepsizes h (but over the same interval) the error of the integration method is proportional, in first approximation, to a certain power of h.

Definition. *A numerical integration method has order N, if the integration error upon integration from x_0 to a fixed point x has the order of magnitude $O(h^N)(^1)$.*

The reason why a knowledge of this order N, which is characteristic for the method in question, is of importance, is that it allows us to tell by how much the results are improved when the step is reduced. In general, one prefers methods with a large N, since then a reduction of h promises a larger gain in accuracy. One should not overlook, however, that such methods also make the error *grow much more rapidly* when h is increased (which, of course, is what one wants, in order to reduce the number of steps).

Our analysis for the differential equation $y' = y$ suggests the following

Theorem 8.1. *The Euler method has order 1, the method (16) order N.*

For the *determination of the order* of a method, we begin, first of all, by considering the local error, that is the error in *one* integration step.

Let $Y(x)$ again be the exact solution of the differential equation (determined by the initial condition $Y(x_0) = y_0$), $y_0, y_1, \ldots$ the numerical solution obtained by stepwise integration (with support points $x_0, x_1, \ldots$), whereas $\bar{y}(x)$ is the *exact* solution of the differential equation determined by the initial condition

$$\bar{y}(x_k) = y_k \quad (k \text{ fixed}) \tag{19}$$

(cf. Fig. 8.4).

---

[1] And this must be true for every differential equation (3) with a right-hand side $f(x,y)$ which is sufficiently often differentiable. (Editors' remark)

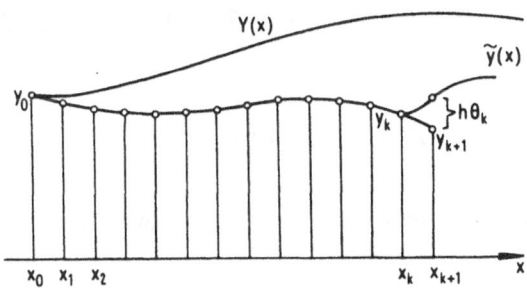

Figure 8.4. *To the definition of the local error* θ_k

Then the quotient

$$\theta_k = \frac{y_{k+1} - \tilde{y}_{k+1}}{h} \tag{20}$$

is called the *local error*[2] of the method at step k (from x_k to x_{k+1}). If, for the moment, one assumes θ_k known and makes use of the fact that $\tilde{y}_{k+1}$ as the solution of the differential equation $\tilde{y}' = f(x,\tilde{y})$ with initial condition (19) can be written in the form

$$\tilde{y}_{k+1} = y_k + \int_{x_k}^{x_{k+1}} f(x,\tilde{y})dx \ ,$$

one obtains

$$y_{k+1} = \tilde{y}_{k+1} + h\theta_k = y_k + \int_{x_k}^{x_{k+1}} (f(x,\tilde{y}) + \theta_k)dx \ . \tag{21}$$

The numerical solution, therefore, is at the same time the exact solution of a differential equation

$$y' = f(x,\tilde{y}) + \theta_k$$

---

[2] In the literature one usually designates the quantity $y_{k+1} - \tilde{y}_{k+1} = h\theta_k$ as local error, whereas θ_k in (20) is called local error per unit step. (Editors' remark)

on the interval $x_k \leq x \leq x_{k+1}$. On that interval, this solution $y(x)$ thus satisfies $y' - \bar{y}' = \theta_k$; therefore, $|y - \bar{y}| \leq h|\theta_k|$, and, existence and uniform boundedness of $\partial f/\partial y$ being assumed, $f(x,y) - f(x,\bar{y}) = O(h\theta_k)$, that is,

$$y' = f(x,y) + \theta_k + O(h\theta_k) .$$

Consequently, for all k and x $(x_k \leq x \leq x_{k+1})$ one has

$$y' - Y' = f(x,y) - f(x,Y) + \theta_k + O(h\theta_k); \tag{22}$$

the error $\varepsilon = y - Y$ thus satisfies *in first approximation* (for small h) the differential equation

$$\varepsilon' = \left. \frac{\partial f}{\partial y} \right|_{y=Y} \cdot \varepsilon + \theta_k, \quad \varepsilon(x_0) = 0, \tag{23}$$

where $k = [(x - x_0)/h]$. The following now holds:

Theorem 8.2. *If there exists a natural number N such that*

$$\lim_{\substack{h \to 0 \\ kh = x - x_0}} \frac{\theta_k}{h^N} = \Phi(x) \tag{24}$$

is a (not identically vanishing) continuous function, then the method in question has order N, and one has in first approximation

$$\varepsilon(x) \approx h^N E(x), \tag{25}$$

where $E(x)$ is the solution of the differential equation

$$\frac{dE}{dx} = \frac{\partial f}{\partial y}(x,Y(x)) E + \Phi(x), \quad E(x_0) = 0 . \tag{26}$$

What is not being said here, however, is the fact that the order N depends only on the method, and not on the differential equation, provided f satisfies a Lipschitz condition([3]).

Example. In the method of Euler, the local error is

$$\theta_k = \frac{(y_k + hy'_k) - (y_k + hy'_k + \frac{1}{2}h^2 y''_k + \cdots)}{h} = -\frac{h}{2}y''_k - \frac{h^2}{6}y'''_k - \cdots .$$

Therefore (considering that convergence has already been proved),

$$\lim_{\substack{h \to 0 \\ kh = x - x_0}} \frac{\theta_k}{h} = -\frac{1}{2}Y''(x) .$$

The method has thus order 1, and

$$\lim_{h \to 0} \frac{y(x) - Y(x)}{h} = E(x),$$

where $E(x)$ is the solution of

$$E' = \frac{\partial f}{\partial y}(x, Y(x)) E - \frac{Y''(x)}{2}, \quad E(x_0) = 0.$$

Note: The statement concerning the order is not valid if Y'' is not continuous, as for example in the differential equation

$$y' = \sqrt[3]{y} + \sqrt{x} ,$$

which has, among others, the solution $Y(x) = 1.415137653 \ldots x^{3/2}$.

---

[3] Actually, one needs sufficient smoothness of f. (Translator's remark)

§8.4. Methods of Runge-Kutta type

A general approach for constructing methods of higher order proceeds as follows:

In each integration interval $[x_k, x_{k+1}]$ a number of auxiliary support points x_A, x_B, x_C , . . . are chosen, whose relative positions within the interval are defined by factors ρ_A, ρ_B, ρ_C , . . . (which for the method in question are fixed once and for all):

$$
\begin{aligned}
x_A &= x_k + \rho_A h, \\
x_B &= x_k + \rho_B h, \\
x_C &= x_k + \rho_C h, \quad \text{etc.}
\end{aligned}
\tag{27}
$$

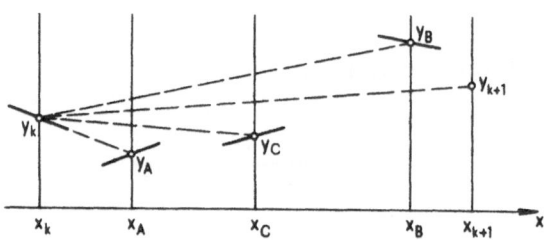

Figure 8.5. *Method of Runge-Kutta type* (Example with 3 auxiliary points A, B, C)

One then defines (cf. Fig. 8.5)

$$
\begin{aligned}
y_A &= y_k + h\sigma_0^A y_k', & y_A' &= f(x_A, y_A), \\
y_B &= y_k + h(\sigma_0^B y_k' + \sigma_A^B y_A'), & y_B' &= f(x_B, y_B), \\
y_C &= y_k + h(\sigma_0^C y_k' + \sigma_A^C y_A' + \sigma_B^C y_B'), & y_C' &= f(x_C, y_C), \quad \text{etc.}
\end{aligned}
\tag{28}
$$

until, finally,

$$
y_{k+1} = y_k + h(\sigma_0 y_k' + \sigma_A y_A' + \sigma_B y_B' + \sigma_C y_C' + \cdots).
\tag{29}
$$

The σ's are determined such that the final value y_{k+1}, the only one used later on, agrees as closely as possible with the exact value, that is, in such a way that the method achieves as high an order as possible. For this, it is necessary, first of all, that

$$\rho_A = \sigma_0^A$$
$$\rho_B = \sigma_0^B + \sigma_A^B$$
$$\rho_C = \sigma_0^C + \sigma_A^C + \sigma_B^C$$

$$\cdot$$
$$\cdot$$
$$\cdot$$

(30)

$$1 = \sigma_0 + \sigma_A + \sigma_B + \sigma_C + \cdots \ ,$$

which is equivalent to the differential equation $y' = 1$ having intermediate values $y_A, y_B, y_C, \ldots$ and final value y_{k+1} that are all exact.

A *method of Runge-Kutta type* is therefore uniquely determined by a triangular matrix([4])

$$
\Sigma =
\begin{bmatrix}
\sigma_0^A & & & & \\
\sigma_0^B & \sigma_A^B & & 0 & \\
\sigma_0^C & \sigma_A^C & \sigma_B^C & & \\
\cdot & & \cdot & & \\
\cdot & & & \cdot & \\
\cdot & & & & \cdot \\
\sigma_0^Z & \sigma_A^Z & \sigma_B^Z & \cdots & \sigma_Y^Z \\
\sigma_0 & \sigma_A & \sigma_B & \cdots & \sigma_Y & \sigma_Z
\end{bmatrix} \quad ;
$$

(31)

the ρ-values are simply the row sums.

Examples. a) The *method of Heun* is given by the matrix

$$
\Sigma_H =
\begin{bmatrix}
1 & 0 \\
\frac{1}{2} & \frac{1}{2}
\end{bmatrix}.
$$

(32)

There is only one auxiliary point (cf. Fig. 8.6),

---

[4] In the literature a row of zero elements is usually added on top of the matrix to indicate that the evaluation of f at (x_k, y_k) utilizes no auxiliary values. (Translator's remark)

$$x_A = x_k + h = x_{k+1}, \quad y_A = y_k + h y_k', \quad y_A' = f(x_A, y_A),$$ (33)

and the final value is computed according to

$$y_{k+1} = y_k + \frac{h}{2}\,(y_k' + y_A').$$ (34)

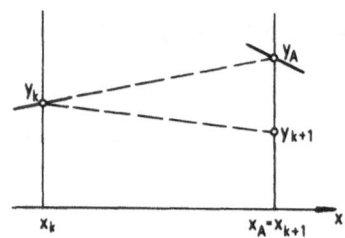

Figure 8.6. *Method of Heun*

Determination of the order of this method: Let $\bar{y}(x)$ again denote the exact solution for the initial condition $\bar{y}(x_y) = y_k$. Then, with $\bar{y}_{k+1} = \bar{y}(x_{k+1})$, one first has

$$y_A \;=\; y_k + h y_k' = \bar{y}_{k+1} - \frac{h^2}{2}\,y_k'' - \cdots ,$$

$$y_A' \;=\; f(x_A, y_A) = f(x_A, \bar{y}_{k+1}) - \frac{h^2}{2}\,y_k''\,\frac{\partial f}{\partial y} - \cdots ,$$

$$y_{k+1} \;=\; y_k + \frac{h}{2}\,(y_k' + y_A') = y_k + \frac{h}{2}\,(y_k' + \bar{y}_{k+1}' - \frac{h^2}{2}\,y_k''\,\frac{\partial f}{\partial y} - \cdots \;) .$$

In comparison, as is easily verified,

$$\bar{y}_{k+1} = y_k + \frac{h}{2}\,(y_k' + \bar{y}_{k+1}') - \frac{h^3}{12}\,y_k''' - \cdots .$$

(Beginning with the h^3-term, the terms of this series, by the way, correspond precisely to the error of the trapezoidal rule, cf. §8.6.) Therefore,

$$\frac{y_{k+1} - \bar{y}_{k+1}}{h} = h^2 \left[\frac{y_k'''}{12} - \frac{\partial f}{\partial y} \frac{y_k''}{4} \right] + O(h^3) .$$

On the right we have the local error θ_k; since

$$\lim_{h \to 0} \frac{\theta_k}{h^2} = \frac{Y'''}{12} - \frac{Y''}{4} \frac{\partial f}{\partial y} \bigg|_{y=Y} , \tag{35}$$

the method of Heun, according to Theorem 8.2, has order 2. By estimating the expressions occurring on the right in (35) one can determine a suitable stepsize.

One occasionally recommends as a criterion for the choice of h the agreement between y_A and y_{k+1}. The fact that one can be taken in, that way, is shown by the example $y' = x^2 + y^2$ with $y(0) = -1$, $h = 1$. Indeed, for $k = 0$ one obtains $y_A = y_{k+1} = 0$, even though the value of the exact solution is $Y(1) = -.23 \ldots$.

b) The *classical Runge-Kutta method* is defined by the matrix

$$\Sigma_{RK} = \begin{bmatrix} \frac{1}{2} & & & 0 \\ 0 & \frac{1}{2} & & \\ 0 & 0 & 1 & \\ \frac{1}{6} & \frac{1}{3} & \frac{1}{3} & \frac{1}{6} \end{bmatrix} . \tag{36}$$

Interpretation (cf. Fig. 8.7):

$$x_A = x_k + \frac{h}{2}, \quad y_A = y_k + \frac{h}{2} y_k', \quad y_A' = f(x_A, y_A),$$

$$x_B = x_A, \quad y_B = y_k + \frac{h}{2} y_A', \quad y_B' = f(x_B, y_B), \tag{37}$$

$$x_C = x_k + h, \quad y_C = y_k + h y_B', \quad y_C' = f(x_C, y_C),$$

$$y_{k+1} = y_k + \frac{h}{6} \left(y_k' + 2y_A' + 2y_B' + y_C' \right). \tag{38}$$

This method has order 4 (without proof([1])).

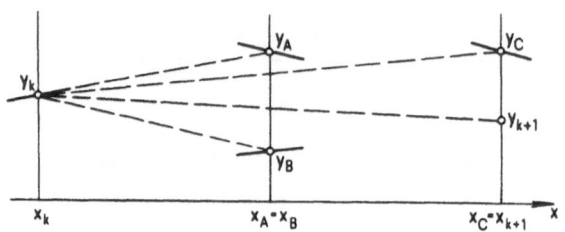

Figure 8.7.   *Classical Runge-Kutta method*

Example. We consider again the differential equation $y' = x^2 + y^2$, $y(0) = -1$, but now choose the step $h = .2$:

$$
\begin{aligned}
x_0 &= 0, & y_0 &= -1, & y_0' &= 1, \\
x_A &= .1, & y_A &= -.9, & y_A' &= .82, \\
x_B &= .1, & y_B &= -.918, & y_B' &= .852724, \\
x_C &= .2, & y_C &= -.8294552, & y_C' &= .727995929, \\
x_1 &= .2, & y_1 &= -.830885202.
\end{aligned}
$$

The exact value would be $Y(.2) = -.830881313772$; the error of y_1 is thus about $-3.9_{10}{-}6$. It is remarkable how much better y_1 is, compared to the auxiliary values y_A, y_B, y_C, whose errors are $9.772_{10}{-}3$, $-9.288_{10}{-}3$, $1.426_{10}{-}3$, respectively.

c) The *method of Nyström*([2]) uses 5 auxiliary points (i.e., the matrix has order 6) and has a local error of order 5:

[1] For a proof see Bieberbach L.: On the remainder of the Runge-Kutta formula in the theory of ordinary differential equations, *Z. Angew. Math. Phys.* **2**, 233–248 (1951).

[2] Nyström E.J.: Über die numerische Integration von Differentialgleichungen, *Acta Soc. Sci. Fenn.* **50**, 13, 1–55 (1925).

$$\Sigma_N = \begin{bmatrix} \frac{1}{3} & & & & & \\ \frac{4}{25} & \frac{6}{25} & & & 0 & \\ \frac{1}{4} & -3 & \frac{15}{4} & & & \\ \frac{6}{81} & \frac{90}{81} & -\frac{50}{81} & \frac{8}{81} & & \\ \frac{6}{75} & \frac{36}{75} & \frac{10}{75} & \frac{8}{75} & 0 & \\ \frac{23}{192} & 0 & \frac{125}{192} & 0 & -\frac{81}{192} & \frac{125}{192} \end{bmatrix} .$$

For the example above, one obtains with this method $y_1 = .830882010$, which is five times as accurate as the result with the Runge-Kutta method.

d) There is a *method of Huta*[3] which uses 7 auxiliary points and has order 6.

Questions of implementation. If the system of differential equations

$$y_j' = f_j(x, y_1, y_2, \ldots, y_n) \quad (j = 1, 2, \ldots, n)$$

with given initial values $y_j(x_0)$ is to be integrated numerically by means of a method determined by the matrix

$$\Sigma = \begin{bmatrix} \sigma_0^A & & & & \\ \sigma_0^B & \sigma_A^B & & 0 & \\ \cdot & & \cdot & & \\ \cdot & & & \cdot & \\ \cdot & & & & \cdot \\ \sigma_0^Z & \sigma_A^Z & \cdots & \sigma_Y^Z & \\ \sigma_0 & \sigma_A & \cdots & \sigma_Y & \sigma_Z \end{bmatrix} ,$$

---

[3] Huta A.: Une amélioration de la méthode de Runge-Kutta-Nyström pour la résolution numérique des équations différentielles du premier ordre, *Acta Fac. Nat. Univ. Comenian. Math.* **1**, 201–224 (1956); Huta A.: Contribution à la formule de sixième ordre dans la méthode de Runge-Kutta-Nyström, *Acta. Fac. Nat. Univ. Comenian. Math.* **2**, 21–24 (1957).

one has to proceed as follows (where y_{kj} will denote the value of the function y_j at the point x_k):

Beginning with the given initial values $y_j(x_0) = y_{0j}$, compute for $k = 0, 1, \ldots$:

1) From $x_k, y_{k1}, y_{k2}, \ldots, y_{kn}$ the derivatives

$$y'_{kj} = f_j(x_k, y_{k1}, \ldots, y_{kn}) \quad \text{(for all } j\text{)}.$$

2) The auxiliary values at the point $x_A = x_k + \rho_A h$:

$$y_{Aj} = y_{kj} + h\sigma_0^A y'_{kj} \quad \text{(for all } j\text{)}.$$

3) The derivatives

$$y'_{Aj} = f_j(x_A, y_{A1}, \ldots, y_{An}) \quad \text{(for all } j\text{)}.$$

4) The auxiliary values at the point $x_B = x_k + \rho_B h$:

$$y_{Bj} = y_{kj} + h(\sigma_0^B y'_{kj} + \sigma_A^B y'_{Aj}) \quad \text{(for all } j\text{)}.$$

5) The derivatives

$$y'_{Bj} = f_j(x_B, y_{B1}, \ldots, y_{Bn}) \quad \text{(for all } j\text{)}.$$

6) Etc., until finally

$$y_{k+1,j} = y_{kj} + h(\sigma_0 y'_{kj} + \sigma_A y'_{Aj} + \sigma_B y'_{Bj} + \cdots + \sigma_Z y'_{Zj}) \quad \text{(for all } j\text{)}.$$

While "for all j" is dealt with by a **for**-statement (**for** $j := 1$ **step** 1 **until** n **do**), the auxiliary points A, B, C must be programmed out explicitly. A loop running also over the auxiliary points would actually be possible, but is not very economical: indeed, by storing

σ_0^A as *sigma*[1,0],

σ_0^B as *sigma*[2,0], σ_A^B as *sigma*[2,1],

.

.

.

σ_0 as *sigma*[m, 0], σ_A as *sigma*[m, 1], . . . ,

ρ_A as *rho*[1], ρ_B as *rho*[2], . . . ,

and furthermore the intermediate values y'_{kj}, y'_{Aj}, y'_{Bj} , . . . (for all j) as an **array** $z[0{:}m - 1,\ 1{:}n]$, an integration step can be described as follows (The procedure *fct* describes the differential equation; upon exit, $z1$ contains the vector $\mathbf{f}(x, \mathbf{y})$.):

```
x := xk;
for j := 1 step 1 until n do y1 [j] := y[j];
for p := 1 step 1 until m do
begin
    fct(n,x,y,z1);
    x := xk + h × rho[p];
    for j := 1 step 1 until n do
    begin
        z [p − 1,j] := z1[j];
        s := 0;
        for q := 0 step 1 until p − 1 do
            s := s + sigma[p,q] × z [q,j];
        y [j] := y1[j] + h × s
    end for j
end for p;
```

§8.5. Error considerations for the Runge-Kutta method when applied to linear systems of differential equations

For a *linear system*

$$\frac{d\mathbf{y}}{dx} = \mathbf{A}(x)\mathbf{y}, \quad \mathbf{y}(0) = \mathbf{y}_0, \tag{39}$$

the method of Runge-Kutta offers no particular advantages; in each step one must compute four times the derivatives, which in this case means 4 multiplications of a matrix by a vector.

This holds true even in the case where $A(x) = A$ is a constant matrix (linear differential equation with constant coefficients), but at least one can then better observe the numerical behavior of the method. If the n components of the solution at the points x_k, x_A , . . . are collected into respective vectors y_k, y_A , . . . , one indeed has for the kth step:

$$y_A = y_k + \frac{h}{2} y_k' = \left[I + \frac{h}{2} A \right] y_k,$$

$$y_B = y_k + \frac{h}{2} y_A' = \left[I + \frac{h}{2} A + \frac{h^2}{4} A^2 \right] y_k,$$

$$y_C = y_k + h y_B' = \left[I + hA + \frac{h^2}{2} A^2 + \frac{h^3}{4} A^3 \right] y_k, \qquad (40)$$

$$y_{k+1} = y_k + \frac{h}{6} \left[y_k' + 2y_A' + 2y_B' + y_C' \right]$$

$$= \left[I + hA + \frac{h^2}{2} A^2 + \frac{h^3}{6} A^3 + \frac{h^4}{24} A^4 \right] y_k .$$

The solution vector y in each step is thus multiplied by the factor

$$I + hA + \frac{h^2}{2} A^2 + \frac{h^3}{6} A^3 + \frac{h^4}{24} A^4$$

(without this matrix being actually computed). In contrast, for the exact solution $Y(x)$ one has

$$\mathbf{Y}_{k+1} = e^{h\mathbf{A}}\mathbf{Y}_k .$$

Therefore, the local error (20), now a vector θ_k, becomes

$$\theta_k = \frac{1}{h}\left[\mathbf{y}_{k+1} - \tilde{\mathbf{y}}_{k+1}\right] = \frac{1}{h}\left[\left[\mathbf{I} + h\mathbf{A} + \frac{h^2}{2}\mathbf{A}^2 + \frac{h^3}{6}\mathbf{A}^3 + \frac{h^4}{24}\mathbf{A}^4\right] - e^{h\mathbf{A}}\right]\mathbf{y}_k,$$

or in first approximation,

$$\theta_k = -\frac{h^4}{120}\mathbf{A}^5\mathbf{y}_k , \tag{41}$$

as is consistent with the order 4 of the method. *Evidently, one must make* $h^4\mathbf{A}^5/120$ *small, if one wishes to keep the error small.*

Further insights are provided by the example of the oscillator equation $y'' + y = 0$, which of course will serve here only as a model, since the exact solution is known. With $z = y'$ one can write it as a system

$$\begin{bmatrix} y \\ z \end{bmatrix}' = \begin{bmatrix} 0 & 1 \\ -1 & 0 \end{bmatrix}\begin{bmatrix} y \\ z \end{bmatrix};$$

thus,

$$\mathbf{A} = \begin{bmatrix} 0 & 1 \\ -1 & 0 \end{bmatrix}.$$

Since $\mathbf{A}^5 = \mathbf{A}$, it is $h^4/120$ that must be made small here. One indeed obtains with

$$h = .1: \quad \frac{10^{-4}}{120} \approx 10^{-6}, \qquad \text{i.e., about 6-digit accuracy},$$

$$h = .3: \quad \frac{81 \times 10^{-4}}{120} \approx 10^{-4}, \quad \text{i.e., about 4-digit accuracy}.$$

Since a complete oscillation amounts to integration from 0 to 2π, there follows: *With the Runge-Kutta method one needs about* 20 *steps per oscillation, if 4-digit accuracy is required,* 60 *steps for* 6 *digits,* 200 *steps for* 8 *digits.*

If the solution is a superposition of different oscillations, then the number of integration steps has to be related to a full oscillation of the highest frequency. For example, if

$$y^{(4)} + 101y'' + 100y = 0$$

(frequencies 1 and 10, i.e., $\sin x + \sin (10x)$ is a solution) is to be integrated with 4-digit accuracy, one must choose $h = .03$.

Actually, this severe requirement can be somewhat alleviated, if the high frequencies contribute only weakly. Thus, $h = .1$ ought to be sufficient for 4-digit accuracy if one wants to integrate the special solution $\sin x + .01 \sin (10x)$ of the above equation. In no case, however, is it permissible to increase h at will, even if the presence of $\sin (10x)$ is arbitrarily weak. This is shown by the following analysis.

Componentwise analysis of the error. The content of formula (40) can be refined if one introduces the eigenvalues λ_j (assumed to be all simple) and the eigenvectors v_j of the matrix $\mathbf{A}$. Then there exists for $y(x)$ a unique representation

$$y(x) = \sum_{k=1}^{n} d_k(x)v_k ,$$

where $d_j(x)v_j$ is called the component of y belonging to the eigenvalue λ_j. Therefore,

$$\mathbf{A}y(x) = \sum_{k=1}^{n} d_k(x)\mathbf{A}v_k = \sum_{k=1}^{n} \lambda_k d_k(x)v_k ,$$

from which it follows, first of all, that $d_k(x) = c_k \exp (\lambda_k x)$, hence that

$$y(x) = \sum_{k=1}^{n} c_k e^{\lambda_k x} v_k \tag{42}$$

is the general solution of $\mathbf{y}' = \mathbf{A}\mathbf{y}$. The coefficients c_k can be obtained by expanding the initial vector $\mathbf{y}(0)$ in the $\mathbf{v}_j$ (for simplicity we assume $x_0 = 0$):

$$\mathbf{y}(0) = \sum_{k=1}^{n} c_k \mathbf{v}_k \ . \tag{43}$$

It then follows further that

$$\mathbf{A}^2 \mathbf{y}(x) = \sum_{k=1}^{n} \lambda_k d_k(x) \mathbf{A} \mathbf{v}_k = \sum_{k=1}^{n} \lambda_k^2 d_k(x) \mathbf{v}_k \ ,$$

etc., and in general for any analytic function F, that

$$F(\mathbf{A})\mathbf{y}(x) = \sum_{k=1}^{n} F(\lambda_k) d_k(x) \mathbf{v}_k \ . \tag{44}$$

When multiplying by $F(\mathbf{A})$, the component of the solution vector belonging to the eigenvalue λ_j (of $\mathbf{A}$) is thus amplified by the factor $F(\lambda_j)$ ($j = 1, \ldots, n$).

If one integrates the differential equation $\mathbf{y}' = \mathbf{A}\mathbf{y}$ by Runge-Kutta, the component of the solution belonging to the eigenvalue λ_j, according to (40), is thus multiplied in each step by the factor

$$F(h\lambda_j) = 1 + h\lambda_j + \frac{h^2}{2} \lambda_j^2 + \frac{h^3}{6} \lambda_j^3 + \frac{h^4}{24} \lambda_j^4 \ , \tag{45}$$

whereas the correct amplification factor would be $e^{h\lambda_j}$. *Numerical integration by Runge-Kutta is therefore as good as the amplification factors (45) agree with $e^{h\lambda_j}$ (for all eigenvalues λ_j of the matrix $\mathbf{A}$).* A comparison of these factors for various values of $h\lambda$ is shown in Table 8.1.

Table 8.1. *Examples for the amplification factor $F(h\lambda)$ of the Runge-Kutta method*

$h\lambda$	$F(h\lambda)$	$e^{h\lambda}$
2	7	7.38905610
.5	1.64843750	1.64872127
−.1	.90483750	.90483742
− 1	.37500000	.36787944
− 2	.33333333	.13533528
− 5	13.70833333	.00673795
.2 i	.98006667 + .19866667 i	.98006658 + .19866933 i
$i\pi/2$	.01996896 + .92483223 i	i
$i\pi$	.12390993 − 2.02612013 i	−1

Now it is true that not all amplification factors $F(h\lambda_j)$ ($j = 1, \ldots, n$) of the components $d_j \mathbf{v}_j$ must agree equally well with $e^{h\lambda_j}$. For an eigenvector which contributes only weakly towards the solution, the deviation may even be relatively large.

Consider, for example, $y'' + 101y' + 100y = 0$, where

$$\mathbf{A} = \begin{bmatrix} 0 & 1 \\ -100 & -101 \end{bmatrix}, \quad \lambda_1 = -1, \quad \lambda_2 = -100 .$$

Here one must first choose h so small that $h\lambda$ remains small for both eigenvalues; for example, $h = .001$, with which the amplification factor for λ_2 becomes $F(-.1) = .9048375$ instead of $e^{-1} = .90483742$. After 100 steps the component of λ_2 is reduced to the fraction $e^{-10} \approx .00005$ of the original value. If one now puts $h = .005$, the amplification factor for λ_2 becomes $.6067708$ instead of $e^{-.5} = .6065307$, which is amply accurate. After an additional 40 steps (i.e., at $x = x_0 + .3$), the component belonging to λ_2 is practically extinguished. If one now continues integrating with $h = .02$, the amplification factors are for

$\lambda_1 = -1$: $.98019867$ (practically exact),

$\lambda_2 = -100$: $.33333333$ instead of $.13533528$.

Since the large deviation for λ_2 can no longer do any damage, one obtains in this way very accurate results. To be noted, however, is the following:

Whereas the components of the solution that are already damped out need no longer be integrated accurately, it is absolutely inadmissible that their amplification factors become larger than 1 in absolute value. This rule imposes severe restrictions on the possibility of enlarging the stepsize in the Runge-Kutta method (and also in the method of Euler and in the other methods of Runge-Kutta type), even if the dominant components of the solution would permit such an enlargement of h.

If in the above example one were to choose $h = .05$, the factors would be for

$$\lambda_1 = -1: \qquad .951229427 \text{ instead of } .951229425,$$
$$\lambda_2 = -100: \qquad 13.70833333 \text{ instead of } .00673795.$$

In this case it would be true that the component belonging to λ_1 is still treated with adequate accuracy, but the component of λ_2 would again be magnified and would poison the solution in a short time.

§8.6. The trapezoidal rule

If $Y(x)$ again denotes the exact solution of the given differential equation (3), and if one puts $Y(x_k) = Y_k$, $Y'(x_k) = Y'_k$, etc., one has (under suitable regularity conditions)([1]):

$$Y_{k+1} - Y_k = \frac{h}{2} (Y'_k + Y'_{k+1}) - \frac{h^3}{12} Y'''(\xi), \text{ where } x_k \leq \xi \leq x_{k+1} \,. (46)$$

By neglecting here the h^3-term, one obtains the *trapezoidal rule*

---

[1] Derivation, e.g., in Krylov V.I: *Approximate Calculation of Integrals*, MacMillan, New York 1962, §6.3. (Translator's remark)

$$y_{k+1} - y_k = \frac{h}{2} \left(y_k' + y_{k+1}' \right), \qquad (47)$$

which is to be supplemented by the relation

$$y_{k+1}' = f(x_{k+1}, y_{k+1}) \qquad (48)$$

in order to have two equations for the two unknowns y_{k+1} and y_{k+1}'. This system of equations, in principle, must be solved in each step.

In the general case, that is, when $f(x,y)$ is nonlinear in y, one conveniently solves it approximately with a predictor-corrector combination, first determining, by means of a *predictor*

$$y_A = y_k + h y_k',$$

an approximate value for y_{k+1}, and then substituting the derivative $y_A' = f(x_{k+1}, y_A)$ in place of y_{k+1}' into the *corrector*, that is, into the trapezoidal rule (47). One easily recognizes in this combination the method of Heun, which thus has arisen from the trapezoidal rule.

In contrast to the method of Runge-Kutta, one indeed gains something here, when the differential equation is linear, since the two equations (47), (48) for y_{k+1} and y_{k+1}' are then also linear, and therefore can be solved without the detour via a predictor. This simplification, in particular, applies also to a system of linear differential equations. Let

$$\mathbf{y}' = \mathbf{A}(x)\mathbf{y} + \mathbf{b}(x), \quad \mathbf{y}(x_0) = \mathbf{y}_0, \qquad (49)$$

be such a system with initial conditions, where $\mathbf{A}(x)$ is a matrix depending on x and $\mathbf{b}(x)$ a vector depending on x. For this system, the trapezoidal rule becomes

$$\mathbf{y}_{k+1} - \mathbf{y}_k = \frac{h}{2} \left(\mathbf{A}_k \mathbf{y}_k + \mathbf{b}_k + \mathbf{A}_{k+1} \mathbf{y}_{k+1} + \mathbf{b}_{k+1} \right)$$

or

$$\left[\mathbf{I} - \frac{h}{2} \mathbf{A}_{k+1} \right] \mathbf{y}_{k+1} = \left[\mathbf{I} + \frac{h}{2} \mathbf{A}_k \right] \mathbf{y}_k + \frac{h}{2} \left(\mathbf{b}_k + \mathbf{b}_{k+1} \right). \qquad (50)$$

The integration step from x_k to x_{k+1} thus requires the solution of a linear system of equations with the coefficient matrix $\mathbf{I} - \frac{1}{2} h \mathbf{A}(x_{k+1})$, which for small h is usually very well-conditioned (cf. §10.7).

The fact that in each step one must solve a linear system of equations should not be held against the trapezoidal rule, since it is precisely in this way that great advantages are realized which other methods do not have. In order to better analyze these advantages, we first examine the special case

$$\mathbf{A}(x) = \mathbf{A} \ (\text{constant}), \ \mathbf{b} = \mathbf{0}.$$

Then

$$y_{k+1} = \left[\mathbf{I} - \frac{h}{2} \mathbf{A} \right]^{-1} \left[\mathbf{I} + \frac{h}{2} \mathbf{A} \right] y_k$$

is the relation for one integration step, while for the exact solution one has

$$\mathbf{Y}_{k+1} = e^{h\mathbf{A}} \mathbf{Y}_k .$$

The method is therefore as good as the matrices

$$\left[\mathbf{I} - \frac{h}{2} \mathbf{A} \right]^{-1} \left[\mathbf{I} + \frac{h}{2} \mathbf{A} \right] \quad \text{and} \quad e^{h\mathbf{A}}$$

agree with one another. The local error is essentially

$$\theta_k = \frac{1}{h} \left[\left[\mathbf{I} - \frac{h}{2} \mathbf{A} \right]^{-1} \left[\mathbf{I} + \frac{h}{2} \mathbf{A} \right] - e^{h\mathbf{A}} \right] y_k = \frac{h^2}{12} \mathbf{A}^3 y_k + \cdots ,$$

so that the order is equal to $2(^2)$.

For the componentwise analysis of the error we can resume our considerations of §8.5: The component of the solution belonging to an eigenvalue λ (of $\mathbf{A}$) in each step of the trapezoidal rule is multiplied by

$$F(h\lambda) = \frac{1 + \dfrac{h}{2}\lambda}{1 - \dfrac{h}{2}\lambda}, \tag{51}$$

the exact amplification factor being $e^{h\lambda}$. But the quantities

$$\left| \frac{1 + \dfrac{h}{2}\lambda}{1 - \dfrac{h}{2}\lambda} \right| \quad \text{and} \quad |e^{h\lambda}| \, ,$$

depending on the value of $h\lambda$, are now either *both* < 1, or *both* $= 1$, or *both* > 1. In other words: *The trapezoidal rule reproduces damping and magnification in a qualitatively correct way.* In particular, therefore, a damped component of the solution of the system of differential equations, when integrated numerically by the trapezoidal rule, is always going to be damped, even if the damping factor is inaccurate. Some examples for the value of the amplification factor (51) are indicated in Table 8.2 and are compared with the exact factor $e^{h\lambda}$. Notice how much more inaccurate these values are, as compared with the Runge-Kutta method (see Table 8.1), but also how the factor remains less than 1 in absolute value even for $h\lambda = -5$.

---

[2] For linear multistep methods (see §8.7), hence in particular for the trapezoidal rule, it suffices for the determination of the order to consider the *linear* differential equation $y' = \lambda y$ (or the system $y' = \mathbf{A}y$). (Editors' remark)

Table 8.2. *Examples for the amplification factor $F(h\lambda)$ of the trapezoidal rule*

$h\lambda$	$F(h\lambda)$	$e^{h\lambda}$
.5	1.66666667	1.64872127
−.1	.90476190	.90483742
−1	.33333333	.36787944
−2	0	.13533528
−5	−.42857143	.00673795
.2 i	.98019802 + .19801980 i = exp(.19933730 i)	.98006658 + .19866933 i
$i\pi/2$	.23697292 + .97151626 i = exp(1.33154750 i)	i
$i\pi$	−.42319912 + .90603670 i = exp(2.00776964 i)	−1

As to the accuracy of reproducing the various components of the solution, we can first of all make the same observation as in the case of the Runge-Kutta method: When choosing h, the components belonging to the various eigenvalues must be taken into consideration within the context of their strengths. Components already damped out need no longer be integrated accurately, so long as the corresponding factor continues to satisfy $|F(h\lambda)| < 1$. This last condition, in case of the trapezoidal rule, however, is fulfilled automatically for all damped components, since for $h > 0$ and Re(λ) < 0, we always have $|F(h\lambda)| < 1$. *In the trapezoidal rule, h can thus be increased as much as the accuracy of the components which still contribute significantly allows it; otherwise, there are no limits set to the increase of h.*

Example. To be solved is the system of differential equations $\mathbf{y}' = \mathbf{A}\mathbf{y}$, $\mathbf{y}(0) = \mathbf{y}_0$, with

$$\mathbf{A} = \begin{bmatrix} 0 & 1 & -1 \\ -1 & -9 & 1 \\ 1 & -1 & -10 \end{bmatrix}, \quad \mathbf{y}_0 = \begin{bmatrix} 1 \\ 1 \\ 1 \end{bmatrix}.$$

The eigenvalues of $\mathbf{A}$ are $\lambda_1 = -.213$ and $\lambda_{2,3} = -9.39 \pm .87i$. The system is to be integrated with an accuracy of 3 to 4 digits.

Since at the beginning all eigenvalues presumably contribute to the solution, and since max $|\lambda| \approx 10$, one must make $h^2 10^3/12 \approx 10^{-4}$, that is, choose $h = 10^{-3}$. After integrating with this stepsize over 200 steps (to $x = .2$), the components of λ_2 and λ_3 are multiplied by the factor $|e^{-200h\lambda}| = e^{-1.878} \approx .16$. Since the order is 2, this permits approximately a doubling of the stepsize, more precisely, a multiplication by $\sqrt{1/.16} = 2.5$. With $h = .0025$, one can for example carry out 120 additional steps (to $x = .5$), reducing to components of $\lambda_{2,3}$ further to $e^{-4.7} \approx .01$; thus, h can be increased to .01 (to 10-times the initial value). After 50 additional steps (to $x = 1$) the components of $\lambda_{2,3}$ already drop to 1/10000 of their initial values. In 4-digit computation they can therefore be neglected, that is, during further integration, h needs to conform only with the eigenvalue $\lambda_1 = -.213$, and this allows $h = .35$ (in 4-digit computation). For example, one can integrate with 100 more steps up to $x = 36$; the total number of steps is only $200 + 120 + 50 + 100 = 470$.

§8.7. General difference formulae

Euler's formula, written in the form

$$y_k - y_{k-1} = hy'_{k-1} \; ,$$

and the trapezoidal rule (47), are special cases of the general class([1])

$$\sum_{j=0}^{m} \alpha_j y_{k-j} = h \sum_{j=0}^{m} \beta_j y'_{k-j} \; , \tag{52}$$

namely with

$$m = 1,$$
$$\alpha_0 = 1, \quad \alpha_1 = -1,$$
$$\beta_0 = 0, \quad \beta_1 = 1$$

and

---

[1] Customary name: *linear multistep methods*. (Editors' remark)

$$m = 1,$$
$$\alpha_0 = 1, \quad \alpha_1 = -1,$$
$$\beta_0 = \beta_1 = \frac{1}{2},$$

respectively.

Generally, a difference formula (52), when $y_{k-m}, y_{k-m+1}, \ldots, y_{k-1}$ (and the derivatives $y'_{k-m} = f(x_{k-m}, y_{k-m}), \ldots, y'_{k-1} = f(x_{k-1}, y_{k-1}))$ are known, is used in such a way that one looks at this formula as a linear equation in the unknowns y_k, y'_k:

$$\alpha_0 y_k - h\beta_0 y'_k = \text{given}, \tag{53}$$

from which, together with the differential equation

$$f(x_k, y_k) - y'_k = 0, \tag{54}$$

one can determine y_k. This is particularly easy in two cases:

a) When, as in the Euler method, $\beta_0 = 0$, one obtains y_k directly from (53) (so-called *explicit methods*).

b) When the differential equation is linear, one has only to solve two linear equations in two unknowns.

On the other hand, when, as for example in the trapezoidal rule, $\beta_0 \neq 0$ (*implicit methods*) and the function $f(x,y)$ is nonlinear in y, then one solves the two equations by *iteration*, alternately determining y_k from (53) and y'_k from (54). For sufficiently small h, this iteration is guaranteed to converge; after its termination, the integration step is completed.

One further speaks of a *predictor-corrector method* if one succeeds with a predictor formula to determine such a good approximation y_k that a single substitution in (54), and subsequently in (53), already yields a sufficiently accurate y_k-value.

Additional examples of such methods:

The *secant rule*([2]) is an explicit method, defined by

$$y_k - y_{k-2} = 2hy'_{k-1} \ . \tag{55}$$

Here,

$$m \ = \ 2,$$
$$\alpha_0 \ = \ 1, \quad \alpha_1 = 0, \quad \alpha_2 = -1,$$
$$\beta_0 \ = \ 0, \quad \beta_1 = 2, \quad \beta_2 = 0.$$

Simpson's rule

$$y_k - y_{k-2} = \frac{h}{3} \left(y'_k + 4y'_{k-1} + y'_{k-2} \right), \tag{56}$$

on the other hand, is implicit:

$$m \ = \ 2,$$
$$\alpha_0 \ = \ 1, \quad \alpha_1 = 0, \quad \alpha_2 = -1,$$
$$\beta_0 \ = \ \tfrac{1}{3}, \quad \beta_1 = \tfrac{4}{3}, \quad \beta_2 = \tfrac{1}{3} \ .$$

One can ask, of course, how such formulae are obtained in general; there is actually a rather simple answer to this.

We apply formula (52) to the differential equation $y' = y$, $y(0) = 1$; then e^x, that is, $y_k = e^{kh}$, should be a solution. One must therefore determine the α_j, β_j at least in such a way that the two sides of (52) agree for $y_k = e^{kh}$ "as much as possible". If we substitute $z = e^h$, this desideratum takes on the form

$$\sum_{j=0}^{m} \alpha_j z^{k-j} \approx \log z \sum_{j=0}^{m} \beta_j z^{k-j} \quad \text{(for all } k\text{)} ,$$

---

[2] Also called *midpoint rule*. (Translator's remark)

$$\log z \approx \frac{\sum\limits_{j=0}^{m} \alpha_j z^{m-j}}{\sum\limits_{j=0}^{m} \beta_j z^{m-j}} = \frac{A(z)}{B(z)} . \tag{57}$$

The problem is thus reduced to the task of approximating $\log z$ as well as possible by a rational function, that is, by the quotient of the two polynomials

$$A(z) = \alpha_0 z^m + \alpha_1 z^{m-1} + \cdots + \alpha_m ,$$
$$B(z) = \beta_0 z^m + \beta_1 z^{m-1} + \cdots + \beta_m . \tag{58}$$

The only question is in which domains of the z-plane this approximation is supposed to be good. If one is interested only in a large order of the method, the approximation must be good in the neighborhood of $h = 0$, thus near $z = 1$.

A very crude approximation at $z = 1$ is

$$\log z \approx z - 1, \tag{59}$$

that is,

$$A(z) = z - 1, \qquad B(z) = 1,$$
$$\alpha_0 = 1, \quad \alpha_1 = -1, \quad \beta_0 = 0, \quad \beta_1 = 1,$$

wherein one recognizes again the Euler method. For an improvement, one averages (59) with

$$\log z = -\log \frac{1}{z} \approx -\left[\frac{1}{z} - 1\right] = 1 - \frac{1}{z} .$$

One so obtains

$$\log z \approx \frac{z - \dfrac{1}{z}}{2} = \frac{z^2 - 1}{2z},$$

$$A(z) = z^2 - 1, \quad B(z) = 2z, \tag{60}$$
$$\alpha_0 = 1, \quad \alpha_1 = 0, \quad \alpha_2 = -1,$$
$$\beta_0 = 0, \quad \beta_1 = 2, \quad \beta_2 = 0,$$

which is the secant rule.

As a further experiment, we take the series of log z and truncate it after the second term:

$$\log z \approx z - 1 - \frac{(z-1)^2}{2} = -\frac{z^2 - 4z + 3}{2}.$$

Then,

$$A(z) = z^2 - 4z + 3, \quad B(z) = -2,$$
$$\alpha_0 = 1, \quad \alpha_1 = -4, \quad \alpha_2 = 3,$$
$$\beta_0 = \beta_1 = 0, \quad \beta_2 = -2,$$

that is,

$$y_k - 4y_{k-1} + 3y_{k-2} = -2hy'_{k-2}. \tag{61}$$

If we substitute in this formula the exact solution $y_k = e^{kh}$ of the differential equation $y' = y$, $y(0) = 1$ ($h = .1$), then the resulting difference between the left-hand and right-hand side is indicated in the 3rd column of Table 8.3. If, on the other hand, one resorts to this equation (61) for the numerical solution of the differential equation, that is, if one uses it as a recurrence formula for y_k, one obtains the y_k noted in the 4th column, which diverge to $-\infty$. The method is thus useless, like all formulas which are produced by series expansion of log z.

Table 8.3. *Application of the difference formula* (61) *to* $y' = y$, $y(0) = 1$

$x_k = kh$	exact solution $Y_k = e^{kh}$	difference in (61) upon substitution of Y_k	numerical solution y_k
0	1		1
.1	1.105171		1.105171
.2	1.221403	.000719	1.220684
.3	1.349859	.000795	1.346188
.4	1.491825	.000878	1.478563
.5	1.648721	.000971	1.606452
.6	1.822119	.001073	1.694406
.7	2.013753	.001186	1.636979
.8	2.225541	.001310	1.125814
.9	2.459603	.001448	−.735075
1.0	2.718282	.001600	−6.542905
.	.	.	↓
.	.	.	
.	.	.	$-\infty$

Useful, however, are those methods in which $B(z)$ is formed through truncation of the series of $1/\log z$. We have, first of all,

$$\frac{t}{\log \dfrac{1}{1-t}} = \frac{-t}{\log(1-t)} = \sum_{k=0}^{\infty} \sigma_k t^k = 1 - \frac{1}{2}t - \frac{1}{12}t^2 - \frac{1}{24}t^3 - \cdots \quad (62)$$

(converge radius 1). Now one substitutes

$$z = \frac{1}{1-t}, \quad \text{i.e.,} \quad t = 1 - \frac{1}{z},$$

and finds

$$\frac{1 - \dfrac{1}{z}}{\log z} = \sum_{k=0}^{\infty} \sigma_k \left(1 - \frac{1}{z}\right)^k,$$

$$\log z = \frac{1 - \dfrac{1}{z}}{\displaystyle\sum_{k=0}^{\infty} \sigma_k \left[1 - \dfrac{1}{z}\right]^k} \cdot \tag{63}$$

$\log z$ can thus be approximated by

$$\log z \approx \frac{1 - \dfrac{1}{z}}{\displaystyle\sum_{k=0}^{m} \sigma_k \left[1 - \dfrac{1}{z}\right]^k} \cdot$$

Multiplying numerator and denominator by z^m, there finally results

$$\log z \approx \frac{A(z)}{B(z)}$$

with

$$A(z) = z^m - z^{m-1},$$
$$B(z) = \sum_{k=0}^{m} \sigma_k z^{m-k}(z-1)^k . \tag{64}$$

Example. For $m = 3$ one obtains

$$\sigma_0 = 1, \quad \sigma_1 = -\frac{1}{2}, \quad \sigma_2 = -\frac{1}{12}, \quad \sigma_3 = -\frac{1}{24},$$

$$A(z) = z^3 - z^2, \quad B(z) = \frac{1}{24}(9z^3 + 19z^2 - 5z + 1),$$

$$\alpha_0 = 1, \quad \alpha_1 = -1, \quad \alpha_2 = \alpha_3 = 0,$$

$$\beta_0 = \frac{9}{24}, \quad \beta_1 = \frac{19}{24}, \quad \beta_2 = -\frac{5}{24}, \quad \beta_3 = \frac{1}{24},$$

thus the integration formula

$$y_k - y_{k-1} = \frac{h}{24} \left(9y'_k + 19y'_{k-1} - 5y'_{k-2} + y'_{k-3} \right). \tag{65}$$

It defines the implicit *method of Adams-Moulton* with $m = 3$. It has order 4.

The order of the method can be deduced directly from the order of approximation of log z: For the polynomials (64), in fact, one has for $z \to 1$,[3]

$$\frac{A(z)}{B(z)} = \log z + O\left(\left[1 - \frac{1}{z} \right]^{m+2} \right). \tag{66}$$

A more detailed analysis shows that $O((1 - 1/z)^{m+2})/h$ corresponds to the local error θ, which, since $1 - 1/z \approx \log z = h$, is thus of the order $O(h^{m+1})$; the order therefore is equal to $m + 1$.

In order to produce an explicit method, one first multiplies numerator and denominator in the representation (63) for log z by $z = 1/(1 - t)$:

$$\log z = \frac{z \left[1 - \dfrac{1}{z} \right]}{\dfrac{1}{1 - \left[1 - \dfrac{1}{z} \right]} \displaystyle\sum_{k=0}^{\infty} \sigma_k \left[1 - \frac{1}{z} \right]^k} \, ,$$

which yields

$$\log z = \frac{z - 1}{\displaystyle\sum_{k=0}^{\infty} \tau_k \left[1 - \frac{1}{z} \right]^k} \tag{67}$$

---

[3] The relation (66) follows readily from (63) and the approximation stated immediately thereafter. (Translator's remark)

with

$$\tau_k = \sum_{j=0}^{k} \sigma_j .$$ (68)

One thus obtains, approximately,

$$\log z \approx \frac{z - 1}{\sum\limits_{k=0}^{m-1} \tau_k \left[1 - \frac{1}{z} \right]^k} ,$$

and, multiplying numerator and denominator by z^{m-1}, finally

$$\log z \approx \frac{A(z)}{B(z)}$$

with

$$A(z) = z^m - z^{m-1} ,$$
$$B(z) = \sum_{k=0}^{m-1} \tau_k z^{m-1-k}(z - 1)^k .$$ (69)

Here, $B(z)$ is a polynomial of degree $m - 1$, so that $\beta_0 = 0$.

Example. For $m = 3$, one gets

$$A(z) = z^3 - z^2, \quad B(z) = \frac{1}{12} (23z^2 - 16z + 5),$$

from which one obtains the integration formula

$$y_k - y_{k-1} = \frac{h}{12} (23y'_{k-1} - 16y'_{k-2} + 5y'_{k-3}) ,$$ (70)

known as the *method of Adams-Bashforth with m = 3*. Because of

$$\frac{A(z)}{B(z)} = \log z + O((z-1)^4),$$

this method has order 3.

The start-up computation. Every difference formula of the type

$$\sum_{j=0}^{m} \alpha_j y_{k-j} = h \sum_{j=0}^{m} \beta_j y'_{k-j}$$

presupposes that the values y_{k-m}, y_{k-m+1} ,...., y_{k-1} are already known. This however, when $k = 1$, is the case only for methods with $m = 1$ (Euler method, trapezoidal rule). Otherwise, the previous history, consisting of the values y_{-1}, y_{-2} ,...., y_{1-m}, is nonexistent. There are two ways out of this dilemma:

1) Integrate with this difference formula only from $k = m$ onward, having previously computed $y_1, \ldots, y_{m-1}$ by means of a method of the Runge-Kutta type.

2) The missing information is made available artificially. Note, in this connection, that in the methods of Adams-Bashforth and Adams-Moulton only the derivatives y'_{-1}, y'_{-2} ,...., y'_{1-m} are actually needed, which facilitates the problem considerably.

We illustrate the second approach with the example of the Adams-Bashforth method with $m = 2$. For this method one has

$$A(z) = z^2 - z, \quad B(z) = \frac{3}{2} z - \frac{1}{2},$$

thus

$$y_k - y_{k-1} = \frac{h}{2} (3y'_{k-1} - y'_{k-2}) . \tag{71}$$

Let the equation to be integrated be again $y' = y$, $y(0) = 1$, with $h = .1$. The missing information here consists soley of y'_{-1}. We first put $y'_{-1} = y'_0 = 1$ and integrate over $m = 2$ steps:

k	x_k	y_k	y'_k	$\Delta y'$	$\Delta^2 y'$	$\Delta^3 y'$
-1	$-.1$		1		extrapolated backwards	
				.085		
0	0	1	1	$-\;-\;-$ .015		
				.1 $-\;-\;-$	0	
1	.1	1.1	1.1		.015 $-\;-\;-\;-\;-$	
				.115		
2	.2	1.215	1.215			

From the values y'_0, y'_1, y'_2 one then extrapolates y'_{-1} in such a way that the third (in general, the $(m+1)$st) difference becomes 0. This yields here the value $y'_{-1} = 1 - .085 = .915$, with which one integrates once more. (In the general case one would have to extrapolate back to y'_{1-m}.)

k	x_k	y_k	y'_k	$\Delta y'$	$\Delta^2 y'$	$\Delta^3 y'$
-1	$-.1$		.915		extrapolated backwards	
				.092862		
0	0	1	1	$-\;-\;-\;-$ .011388		
				.10425 $-\;-\;-\;-$	0	
1	.1	1.10425	1.10425		.011388 $-\;-\;-\;-$	
				.115638		
2	.2	1.219888	1.219888			

From these values one obtains, again by backward extrapolation, $y'_{-1} = .907138$, whereupon one integrates forward once more, etc. (The exact value would be $Y'(-.1) = e^{-.1} = .90483742 \ldots .$)

§8.8. The stability problem

As we have seen, some difference formulae derived from rational approximations of log z, for example (61), are useless. We thus turn to the question of stability of an integration method of type (52).

We apply the integration method to be examined to the differential equation $y' = \lambda y$, where λ can be arbitrary complex. This covers also the behavior of a system $\mathbf{y}' = \mathbf{A}\mathbf{y}$, because the solution component of this system belonging to the eigenvalue λ behaves in every respect like the solution of $y' = \lambda y$. The difference formula (52) then becomes

$$\sum_{j=0}^{m} \alpha_j y_{k-j} = h\lambda \sum_{j=0}^{m} \beta_j y_{k-j} , \qquad (72)$$

from which one infers that for the behavior of the numerical solution only the product $h\lambda$ is important. We therefore may as well integrate the differential equation $y' = y$ with the "reduced steplength" $s = h\lambda$, where s, however, can be complex. The difference formula then becomes

$$\sum_{j=0}^{m} (\alpha_j - s\beta_j)y_{k-j} = 0. \qquad (73)$$

This is a linear difference equation with constant coefficients, whose general solution can be sought in the form

$$y_k = \sum_{v=1}^{m} g_v z_v^k . \qquad (74)$$

Substitution into the preceding equation (73) yields at once

$$\sum_{j=0}^{m} \sum_{v=1}^{m} (\alpha_j - s\beta_j)g_v z_v^{k-j} = 0 ,$$

or

$$\sum_{v=1}^{m} g_v z_v^{k-m} \left\{ \sum_{j=0}^{m} \left[\alpha_j - s\beta_j \right] z_v^{m-j} \right\} = 0 ,$$

where the expression in braces is independent of k. Since this relation must hold for all k, there follows

$$\sum_{j=0}^{m} (\alpha_j - s\beta_j)z_v^{m-j} = 0 ,$$

or, introducing again the polynomials (58),

$$A(z) - sB(z) = 0 \quad \text{for} \quad z = z_\nu, \; \nu = 1, \ldots, m \; . \tag{75}$$

The basic numbers $z_1, z_2, \ldots, z_m$ thus are solutions of this algebraic equation, and the numerical solution has the form (74) (with certain coefficients g_ν), while for the exact solution $Y(x) = ce^x$ with $x = sk$ one has

$$Y_k = c(e^s)^k \; .$$

Now for large k, however, the numerical solution (74) consists practically only of the term $g_1 z_1^k$, where z_1 denotes the largest in modulus of the roots $z_1, \ldots, z_m$. *Thus, if the numerical solution is more or less to follow the exact solution, the dominant root z_1 of the equation (75) must lie in the vicinity of e^s.*

There is *one* root of (75) which always lies near e^s, because the coefficients α_j, β_j were determined in §8.7 such that

$$\frac{A(z)}{B(z)} \approx \log z \; ,$$

so that, with $s = \log z$, also $A(z) - sB(z) \approx 0$. However, this is not enough; this root near e^s must also be the largest in absolute value.

Example. For the secant rule (55) we have $A(z) = z^2 - 1$, $B(z) = 2z$; the equation (75) thus becomes

$$A(z) - sB(z) = z^2 - 2sz - 1 = 0$$

and has the solution

$$z = s \pm \sqrt{s^2 + 1} \; . \tag{76}$$

The product of the two roots is -1; for the larger in absolute value one therefore has $|z_1| > 1$. This larger root, in fact, defines a conformal mapping of the s-plane, cut along the segment from $s = i$ to $s = -i$, onto $|z| > 1$:

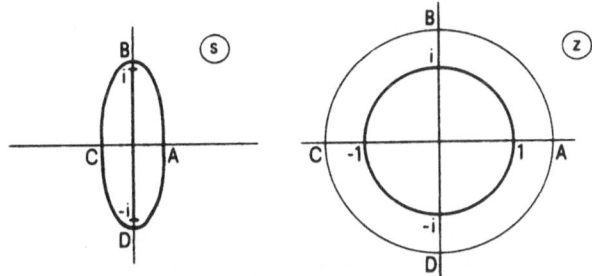

Figure 8.8. *Mapping of the z-plane to the w-plane defined
by the larger of the two solutions (76)*

while $w = e^s$ yields the following picture:

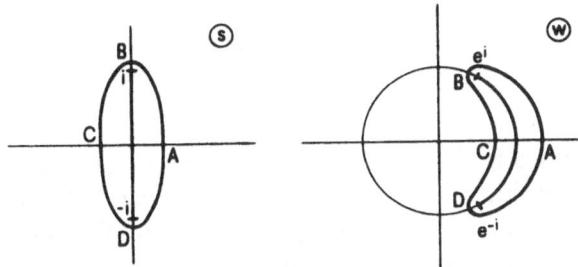

Figure 8.9. *Mapping of the z-plane to the w-plane by means of $w = e^s$*

Only for the part of the neighborhood of the origin lying in the right half
of the s-plane is the root

$$z = s + \sqrt{s^2 + 1} = 1 + s + \frac{1}{2} s^2 - \frac{1}{8} s^4 - \cdots$$

near e^s the larger one in absolute value. In the left half-plane, however,
one has $z_1 \approx e^{-s}$ for the larger root. The secant rule is therefore unstable
for Re $s < 0$ (though only weakly unstable).

Now in order to obtain a measure for the error, we note that for
large k one has approximately $y_{k+1} = y_k z_1$, while in the notations of §8.3,

$$\bar{y}_{k+1} = y_k e^s,$$

hence for the local error (20),

$$\theta_k = \frac{z_1 - e^s}{s} \, y_k \, .$$

For us, however, it is more meaningful to consider the local relative error, that is, the local absolute error of the logarithm:

$$\frac{\log y_{k+1} - \log \bar{y}_{k+1}}{s} = \frac{\log z_1 - s}{s} \, .$$

Its absolute value is introduced as universal error measure:

$$\psi(s) = \left| \frac{\log z_1 - s}{s} \right| \, . \tag{77}$$

Here, z_1 continues to denote the dominant root of (75), and $\log z_1$ is that value of the logarithm which lies closest to s. $\psi(s)$ is a real function of the complex variable s, which for each value of the reduced stepsize s indicates the relative error per unit integration step.

In order now to arrive at a criterion for stability, we first require that $\psi(0) = 0$ and $\psi(s)$ is small for $|s|$ small (i.e., for small h one should nearly obtain the exact solution). This requires that for all sufficiently small $|s|$ the root z near e^s is the largest in absolute value, that is, for $s \to 0$ we must have

$$\log z_1(s) - s = o(s) \tag{78}$$

and in particular, $\log z_1(0) = 0$, thus $z_1(0) = 1$, or

$$A(1) = 0. \tag{79}$$

Furthermore, by (78),

$$\frac{d \log z_1}{ds}\bigg|_{s=0} - 1 = 0, \quad \text{i.e.,} \quad \frac{dz_1}{ds}\bigg|_{s=0} = 1.$$

From $A(z_1) - sB(z_1) = 0$, however, there follows

$$\frac{dA}{dz_1}\frac{dz_1}{ds} - B(z_1) - s\frac{dB}{dz_1}\frac{dz_1}{ds} = 0,$$

thus, for $s = 0$, $z_1 = 1$,

$$A'(1) - B(1) = 0. \tag{80}$$

(79) *and* (80) *are merely necessary conditions for the stability of a method (so-called consistency conditions).*

Now since the roots of an algebraic equation, as is well known, depend continuously on the coefficients, the root z of $A(z) - sB(z) = 0$ lying in the vicinity of e^s is for all sufficiently small $|s|$ the largest in absolute value, if this is the case for $s = 0$, that is, if $z = 1$ is the root of maximum modulus of $A(z) = 0$, and besides is simple. Consequently, as a condition for the stability of the difference formula (52) we have

$$\frac{A(z)}{z-1} \neq 0 \quad \text{for} \quad |z| \geq 1. \tag{81}$$

This condition guarantees the so-called strong stability of the method.

Examples. 1) The method of Adams-Bashforth with $m = 3$ is defined by the difference formula (70). One has

$$A(z) = z^3 - z^2, \quad B(z) = \frac{1}{12}(23z^2 - 16z + 5),$$

and therefore

$$A(1) = 0, \quad A'(1) = 1 = B(1).$$

Furthermore,

$$\frac{A(z)}{z-1} = z^2 ,$$

which is certainly different from 0 for $|z| \geq 1$. The method is thus stable for all sufficiently small $|s|$. The same is true for all Adams-Bashforth methods, which according to §8.7 are characterized by the polynomials (69). If, however, s is made more and more negative, then sooner or later the root lying in the "vicinity" of e^s is no longer the largest in absolute value, and the method becomes unstable. This happens the sooner (that is, already for smaller $|s|$) the larger m. For example, if $m = 16$, the method is unstable already for $s = -.05$.

2) As a second example we consider the fantasy method defined by (61), that is, by the polynomials

$$A(z) = z^2 - 4z + 3, \quad B(z) = -2 .$$

Here, $A(1) = 0$, $A'(1) = -2 = B(1)$, but $A(z)/(z-1) = z - 3$ has a root outside the unit circle; the method is therefore unstable. The solution for small $|s|$ and large k behaves about like $y_k = 3^k$, and this almost independently of the differential equation.

3) For the secant method we have

$$A(z) = z^2 - 1, \quad B(z) = 2z,$$
$$A(1) = 0, \quad\quad A'(1) = 2 = B(1).$$

But

$$\frac{A(z)}{z-1} = z + 1$$

has a root precisely on the unit circle. While the stability condition is not satisfied, we have here a limit case which must be examined more carefully. To this end, we consider the differential equation $y' = -y$, $y(0) = 1$, which is to be integrated numerically by the secant rule, whereby the steplength h is assumed to be positive. The reduced steplength is here $s = -h$. The numerical solution therefore has the form

$$y_k = g_1 z_1^k + g_2 z_2^k = g_1 \left[-h - \sqrt{1+h^2} \right]^k + g_2 \left[-h + \sqrt{1+h^2} \right]^k ,$$

where g_1 and g_2 remain to be determined. By means of the substitution $h = \sinh \eta$ one obtains

$$z_1 = - \sinh \eta - \cosh \eta = - e^{\eta}, \quad z_2 = - \sinh \eta + \cosh \eta = e^{-\eta},$$

and thus

$$y_k = g_1(-1)^k e^{\eta k} + g_2 e^{-\eta k}.$$

To construct the second starting value y_1, we may first carry out one step according to Euler, with $y_0 = 1$. Then,

$$y_0 = 1 = g_1 + g_2, \quad y_1 = 1 - h = - g_1 e^{\eta} + g_2 e^{-\eta},$$

from which there follows

$$g_1 = \frac{\cosh \eta - 1}{2 \cosh \eta}, \quad g_2 = \frac{\cosh \eta + 1}{2 \cosh \eta}.$$

For small h, however, one now has

$$\frac{\cosh \eta - 1}{2 \cosh \eta} = \frac{\sqrt{1 + h^2} - 1}{2\sqrt{1 + h^2}} \approx \frac{h^2}{4},$$

$$\frac{\cosh \eta + 1}{2 \cosh \eta} = \frac{\sqrt{1 + h^2} + 1}{2\sqrt{1 + h^2}} \approx 1 - \frac{h^2}{4}$$

and furthermore,

$$e^{\eta k} = \exp\left[\frac{\eta h k}{\sinh \eta}\right] = \exp\left[\frac{\eta}{\sinh \eta} x_k\right].$$

For small η, on the other hand, $\eta/\sinh \eta \approx 1$, hence $e^{\eta k} \approx e^{x_k}$,

$e^{-\eta k} \approx e^{-x_k}$, so that for the numerical solution one obtains approximately

$$y_k \approx (-1)^k \frac{h^2}{4} e^{x_k} + \left[1 - \frac{h^2}{4} \right] e^{-x_k} .$$

One sees that the solution is made up of an oscillatory increasing, and a decreasing term. The first, to be sure, is small, but if one integrates long enough it will eventually dominate, and further integration becomes illusory (cf. Fig. 8.10).

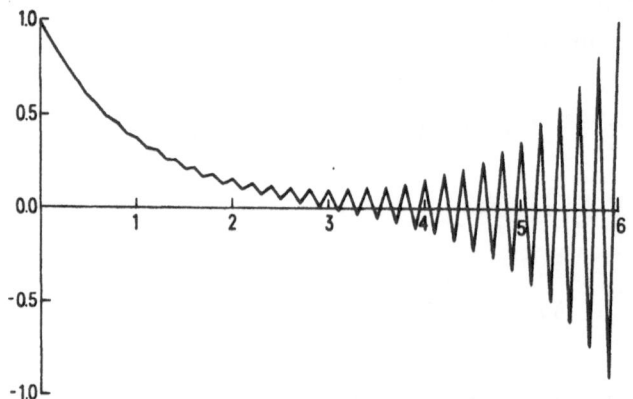

Figure 8.10. *Integration of $y' = -y$, $y(0) = 1$, with the secant rule and $h = .1$*

The point x_k at which a negative y-value is to be expected for the first time is approximately given by

$$e^{-x_k} = \frac{h}{2}, \quad \text{i.e.,} \quad x_k = \ln \frac{2}{h} .$$

The onset of the oscillation, therefore, is further and further delayed upon decreasing h. It is even true that in every finite interval the numerical solution converges to the exact solution as $h \to 0$, that is, the instability – provided one considers only a finite interval – can be eliminated by decreasing h.

One calls this phenomenon *weak instability*. It is characterized, according to Dahlquist, by the following two conditions:

a) $\dfrac{A(z)}{z-1} \neq 0$ for $|z| > 1$.

b) The zeros of $A(z)/(z-1)$ located on the circumference of the unit circle are simple.

4) As a further example we consider the difference formula

$$y_k - y_{k-3} = \frac{h}{4}\,(9y'_{k-1} + 3y'_{k-3})\,.$$

Here,

$$A(z) = z^3 - 1\,, \qquad B(z) = \frac{3}{4}\,(3z^2 + 1)\,,$$

$$A(1) = 0\,, \qquad\qquad A'(1) = 3 = B(1)\,,$$

$$\frac{A(z)}{z-1} = z^2 + z + 1\,.$$

Both roots of $A(z)/(z-1)$ lie on the unit circle and are simple. The method is thus weakly unstable.

A case study for stability. For a more detailed discussion of stability we select the method of Adams-Bashforth with $m = 2$:

$$y_k - y_{k-1} = \frac{h}{2}\,(3y'_{k-1} - y'_{k-2})\,,$$

which has order 2. Here,

$$A(z) = z^2 - z\,, \qquad B(z) = \frac{3}{2}\,z - \frac{1}{2}\,;$$

the roots of the equation $A(z) - sB(z)$ are thus

$$z = \frac{1}{2} + \frac{3}{4}\,s \pm \sqrt{\left[\frac{1}{2} + \frac{3}{4}\,s\right]^2 - \frac{s}{2}}\,, \tag{82}$$

and this function $z(s)$ has the branch points $(-2 \pm 4\sqrt{2}\,i)/9$.

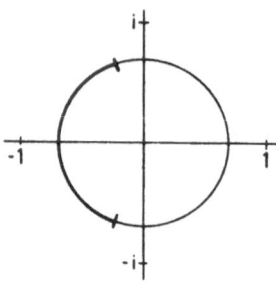

Figure 8.11. *To the definition of z(s) in (82)*

Now cutting open the Riemann surface of this function along the circular arc $|s| = \frac{2}{3}$ between the branch points (see Fig. 8.11), one finds that $z(s)$ in the sheet belonging to the value $z_1(0) = 1$ is larger in absolute value than in the other sheet determined by $z_2(0) = 0$. (On the cut one has $|z_1(s)| = |z_2(s)|$.) To the right of the circular arc, thus in particular for $|s| < \frac{2}{3}$, the root lying in the vicinity of e^s is therefore the larger one in absolute value; it of course deviates more and more from e^s for larger $|s|$. Upon crossing the circular arc, the other root (which is quite different from e^s) suddenly becomes larger in absolute value. The method is unstable as soon as $s = h\lambda$ lies in that domain. Note, however, that along a path around the cut one can pass continuously from the stable to the unstable domain. The error then increases from tiny to huge values. In Table 8.4 a few function values are given for illustration.

Table 8.4. *Some values $z_1(s)$ and $z_2(s)$ of the function (82)*

s	$z_1(s)$	$z_2(s)$	e^s
2	3.7321	.2679	7.3891
.5	1.5931	.1569	1.6487
0	1	0	1
−.5	.6404	−.3904	.6065
−.66	.5795	−.5695	.5169
−.68	−.5932	.5732	.5066
−2	−2.4142	.4142	.1353
$i\,2/3$	.7887 +	.2113 +	.7859 +
	.7887 i	.2113 i	.6184 i

§8.9. Special cases

A) *Treatment of extremely strong damping.* If a differential equation such as

$$y' + 10000\,y = e^x - y^2, \quad y(0) = 0, \tag{83}$$

must be integrated numerically, then the large coefficient 10000 requires an extremely small integration step (about 10^{-5}). In order to discuss how one can escape from this restriction, we consider more generally the case

$$y' + my = g(x,y), \tag{84}$$

where we assume m to be a large positive constant, but the function values of g to be "normal".

Let us learn from the way we integrated $y' = f(x,y)$ numerically by Euler. The formula $y_{k+1} = y_k + hy'_k$, indeed, can be interpreted by saying that one integrates *exactly* the differential equation $y' = f_k$, in which $y'_k = f_k = f(x_k,y_k)$ is held fixed, from x_k to x_{k+1}.

An analogous method can be obtained for the differential equation (84), if for fixed $g_k = g(x_k,y_k)$ one integrates exactly the equation

$$y' + my = g_k \tag{85}$$

from x_k to x_{k+1}. The exact general solution of (85) is

$$y = ce^{-m(x-x_k)} + \frac{g_k}{m}.$$

With $y(x_k) = y_k$ one obtains $c = y_k - g_k/m$ and thus the formula of integration

$$y_{k+1} = y_k e^{-mh} + g_k \frac{1 - e^{-mh}}{m}. \tag{86}$$

This is a generalization of Euler's formula and reduces to it when $m \to 0$.
In the above example (83) we first obtain, with $h = .01$,

$$e^{-mh} = e^{-100} \approx 4_{10}\text{--}44 \approx 0 ,$$

$$\frac{1 - e^{-mh}}{m} = \frac{1 - e^{-100}}{10000} \approx 10\text{--}4$$

for the constants in (86), and then in turn,

$$
\begin{array}{lll}
x_0 = 0, & y_0 = 0, & g_0 = 1, \\
x_1 = .01, & y_1 = 10\text{--}4, & g_1 = 1.01005, \\
x_2 = .02, & y_2 = 1.01005_{10}\text{--}4, & g_2 = 1.02020, \\
x_3 = .03, & y_3 = 1.02020_{10}\text{--}4, & \text{etc.}
\end{array}
$$

This is not very accurate, but the usual Euler method immediately incurs
heavy instability, whenever h is not $< 10^{-4}$.

Similarly, one can adapt the method of Heun to the given problem:
its formulae

$$y_A = y_k + hf_k, \quad y_{k+1} = y_k + \frac{h}{2} (f_k + f_A)$$

mean that, having determined the predicted value y_A (by Euler), the
differential equation

$$y' = f_k + \frac{f_A - f_k}{h} (x - x_k)$$

is integrated *exactly* from x_k to x_{k+1}.

For the differential equation (84) the analogue of Heun's method
therefore is: having computed

$$y_A = y_k e^{-mh} + g_k \frac{1 - e^{-mh}}{m} , \quad g_A = g(x_{k+1}, y_A) , \tag{87}$$

the differential equation

$$y' + my = g_k + \frac{g_A - g_k}{h}(x - x_k)$$

is integrated exactly from x_k to x_{k+1}. The general solution is

$$y = ce^{-m(x - x_k)} + \frac{g_k}{m} + \frac{g_A - g_k}{m^2 h}(m(x - x_k) - 1),$$

where, on account of $y(x_k) = y_k$,

$$c = y_k - \frac{g_k}{m} + \frac{g_A - g_k}{m^2 h}.$$

The corrector formula therefore becomes

$$y_{k+1} = y_k e^{-mh} + g_k \frac{1 - e^{-mh}}{m} + (g_A - g_k)\frac{mh - 1 + e^{-mh}}{m^2 h}$$

$$= y_k e^{-mh} + g_k \frac{1 - e^{-mh} - mhe^{-mh}}{m^2 h} + g_A \frac{mh - 1 + e^{-mh}}{m^2 h},$$

or

$$y_{k+1} = y_k e^{-mh} + h(c_0 g_k + c_1 g_A), \qquad (88)$$

where

$$c_0 = \frac{1 - (1 + mh)e^{-mh}}{(mh)^2}, \quad c_1 = \frac{e^{-mh} - 1 + mh}{(mh)^2}. \qquad (89)$$

Since for $mh \to 0$ both c_0 and c_1 tend to 1/2, one obtains also here as limit case indeed the method of Heun.

In the example (83), one now first obtains, with $h = .01$,

$$c_0 = \frac{1 - (1 + 100)e^{-100}}{10000} \approx 10^{-4},$$

$$c_1 = \frac{e^{-100} - 1 + 100}{10000} \approx 99_{10}{-4}.$$

The first 10 integration steps are summarized in Table 8.5. The y_k agree with the exact solution up to one unit in the last decimal digit given.

Table 8.5. *Integration of the differential equation (83) with a special predictor-corrector method*

k	x_k	y_k	g_k	y_A	g_A
0	0	0	1	$1.000000_{10}{-4}$	1.010050
1	.01	$1.009950_{10}{-4}$	1.010050	$1.010050_{10}{-4}$	1.020201
2	.02	$1.020100_{10}{-4}$	1.020201	$1.020201_{10}{-4}$	1.030455
3	.03	$1.030352_{10}{-4}$	1.030455	$1.030455_{10}{-4}$	1.040811
4	.04	$1.040707_{10}{-4}$	1.040811	$1.040811_{10}{-4}$	1.051271
5	.05	$1.051166_{10}{-4}$	1.051271	$1.051271_{10}{-4}$	1.061837
6	.06	$1.061731_{10}{-4}$	1.061837	$1.061837_{10}{-4}$	1.072508
7	.07	$1.072401_{10}{-4}$	1.072508	$1.072508_{10}{-4}$	1.083287
8	.08	$1.083179_{10}{-4}$	1.083287	$1.083287_{10}{-4}$	1.094174
9	.09	$1.094065_{10}{-4}$	1.094174	$1.094174_{10}{-4}$	1.105171
10	.10	$1.105061_{10}{-4}$			

B) *Treatment of a differential equation of the form*

$$y'' + f(x)\, y = 0, \tag{90}$$

where $f(x)$ for all x is very large (positive) and slowly varying. With numerical integration in the usual style one does not get very far, since per unit length about $10\sqrt{f(x)}$ integration steps would be required (for 6-digit accuracy). It is much better to introduce a new function

$$z(x) = y(x) - i \frac{y'(x)}{\sqrt{f(x)}}, \tag{91}$$

which may be thought of as a curve in the complex plane, where x serves as the parameter. With $y(x)$ a solution of the differential equation (90), one clearly has

$$\frac{dz}{dx} = y'(x) - i \frac{y''(x)}{\sqrt{f(x)}} + i \frac{y'(x)f'(x)}{2(f(x))^{3/2}}$$

$$= y'(x) + iy(x)\sqrt{f(x)} + i \frac{y'(x)f'(x)}{2(f(x))^{3/2}},$$

hence, because of $\overline{z(x)} - z(x) = 2i\, y'(x)/\sqrt{f(x)}$,

$$\frac{dz}{dx} = i \sqrt{f(x)}\,z(x) + \frac{\overline{z(x)} - z(x)}{4} \frac{f'(x)}{f(x)}. \tag{92}$$

Since by assumption, f'/f is to be small, one obtains as first approximation in the neighborhood of $x = x_0$ (with $t = x - x_0$, $z_0 = z(x_0)$, $f_0 = f(x_0)$)

$$z(x_0 + t) \approx z_0 e^{it\sqrt{f_0}},$$

that is, the point $z(x)$ describes in first approximation a circular path with radius $|z_0|$ and circular frequency $\sqrt{f_0}$. An improvement is obtained on the basis of the approximation

$$\sqrt{f} = \sqrt{f_0} + \frac{t f_0'}{2\sqrt{f_0}},$$

if, in addition, in all terms in which z is multiplied by small coefficients, z is replaced by $z_0 \exp(it\sqrt{f_0})$. Then the differential equation (92) becomes

$$\frac{dz}{dt} = i\sqrt{f_0}\,z + \frac{itf_0'}{2\sqrt{f_0}}\,e^{it\sqrt{f_0}}\,z_0 + \frac{\bar{z}_0 e^{-it\sqrt{f_0}} - z_0 e^{it\sqrt{f_0}}}{4}\,\frac{f_0'}{f_0}\,.$$

This differential equation can be solved exactly by the method of variation of constants; one obtains

$$z(x_0 + t) = z_0 e^{it\sqrt{f_0}} + \frac{\bar{z}_0 f_0'}{4 f_0^{3/2}}\,\sin(t\sqrt{f_0})$$

$$+ iz_0 \frac{f_0'}{4\sqrt{f_0}}\,t^2 e^{it\sqrt{f_0}} - z_0 \frac{f_0'}{4 f_0}\,t e^{it\sqrt{f_0}}\,. \qquad (93)$$

Apart from the periodic perturbation caused by the sine term, one has the following terms which deviate from the harmonic oscillation:

$$it^2 \frac{f_0'}{4\sqrt{f_0}}\,z \qquad \text{(phase shift due to tangential acceleration)},$$

$$-t \frac{f_0'}{4 f_0}\,z \qquad \text{(decrease of amplitude due to radial acceleration)}\,.$$

Now at time $t = 2\pi k/\sqrt{f_0}$ (k an integer), however, one evidently has

$$z(x_0 + t) = z_0 \left[1 + \frac{if_0'}{4\sqrt{f_0}}\,\frac{4\pi^2 k^2}{f_0} - \frac{f_0'}{4 f_0}\,\frac{2\pi k}{\sqrt{f_0}} \right],$$

that is,

$$z_k^* = z\left[x_0 + \frac{2\pi k}{\sqrt{f_0}} \right] = z_0 + z_0 \frac{f_0'}{f_0^{3/2}}\left[i\pi^2 k^2 - \frac{\pi}{2}\,k \right], \qquad (94)$$

so that the points z_1^*, z_2^*, z_3^*, ... assume the positions depicted in Fig. 8.12 (z_0 is assumed real).

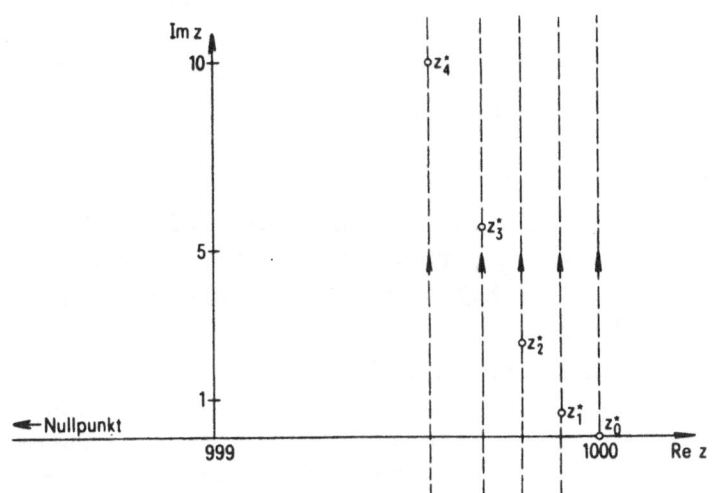

Figure 8.12. *Position of the points z_k^**

Determination of the crossings of the real axis of the z-plane: Since in first approximation, $dz/dt = i\sqrt{f_0}\,z$, and since at time $t = 2\pi k/\sqrt{f_0}$ one has overshot the real axis by the amount

$$z_0 \, \frac{if_0'}{f_0^{2/3}} \, \pi^2 k^2 \,,$$

the crossings take place, in first approximation, at the times

$$t_k = \frac{2\pi k}{\sqrt{f_0}} - \frac{\pi^2 k^2 f_0'}{f_0^2} \,, \tag{95}$$

and one has approximately

$$z_k = z(x_0 + t_k) = z_0 \left[1 - \frac{\pi}{2} \, k \, \frac{f_0'}{f_0^{3/2}} \right] \,. \tag{96}$$

If f' is relatively large, one perhaps merely manages in this way to compute from z_0 the next crossing z_1, and then must choose

$$x_1 = x_0 + \frac{2\pi}{\sqrt{f_0}} - \frac{\pi^2 f_0'}{f_0^2}$$

as new initial point (and there again compute $f(x_1)$, $f'(x_1)$) , etc. If, however, f' is very small, one can in one stroke carry out a great many revolutions, and in this way, for example, compute from z_0 directly z_{100} and $x_{100} = x_0 + t_{100}$.

Now a crossing of the real axis by $z(x)$, however, means that $y'(x) = 0$, i.e., that we are at a maximum of the function $y(x)$. Our method, therefore, allows us to compute from one maximum of the function directly the next, or even – provided $f(x)$ varies sufficiently slowly – to determine only every hundredth maximum. In a similar manner one could also jump from zero to zero (crossings of the imaginary axis by $z(x)$). In view of the fact that one always fixes attention only to the crossing of the point $z(x)$ through a ray, one calls this procedure the *stroboscopic method*.

Numerical example. For the differential equation

$$y'' + (10000 + x)y = 0, \quad y(0) = 1, \quad y'(0) = 0,$$

we have at the beginning $x_0 = 0$, $f_0 = 10000$, $f_0' = 1$, $z_0 = 1$, and therefore, according to (95), (96),

$$z_k = z \left[\frac{2\pi k}{100} - \frac{\pi^2 k^2}{10^8} \right] = 1 - \frac{\pi}{2} \frac{k}{10^6} .$$

Here, one can easily choose $k = 100$, and finds

$$x_{100} = 6.282198, \quad z_{100} = .999843 .$$

From now on, one continues with $f_{100} = 10006.282$, $f_{100}' = 1$:

$$x_{200} = x_{100} + \frac{200\pi}{\sqrt{f}_{100}} - \frac{10^4\pi^2}{f_{100}^2} = 12.562425 \,,$$

$$z_{200} = z_{100} \left[1 - \frac{50\pi}{f_{100}^{3/2}} \right] = .999686 \,.$$

Notes to Chapter 8

§8.1 Most computer codes for solving ordinary differential equations assume the initial value problem given in standard form, as in Eq. (5), or in autonomous form (without the independent variable appearing explicitly), to which it can easily be transformed by adding an equation $dy_0/dx = 1$. In certain applications, for example in network analysis and simulation, one encounters also systems of *implicit differential equations*, or differential equations coupled with nonlinear algebraic equations. The numerical treatment of such problems is briefly discussed in Gear [1971, §11.2] and has received considerable attention in recent years; see , e.g., Gear & Petzold [1984], Lötstedt & Petzold [1986], Griepentrog & März [1986], and Hairer, Lubich & Roche [1989].

The user should always be mindful of the fact that differential equations can often be rendered more manageable, hence their numerical integration more effective, if one applies a preliminary transformation of variables. See §8.9 B) for an elementary example and Daniel & Moore [1970, Part 3] for further examples. Interesting transformations are those of Levi-Civita and of Kustaanheimo and Stiefel used in celestial mechanics to regularize and linearize the Newtonian equations of motion. An extensive treatment of these, including numerical aspects, can be found in Stiefel & Scheifele [1971].

Among the textbooks specifically devoted to the numerical solution of ordinary differential equations we mention the classical work of Henrici [1962], [1963], the book by Gear [1971], the very readable treatment of Lambert [1973], as well as recent monographs by Hairer, Nørsett & Wanner [1987] and Butcher [1987]. (Butcher's book contains a comprehensive list of references up to 1983; a planned sequel is to continue this list from 1983 onwards.) Surveys of developments during the last 15 years can be found in Hall & Watt [1976], Gladwell & Sayers [1980], and Iserles & Powell [1987, Chs. 14–16].

§8.2 Euler proposed his method in the *Institutiones Calculi Integralis* (Euler [1768, §650]), where it is presented as a purely computational procedure. The convergence of the method, and its use for establishing existence and uniqueness theorems for ordinary differential equations, is due to Cauchy (1844) and Lipschitz (1880).

§8.3 The Taylor's series method for solving ordinary differential equations was proposed already by Euler [1768, §656]. Rutishauser's perception of the method as too cumbersome for practical use is still widely held. The required analytic differentiations, at least in the case of differential equations with arithmetic expressions as right-hand functions, nevertheless can be carried out systematically by recursion, and programs have been written to implement this; see Moore [1979, §3.4] and Butcher [1987, §24] for relevant remarks.

A proof of Theorem 8.2, for single differential equations as well as for systems, can be found in Henrici [1962, §§2.2–6, 3.3–6].

§8.4 The derivation of Runge-Kutta type methods by conventional means requires extremely tedious computations. Using appropriate notational devices, in particular, the formalism of rooted trees, Butcher [1965], [1975] succeeds in carrying through these computations in a transparent manner and obtains important results concerning the attainable order of Runge-Kutta formulae (compare also Butcher [1987, Ch.3] and Hairer, Nørsett & Wanner [1987, Ch. III] for detailed expositions). If s denotes the number of stages of an explicit Runge-Kutta formula [that is, the number of rows in the matrix $\sum$ of Eq. (31)] and $p^*(s)$ the maximum attainable order for arbitrary smooth differential equations, it has been known for some time that $p^*(s) = s$ when $1 \leq s \leq 4$. For more than four stages, J.C. Butcher proves, among other things, that $p^*(s) = s - 1$ for $5 \leq s \leq 7$, $p^*(s) = s - 2$ for $8 \leq s \leq 9$, and $p^*(s) \leq s - 2$ for $s \geq 10$. The longstanding question of whether 10 or 11 stages are needed to attain order 8 was settled by Butcher [1985]: ten-stage explicit Runge-Kutta methods of order 8 do not exist; indeed, $p^*(10) = 7$.

Eqs. (28), (29) define what are known as *explicit* Runge-Kutta methods. One can also define *implicit* methods, for which the matrix $\sum$ in Eq. (31) is no longer triangular. The advantages of implicit Runge-Kutta methods are two-fold. In the first place, they can be made to have exceptionally high order. For each s, there exists, in fact, an s-stage method of maximum order $p^*(s) = 2s$ (Butcher [1964]). Secondly, this maximum-order implicit method also possesses favorable stability properties (cf. the notes to §8.6). On the other hand, implicit Runge-Kutta methods require the solution of nonlinear equations at each step, typically by Newton's method, which makes them more costly to use and of interest only in special situations. For a discussion of reasons why in recent years implicit Runge-Kutta methods have received serious attention, see Butcher [1987, §34].

Further developments of Runge-Kutta type formulas have been prompted by a desire to economically estimate the local discretization error and thus control the stepsize. In particular, Fehlberg [1969], [1970] develops what are now called *embedded* Runge-Kutta methods. These consist of pairs of (explicit) Runge-Kutta formulae, one of order p with s stages, the other of order $p^* = p + 1$ with $s^* > s$ stages, both having identical first s stages. They require therefore s^* evaluations of the differential equation per step, which is more than in conventional methods, but they allow an easy estimation of the local discretization error. The formulas developed by Fehlberg correspond to the following values of the parameters p, s and s^*:

p	3	4	5	6	7	8
s	4	5	6	8	11	15
s^*	5	6	8	10	13	17

Embedded Runge-Kutta methods, including more recent methods by Verner and by Prince and Dormand, are studied in Butcher [1987, §37] and in Hairer, Nørsett & Wanner [1987, §§II.4, II.6] (this book contains Fortran listings of two of the Prince-Dormand methods.) An illuminating comparison of embedded Runge-Kutta methods is given in Shampine [1986].

Additional information on various aspects of Runge-Kutta methods is provided in Dekker & Verwer [1984] (where the focus is on the stability of implicit Runge-Kutta methods; see also the notes to §8.6), Hairer, Nørsett & Wanner [1987], and Butcher [1987].

Gear [1971, pp. 83–84] lists a Fortran program implementing the classical fourth-order Runge-Kutta method, with stepsize and error control. Further codes can be found in standard software libraries such as IMSL or NAG.

§8.6 Considerations of the type indicated in §§8.5–8.6 have given rise to a number of stability concepts and related theories of numerical stability. Basically, one wants to make sure that a problem which is Lyapunov stable remains so when approximated by a numerical method. A simple, but instructive, *model problem* is the linear system (39) with constant coefficient matrix A. This problem is Lyapunov stable if all eigenvalues of A have negative real part. If A is diagonalizable, the problem can be reduced to a system of uncoupled equations of the form $dy/dx = \lambda y$, where λ runs through the eigenvalues of A. Assuming, therefore, Re $\lambda < 0$, the exact solution on an interval of length h is damped by the factor $e^{\lambda h}$, whereas the corresponding factor for the numerical method is $F(\lambda h)$, where the function F is characteristic of the method (for example, the function F in (45) for any explicit fourth-order Runge-Kutta method, or the function F in (51) for the trapezoidal rule). The method is said to be *A-stable* (Dahlquist [1963]) if $|F(z)| < 1$ for all complex z with Re $z < 0$. Somewhat less restrictive is the notion of $A(\alpha)$-*stability*, $0 < \alpha < \pi/2$ (Widlund [1967]), which requires $|F(z)| < 1$ for all complex z in the angular opening $|\arg(-z)| < \alpha$, $z \neq 0$. Clearly, explicit Runge-Kutta methods cannot be $A(\alpha)$-stable, for any α, let alone A-stable, since $F(z)$ is a polynomial. If F is a rational function, on the other hand, and in particular a Padé approximant of e^z on the diagonal of the Padé table, the corresponding method is A-stable (Birkhoff & Varga [1965]). The trapezoidal rule, see Eq. (51), is an example of this, and so is the implicit s-stage Runge-Kutta method of maximum order $2s$ (cf. the notes to §8.4). Other Padé approximants of e^z also give rise to A-stable methods, viz. those with denominator degree exceeding the numerator degree by one or two (Ehle [1973]). Then, in fact, one has not only $|F(z)| < 1$ for Re $z < 0$, but also $F(z) \to 0$ as Re $z \to -\infty$, a useful property referred to as *L-stability*.

A powerful and elegant tool for analyzing the A-stability property of numerical methods for ODE's, the so-called *order star*, was introduced by Hairer, Nørsett and Wanner in 1978; a good account of this theory may be found in Wanner [1987].

The mid-1970's saw the generalization of the concept of A-stability to nonlinear problems. A detailed study of nonlinear stability properties, such as *B-stability* and *algebraic stability*, of implicit Runge-Kutta methods can be found in Dekker & Verwer [1984]; see also Butcher [1987] and the survey paper of Lambert [1987].

A differential equations problem is called *stiff* if there are solution components with widely varying decay rates. In the model problem mentioned above, this would mean that all eigenvalues of A have negative real parts, some, but not all, having very large absolute values. For the numerical integration of stiff problems, it is essential that the method be A-stable, or at least $A(\alpha)$-stable for some appropriate α. The survey articles of Lambert [1980] and Curtis [1987], as well as §1.1 in Dekker & Verwer [1984], provide much insight into the nature of stiffness. As far as stiff ODE solvers are concerned, the reader is referred to the comprehensive review article by Byrne & Hindmarsh [1987]. It describes in great detail current software for stiff ODE's (and for differential-algebraic systems), and it contains numerous computational results and an extensive bibliography.

§8.7 Adams predictor and corrector formulae, properly implemented, are among the most effective methods for integrating nonstiff differential equations, particularly in cases where the equations are expensive to evaluate. The full potential of these methods, however, is only realized when one allows for variable stepsize and variable order. A great deal of thought must go into devising sound strategies for controlling the stepsize and the order of the method so as to meet given error criteria. Once such strategies have been designed, there will no longer be any need for special starting procedures. One simply starts with Euler's method and a sufficiently small step, and from then on lets the control mechanisms take over to arrive at a proper order and proper step size. Such matters, in the context of Adams predictor-corrector methods, are thoroughly discussed in Shampine & Gordon [1975], which contains also a computer program.

Another useful class of difference formulae are the so-called *backward differentiation formulae*, which are of the form $\sum_{j=0}^{m} \alpha_j y_{k-j} = h\beta_0 y_k'$ [i.e., $\beta_j = 0$, all $j > 0$, in Eq. (52)]. They are usually chosen to have order m, and then have the useful property of being $A(\alpha)$-stable for appropriate $\alpha = \alpha_m$, at least when $m \leq 6$. This makes these methods attractive for integrating stiff differential equations, and they are in fact used for this purpose in a program written by Gear [1971, p. 159] and in the respective codes of the IMSL and NAG libraries.

In addition to Runge-Kutta methods (§§8.4–8.5) and difference methods (§§8.7–8.8), the *extrapolation methods* of Gragg and of Bulirsch and Stoer form another class of competitive integration methods. The basic idea here is to extend Richardson extrapolation, used for example in Romberg integration (cf. §6.10), to differential equations. A good exposition can be found in Stoer & Bulirsch [1980, §7.2.14], and computer programs in Gear [1971, p. 96] and in the IMSL and NAG libraries.

§8.8 According to a theory of Dahlquist [1956], the conditions a) and b), also referred to as *zero-stability*, together with the conditions of consistency (79), (80), are necessary and sufficient for convergence of the m-step method associated with the polynomials A and B. By convergence one means that whenever the starting values at $x = a$, $a + h, \ldots, a + (m - 1)h$ tend to $y(a)$ as $h \to 0$, one has $y_n \to y(x)$ for any x in some finite interval $[a, b]$, where $n \to \infty$ and $h \to 0$ such that $x = a + nh$. Moreover, this is to hold for all differential equations satisfying a uniform Lipschitz condition. An important result of Dahlquist states that the order of a zero-stable m-step method is at most equal to $m + 1$ if m is odd, and at most equal to $m + 2$ if m is even. The latter holds precisely if all zeros of $A(z)$ lie on the unit circle and are simple. For an exposition of Dahlquist's theory, see Henrici [1962, §5.2].

Considerably more restrictive is the requirement of A-stability (cf. the notes to §8.6), which in fact limits the maximum possible order of a linear multistep method to 2 (Dahlquist [1963]). For all α with $0 < \alpha < \pi/2$ there exist, however, $A(\alpha)$-stable m-step methods of order p with $m = p = 3$ and $m = p = 4$ (Widlund [1967]). The backward differentiation m-step formulae (cf. the notes to §8.7) are also $A(\alpha)$-stable for m as large as 6, but only for restricted values α_m of α.

Nevanlinna & Liniger [1978] study standard stability properties of linear multistep methods as they relate to the (stronger) concept of *contractivity*; the latter allows the derivation of stability results for nonlinear ODE's. See also the survey article of Lambert [1987] for more recent developments in nonlinear stability.

References

Birkhoff, G. and Varga, R.S. [1965]: Discretization errors for well-set Cauchy problems I, *J. Math. and Phys.* **44**, 1–23.

Butcher, J.C. [1964]: Implicit Runge-Kutta processes, *Math. Comp.* **18**, 50–64.

Butcher, J.C. [1965]: On the attainable order of Runge-Kutta methods, *Math. Comp.* **19**, 408–417.

Butcher, J.C. [1975]: An order bound for Runge-Kutta methods, *SIAM J. Numer. Anal.* **12**, 304–315.

Butcher, J.C. [1985]: The non-existence of ten stage eighth order explicit Runge-Kutta methods, *BIT* **25**, 521–540.

Butcher, J.C. [1987]: *The Numerical Analysis of Ordinary Differential Equations. Runge-Kutta and General Linear Methods*, Wiley, Chichester.

Byrne, G.D. and Hindmarsh, A.C. [1987]: Stiff ODE solvers: a review of current and coming attractions, *J. Comput. Phys.* **70**, 1–62.

Curtis, A.R. [1987]: Stiff ODE initial value problems and their solution, in *The State of the Art in Numerical Analysis* (A. Iserles and M.J.D. Powell, eds.), pp. 433–450. Clarendon Press, Oxford.

Dahlquist, G. [1956]: Convergence and stability in the numerical integration of ordinary differential equations, *Math. Scand.* **4**, 33–53.

Dahlquist, G. [1963]: A special stability problem for linear multistep methods, *BIT* **3**, 27–43.

Daniel, J.W. and Moore, R.E. [1970]: *Computation and Theory in Ordinary Differential Equations*, W.H. Freeman, San Francisco.

Dekker, K. and Verwer, J.G. [1984]: *Stability in Runge-Kutta Methods for Stiff Nonlinear Differential Equations*, CWI Monograph **2**, North-Holland, Amsterdam.

Ehle, B.L. [1973]: A-stable methods and Padé approximations to the exponential, *SIAM J. Math. Anal.* **4**, 671–680.

Euler, L. [1768]: *Institutiones Calculi Integralis*, Vol. 1, Part 2, Ch. 7, Petersburg. [Opera Omnia, Ser. 1, Vol. 11, B.G. Teubner, Leipzig and Berlin, 1913.]

Fehlberg, E. [1969]: Klassische Runge-Kutta-Formeln fünfter und siebenter Ordnung mit Schrittweiten-Kontrolle, *Computing* **4**, 93–106.

Fehlberg, E. [1970]: Klassische Runge-Kutta-Formeln vierter und niedrigerer Ordnung mit Schrittweiten-Kontrolle und ihre Anwendung auf Wärmeleitungsprobleme, *Computing* **6**, 61–71.

Gear, C.W. [1971]: *Numerical Initial Value Problems in Ordinary Differential Equations*, Prentice-Hall, Englewood Cliffs, New Jersey.

Gear, C.W. and Petzold, L.R. [1984]: ODE methods for the solution of differential / algebraic systems, *SIAM J. Numer. Anal.* **21**, 716–728.

Gladwell, I. and Sayers, D.K. (eds.) [1980]: *Computational Techniques for Ordinary Differential Equations*, Academic Press, London.

Griepentrog, E. and März, R. [1986]: *Differential-Algebraic Equations and Their Numerical Treatment*, Teubner, Leipzig.

Hairer, E., Lubich, Ch. and Roche, M. [1989]: *The Numerical Solution of Differential-algebraic Systems by Runge-Kutta Methods*, Lecture Notes Math., v. 1409, Springer, Berlin.

Hairer, E., Nørsett, S.P. and Wanner, G. [1987]: *Solving Ordinary Differential Equations. I. Nonstiff Problems*, Springer, New York.

Hall, G. and Watt, J.M. (eds.) [1976]: *Modern Numerical Methods for Ordinary Differential Equations*, Clarendon Press, Oxford.

Henrici, P. [1962]: *Discrete Variable Methods in Ordinary Differential Equations*, Wiley, New York.

Henrici, P. [1963]: *Error Propagation for Difference Methods*, Wiley, New York.

Iserles, A. and Powell, M.J.D. (eds.) [1987]: *The State of the Art in Numerical Analysis*, Clarendon Press, London.

Lambert, J.D. [1973]: *Computational Methods in Ordinary Differential Equations*, Wiley, London.

Lambert, J.D. [1980]: Stiffness, in *Computational Techniques for Ordinary Differential Equations* (I. Gladwell and D.K. Sayers, eds.), pp. 19–24. Academic Press, London.

Lambert, J.D. [1987]: Development in stability theory for ordinary differential equations, in *The State of the Art in Numerical Analysis* (A. Iserles and M.J.D. Powell, eds.), pp. 409–431. Clarendon Press, Oxford.

Lötstedt, P. and Petzold, L. [1986]: Numerical solution of nonlinear differential equations with algebraic constraints I: Convergence results for backward differentiation formulas, *Math. Comp.* **46**, 491–516.

Nevanlinna, O. and Liniger, W. [1978]: Contractive methods for stiff differential equations I, *BIT* **18**, 457–474; II, *BIT* **19**, 53–72.

Moore, R.E. [1979]: *Methods and Applications of Interval Analysis*, SIAM Studies in Applied Mathematics **2**, SIAM, Philadelphia.

Shampine, L.F. [1986]: Some practical Runge-Kutta formulas, *Math. Comp.* **46**, 135–150.

Shampine, L.F. and Gordon, M.K. [1975]: *Computer Solution of Ordinary Differential Equations*, W.H. Freeman, San Francisco.

Stiefel, E.L. and Scheifele, G. [1971]: *Linear and Regular Celestial Mechanics*, Springer, Berlin.

Stoer, J. and Bulirsch, R. [1980]: *Introduction to Numerical Analysis*, Springer, New York.

Wanner, G. [1987]: Order stars and stability, in *The State of the Art in Numerical Analysis* (A. Iserles and M.J.D. Powell, eds.), pp. 451–471. Clarendon Press, Oxford.

Widlund, O.B. [1967]: A note on unconditionally stable linear multistep methods, *BIT* **7**, 65–70.

CHAPTER 9

Boundary Value Problems For Ordinary Differential Equations

For a differential equation of order n, or a system of differential equations whose orders add up to n, one needs n *conditions* in order to single out one solution from among a family of ∞^n. If these n conditions refer to a single point x_0, one speaks of an *initial value problem*, since — apart from singular cases — one has enough information to integrate away from x_0.

If, on the other hand, these n conditions refer to more than one point, then at no point x does one have sufficient information to start the integration. One then speaks of a *boundary value problem*. A typical case is one in which for a differential equation of the form $y'' = f(x,y,y')$ one seeks a solution $y(x)$ on the interval $a \leq x \leq b$ which at each of the two end points a and b of the interval must satisfy a condition involving y and y'. An example is

$$y'' + y = 0, \quad \text{with} \quad y(a) = 0, \quad y(b) = 1.$$

A somewhat more general case is a linear differential equation of order $2m$,

$$(a_m(x)y^{(m)})^{(m)} + (a_{m-1}(x)y^{(m-1)})^{(m-1)} + \cdots + (a_1(x)y')' + a_0(x)y = 0,$$

with m conditions involving $y, y', \ldots, y^{(2m-1)}$ at each of the points a and b.

§9.1. The shooting method

As an example, let us consider a flexible beam clamped on a rotating shaft parallel to it, which under the influence of the rotation bends away from the shaft (cf. Fig. 9.1).

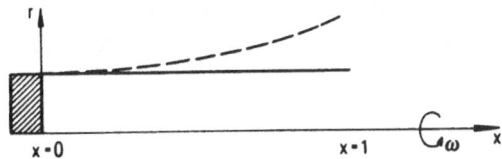

Figure 9.1. *Flexible beam mounted on a rotating shaft, in resting position and during rotation (dotted)*

If m is the mass of the beam per unit length, JE its bending stiffness, ω the circular frequency of the shaft rotation, then the corresponding differential equation is

$$r^{(4)}(x) = \frac{m\omega^2}{JE}\, r(x), \tag{1}$$

and the boundary conditions are

$$r(0) = 1, \quad r'(0) = 0, \quad r''(1) = 0, \quad r'''(1) = 0. \tag{2}$$

Neither at $x = 0$, nor at $x = 1$, does one have sufficient information to start the integration. There still are, for example, ∞^2 solutions which satisfy the boundary conditions at left.

The *shooting method* now simply consists in finding, through systematic trials, that one of the ∞^2 solutions which also satisfies the boundary conditions at $x = 1$. The method can be very tedious, but still succeeds with a tolerable amount of effort for

1) linear differential equations, and

2) differential equations of the 2nd order.

a) *Differential equations of the 2nd order.* If for a differential equation of the form

$$y'' = f(x, y, y')$$ (3)

there is one boundary condition at $x = a$, then only a one-dimensional family of solutions remains that satisfies this condition. We denote this solution family by $y(t, x)$; all functions $y(t, x)$ thus satisfy (3) and the boundary condition at $x = a$. Let the boundary condition at $x = b$ be

$$R[y(b), y'(b)] = 0.$$ (4)

Upon integrating all solutions $y(t, x)$ of the family from a to b, we obtain through substitution in this boundary condition the equation

$$R[y(t, b), y'(t, b)] = 0.$$ (5)

The left-hand side of this equation is now a function $H(t)$ of the variable t, whose zeros determine the desired solutions among the family $y(t, b)$.

In computational practice one actually integrates only a few selected solutions of the family from a to b; through substitution of the corresponding values $y(b)$ and $y'(b)$ into the function R one obtains a few points of the curve $z = H(t)$ from which one tries to determine zeros of $H(t)$ approximately by interpolation.

Example. To be solved is the boundary value problem

$$y'' + 2y'^3 + 1.5y + .5y^2 = 0.05x, \quad y(-5) = -1, \quad y(5) = 0.$$

All solutions $y(t, x)$ satisfying the boundary condition at left can be characterized by

$$y(t, -5) = -1, \quad y'(t, -5) = t;$$ (6)

to each value of t there corresponds a different solution through the point

$x = -5$, $y = -1$. If one imagines all solutions of this family integrated from -5 to 5, one obtains at the point $x = 5$ the condition

$$H(t) = y(t, 5) = 0,$$

which thus must be solved.

Now, however, the function $y(t, 5)$ can only be constructed pointwise; each function value $y(t, 5)$ requires a numerical integration of the differential equation from -5 to 5 with the initial values (6). One so obtains, for example,

t	$y(t, 5)$
0	.049115
.25	.002123
.5	−.021683

The interpolation polynomial of degree 2 for these three points is $.049115 - .23434t + .185488t^2$, from which there results as first approximation $t = .2653$. (The second root $t = .998$ is useless, since it does not lie in the interval $0 \leq t \leq .5$.) Of course, $t = .2653$ is still inaccurate since, for one, the parabola represents only an approximation to $H(t)$, and then also the numerical integration, after all, is only an approximation. One should therefore repeat the integration for additional t-values in the vicinity of $.2653$ and with smaller stepsize h.

b) *Differential equations of order 4.* If we consider the example (1), (2), we find that ∞^2 solutions of the differential equation $r^{(4)}(x) = cr(x)$ satisfy the boundary conditions at the point $x = 0$, namely all solutions with the initial conditions

$$r(0) = 1, \quad r'(0) = 0, \quad r''(0) = s, \quad r'''(0) = t.$$

This solution family therefore has the form $r(s,t,x)$, and, in particular, the boundary values at $x = 1$,

$$H_1(s,t) = r''(s,t, 1), \quad H_2(s,t) = r'''(s,t, 1),$$

remain still functions of s,t.

The solution which also satisfies the boundary conditions at $x = 1$ is determined by the equations $H_1 = H_2 = 0$, and therefore corresponds to that point of the (s,t)-plane which is mapped by

$$H_1 = H_1(s,t), \quad H_2 = H_2(s,t) \tag{7}$$

into the origin of the (H_1, H_2)-plane. Having determined 3 points A, B, C of this mapping (by 3 numerical integrations), one can on the basis of the construction indicated in Fig. 9.2 already find an approximate solution $P = (s,t)$ of (7). One thereby exploits the fact that the mapping $(s,t) \to (H_1,H_2)$ is locally affine.

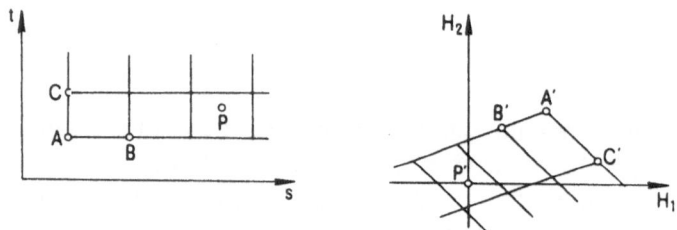

Figure 9.2. *Approximate determination of a common zero P of the functions* $H_1(s,t)$ *and* $H_2(s,t)$

Subsequently, further integrations are carried out, with s,t in the vicinity of the point P thus determined, which usually lead quickly to the desired goal.

§9.2. Linear boundary value problems

For linear differential equations one can apply the same principle. It is found, then, that the mapping in question from the (s,t)-plane to the (H_1,H_2)-plane in the above example (1), (2) (which is indeed linear) is an affinity not only locally, but globally, provided that also the boundary conditions are linear. Thanks to this circumstance, we are in a position to treat rather more general cases.

We consider the general linear differential equation of order n with n boundary conditions:

$$Ly(x) = f(x), \tag{8}$$

where L is a linear homogeneous differential operator of order n,

$$L = \sum_{k=0}^{n} a_k(x)D^k \quad \text{with} \quad D = \frac{d}{dx}, \tag{9}$$

so that, for example, with $L = x^2D^2 + 1$, the differential equation reads

$$x^2y'' + y = f(x).$$

The boundary conditions are likewise assumed to be linear; they may refer to an arbitrary number of points $x_1, \ldots, x_m$:

$$R_j[y(x)] = a_j \quad (j = 1,2, \ldots, n), \tag{10}$$

each R_j being a linear combination of y and its derivatives up to order $n - 1$ at the m points x_i, thus

$$R_j[y(x)] = \sum_{i=1}^{m} \sum_{k=0}^{n-1} r_{ik}^{(j)} y^{(k)}(x_i). \tag{11}$$

These sums, with increasing m, can even become integrals; for example, the following "boundary condition" would be conceivable:

$$\int_a^b y(x)dx = 1.$$

On the other hand, since the differential equation is linear, we can write down the general solution; it is given by

$$y(x) = y_0(x) + \sum_{k=1}^{n} t_k y_k(x),$$

where y_0 is a particular solution of the inhomogeneous equation, while $y_1, \ldots, y_n$ are n independent solutions of the homogeneous equation. Such n independent solutions $y_k(x)$ are characterized by the fact that their Wronskian matrix

$$\mathbf{W}(x) = \{w_{ik}(x) \mid w_{ik}(x) = D^{i-1} y_k(x), \ i,k = 1, \ldots, n\}$$

is nonsingular for all x. For this, it suffices that it be nonsingular for one x. A system of independent solutions can thus be obtained by fixing initial conditions for the y_k such that $\mathbf{W}(x_0)$ is the unit matrix. *Since these $n + 1$ functions can be determined by numerical integration, the general solution of the differential equation in this special case can therefore also be obtained numerically (in tabular form).*

We now substitute the general solution so obtained into the boundary conditions (10) and get

$$R_j[y(x)] = R_j[y_0] + \sum_{k=1}^{n} t_k R_j[y_k] = a_j \quad (j = 1,2, \ldots, n).$$

One thus obtains for the unknowns $t_1, t_2, \ldots, t_n$ the linear system of equations

$$\sum_{k=1}^{n} t_k R_j[y_k] = a_j - R_j[y_0] \quad (j = 1,2, \ldots, n). \tag{12}$$

It is worth noting that all quantities that enter into the coefficient matrix and the right-hand side of this system can be determined by numerical integration, since through an appropriate arrangement of the numerical integration all derivatives of the integrated solution up to the $(n-1)$st become automatically available at each support point. It suffices, therefore, to see to it that all "boundary points" $x_1, x_2, \ldots, x_m$ become support points.

Example.

$$y'' + xy = 1, \quad y(0) + y'(0) = 1, \quad y(1) - y'(1) = 0.$$

One first has to compute by numerical integration a particular solution $y_0(x)$, say the one with $y_0(0) = y_0'(0) = 0$, and then two independent solutions $y_1(x)$, $y_2(x)$ of the homogeneous equation, for example those with $y_1(0) = 1$, $y_1'(0) = 0$ and $y_2(0) = 0$, $y_2'(0) = 1$. One so obtains the functions depicted in Fig. 9.3, with boundary values

$$y_0(1) = .476199, \quad y_0'(1) = .878403,$$

$$y_1(1) = .840057, \quad y_1'(1) = -.468088,$$

$$y_2(1) = .919273, \quad y_2'(1) = .678265.$$

(For these functions one first gets of course only the values (and the derivatives) at the points $x_k = x_0 + kh = .1k$; in the figure these were connected by a curve.)

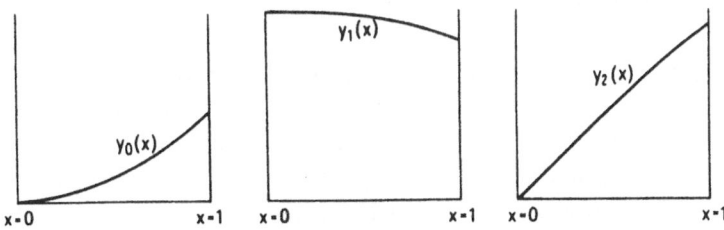

Figure 9.3. *The functions* $y_0(x)$, $y_1(x)$, $y_2(x)$

The boundary operators

$$R_1[y(x)] = y(0) + y'(0), \quad R_2[y(x)] = y(1) - y'(1)$$

for these special solutions take on the values

$$R_1[y_0] = 0, \quad R_2[y_0] = -.402204,$$

$$R_1[y_1] = 1, \quad R_2[y_1] = 1.308145,$$

$$R_1[y_2] = 1, \quad R_2[y_2] = .241008,$$

so that the equations (12) become:

$$
\begin{array}{cc}
t_1 & t_2
\end{array}
$$

1	1	$=$	1
1.308145	.241008	$=$	.402204

and have the solution $t_1 = .151055$, $t_2 = .848945$. Therefore,

$$y(x) = y_0(x) + .151055\, y_1(x) + .848945\, y_2(x)$$

is the desired solution, which is thus determined by the initial conditions

$$y(0) = .151055, \quad y'(0) = .848945.$$

The correctness of the solution can be checked by integrating once more with these initial conditions. One then finds indeed $y(1) = y'(1) = 1.38351$.

It must not be concealed, however, that the coefficient matrix $\{R_j[y_k]\}$ of the system of linear equations (12) for the t_k can be ill-conditioned, or even practically singular. This could be due to an inept choice of the particular solution y_0 and of the independent solutions $y_1, \ldots, y_n$, but may possibly also lie in the nature of the problem. An appropriate choice of the y_i or, what is the same, of the initial conditions defining them, may remedy the predicament. Not infrequently, it is even possible, in this way, to reduce the very number of unknowns t_k that occur.

In the above example one could, say, prescribe for the particular solution y_0 of the inhomogeneous system the initial conditions $y_0(0) = 1$, $y_0'(0) = 0$, in which case the initial condition at left is already satisfied. One then gets $y_0(1) = 1.31625$, $y_0'(1) = .410315$ (cf. Fig. 9.4).

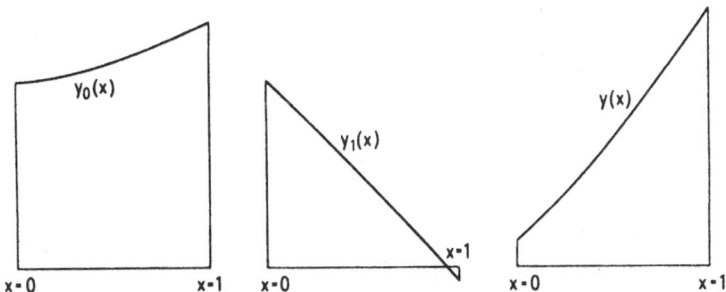

Figure 9.4. *Better selected functions* $y_0(x)$, $y_1(x)$ *and the solution* $y(x)$ *of the boundary value problem*

If one further subjects $y_1(x)$ to the initial conditions $y_1(0) = 1$, $y_1'(0) = -1$, which yields $y_1(1) = -.079216$, $y_1'(1) = -1.14635$, then $y = y_0 + t_1 y_1$ is already the one-parameter family of all solutions which satisfy $y(0) + y'(0) = 1$. In this case one obtains through substitution in the other boundary condition

$$R_2[y_0] = .905940, \qquad R_2[y_1] = 1.06713,$$

and thus only one equation,

$$R_2[y_0 + t_1 y_1] = .905940 + 1.06713 t_1 = 0,$$

which has the solution $t_1 = -.848945$. The desired solution of the boundary value problem can now be written in the simpler form

$$y(x) = y_0(x) - .848945 y_1(x).$$

It is depicted on the right in Fig. 9.4.

Postprocessing. We consider the problem

$$y'' - y + 1 = 0, \quad y(0) = y(10) = 0. \tag{13}$$

Here the differential equation can be solved exactly, but the characteristic difficulties inherent in satisfying the boundary conditions are essentially the same.

The particular solution of the inhomogeneous equation with $y_0(0) = y_0'(0) = 0$ is

$$y_0(x) = 1 - \cosh x.$$

Furthermore, $y_1(x) = \cosh x$ and $y_2(x) = \sinh x$ are two independent solutions of the homogeneous equation, thus

$$y(x) = 1 - \cosh x + t_1 \cosh x + t_2 \sinh x$$

the general solution.

Because of $y(0) = 0$, there follows at once $t_1 = 0$; then from $y(10) = 0$,

$$1 - \cosh 10 + t_2 \sinh 10 = 0,$$

or

$$t_2 = \frac{\cosh 10 - 1}{\sinh 10} = \tanh 5 = .999909 \text{ (to 6 digits)}.$$

Thus, in 6-digit precision,

$$y(x) = 1 - \cosh x + .999909 \sinh x$$

would be the desired solution. Now, regardless of whether this formula is evaluated for "all" x, or whether one integrates again from $x = 0$ to $x = 10$ with the corresponding initial conditions $y(0) = 0$, $y'(0) = .999909$, the boundary value $y(10)$ obtained at $x = 10$ will in fact deviate considerably from 0. With numerical integration, for example, one obtains $y(10) = -.002254$, which is explained by the fact that $y(x)$ for large x depends very sensitively on $y'(0)$.

In order to arrive at a more accurate solution, the solution thus far obtained is denoted by $\bar{y}_0(x)$, and we seek the improved solution $y(x)$ in the form

$$y(x) = \tilde{y}_0(x) + \tilde{\tau}_1 \cosh x + \tilde{\tau}_2 \sinh x.$$

As above, one gets $\tilde{\tau}_1 = 0$, since the boundary condition at the left boundary then is indeed satisfied; thereafter,

$$y(10) = \tilde{y}_0(10) + \tilde{\tau}_2 \sinh 10,$$

hence

$$\tilde{\tau}_2 = -\frac{\tilde{y}_0(10)}{\sinh 10} = \frac{.002254}{11013} \approx 2.047_{10}-7.$$

In this way, one now gets for the initial condition at $x = 0$

$$y'(0) = \tilde{y}'(0) + \tilde{\tau}_2 = .999909 + 2.047_{10}-7,$$

but in 6-digit computation one again obtains $y'(0) = .999909$.

The boundary condition at the point $x = 10$, therefore, was poorly satisfied by $\tilde{y}_0(x)$ not because $\tilde{y}_0(x)$ was determined incorrectly, but because within 6-digit accuracy no value $y'(0)$ exists which would produce a sufficiently small $y(10)$. One therefore cannot, in this case, obtain a better $y(x)$ by numerical integration, but only through linear combination (which is to be carried out for each x):

$$y(x) = \tilde{y}_0(x) + 2.047_{10}-7 \sinh x.$$

This then yields, say,

$$y(1) = .632014 + 2.047_{10}-7 \times 1.17 = .632014,$$

$$\vdots$$

$$y(9) = .631187 + 2.047_{10}-7 \times 4051.54 = .632016,$$
$$y(10) = -.002254 + 2.047_{10}-7 \times 11013 = .000000.$$

It remains to observe, though, that also in this example a more skillful choice of the particular solution could have led us to the correct solution without postprocessing (this is, however, not always possible). If one puts, in fact,

$$y_0(x) \equiv 1, \quad y_1(x) = e^x, \quad y_2(x) = e^{-x},$$

one first obtains for the two boundary operators $R_1[y] = y(0)$, $R_2[y] = y(10)$:

$$R_1[y_0] = 1, \qquad R_2[y_0] = 1,$$
$$R_1[y_1] = 1, \qquad R_2[y_1] = e^{10},$$
$$R_1[y_2] = 1, \qquad R_2[y_2] = e^{-10},$$

so that the equations (12) now become

$$t_1 + t_2 = -1,$$

$$e^{10}t_1 + e^{-10}t_2 = -1.$$

Their solution is $t_1 = -4.53979_{10}-5$, $t_2 = -.999955$, giving

$$y(x) = 1 - 4.53979_{10}-5 \, e^x - .999955e^{-x},$$

a function which satisfies both boundary conditions with an error of about 10^{-6}.

§9.3. The Floquet solutions of a periodic differential equation

A special type of boundary value problem arises in connection with differential equations of the form

$$y'' + \phi(x)y = 0, \tag{14}$$

where $\phi(x)$ is a periodic function with $\phi(x + 2\pi) = \phi(x)$. The differential equation need not necessarily have periodic solutions, but it always possesses solutions of a special type. If, in fact,

$$
\begin{aligned}
y(2\pi) &= ky(0), \\
y'(2\pi) &= ky'(0),
\end{aligned} \tag{15}
$$

where k is a constant, then for all x,

$$
y(x + 2\pi) = ky(x). \tag{16}
$$

(One calls this a *Floquet solution*.) Indeed, $z(x) = y(x + 2\pi)$ is defined as solution of $z'' + \phi(x + 2\pi)z = 0$, with the initial conditions $z(0) = y(2\pi)$, $z'(0) = y'(2\pi)$. Since, then, $z'' + \phi(x)z = 0$, and in addition, $z(0) = ky(0)$, $z'(0) = ky'(0)$, one concludes at once from the homogeneity of the differential equation that $z(x) \equiv ky(x)$, q.e.d.

In order to obtain a Floquet solution, one must evidently find a solution of the differential equation which satisfies the boundary conditions (15). Letting

$$
y = c_1 y_1(x) + c_2 y_2(x),
$$

where y_1, y_2 are defined by the initial conditions

$$
\begin{aligned}
y_1(0) &= 1, & y_1'(0) &= 0, \\
y_2(0) &= 0, & y_2'(0) &= 1,
\end{aligned}
$$

it follows from (15) that

$$
\begin{aligned}
c_1 y_1(2\pi) + c_2 y_2(2\pi) &= ky(0) = kc_1, \\
c_1 y_1'(2\pi) + c_2 y_2'(2\pi) &= ky'(0) = kc_2.
\end{aligned}
$$

Therefore, k and $[c_1, c_2]^T$ are eigenvalue and corresponding eigenvector of the matrix

$$\begin{bmatrix} y_1(2\pi) & y_2(2\pi) \\ y_1'(2\pi) & y_2'(2\pi) \end{bmatrix}, \tag{17}$$

which allows us to construct $y(x)$. We note that the determinant of this matrix is 1, since

$$\frac{d}{dx}\begin{vmatrix} y_1(x) & y_2(x) \\ y_1'(x) & y_2'(x) \end{vmatrix} = 0 \text{ and } \begin{vmatrix} y_1(0) & y_2(0) \\ y_1'(0) & y_2'(0) \end{vmatrix} = 1.$$

This means that the product of the two eigenvalues is 1; only $\lambda = 1$ or $\lambda = -1$ can thus be a double eigenvalue.

Example. If

$$\phi(x) = \frac{1}{4} - \frac{1}{8}\cos x, \tag{18}$$

one obtains for $y_1(x)$, with $y_1(0) = 1$, $y_1'(0) = 0$:

$$y_1(2\pi) = -1.07694, \quad y_1'(2\pi) = -.197774,$$

and for $y_2(x)$, with $y_2(0) = 0$, $y_2'(0) = 1$:

$$y_2(2\pi) = -.808140, \quad y_2'(2\pi) = -1.07694.$$

The eigenvalues and eigenvectors of the matrix

$$\begin{bmatrix} -1.07694 & -.808140 \\ -.197774 & -1.07694 \end{bmatrix}$$

are

$$k_1 = -1.47673, \quad k_2 = -.677154,$$

$$\begin{bmatrix} c_1 \\ c_2 \end{bmatrix} = \begin{bmatrix} 1 \\ .49470 \end{bmatrix}, \quad \begin{bmatrix} c_1 \\ c_2 \end{bmatrix} = \begin{bmatrix} 1 \\ -.49470 \end{bmatrix}.$$

Thus, $y = y_1 - .4947y_2$ is a Floquet solution. It is depicted in Fig. 9.5; Table 9.1, in addition, contains some function values.

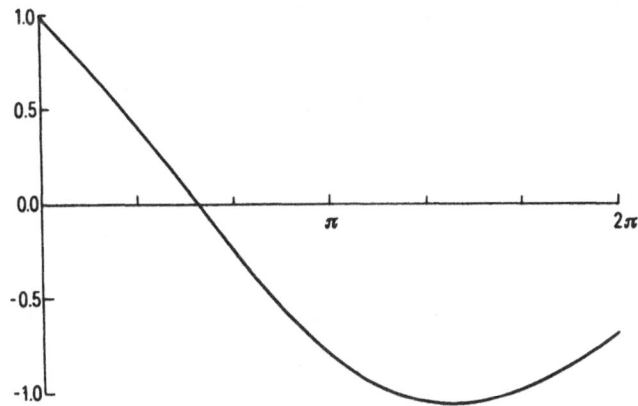

Figure 9.5. *The Floquet solution for the example* (18)

Table 9.1. *Values of the Floquet solution for* (18)

x	y	y'
0	1	−.49470
$\pi/3$	.42192	−.60207
$2\pi/3$	−.23189	−.62015
π	−.79768	−.41995
$4\pi/3$	−1.05532	−.06688
$5\pi/3$	−.96890	.20530
2π	−.67717	.33499

§9.4. Treatment of boundary value problems with difference methods

We consider, as a model, the problem

$$y'' - y + 1 = 0, \quad y(0) = y(10) = 0, \tag{19}$$

which has the exact solution $Y(x) = (e^{10} - e^x)(1 - e^{-x})/(e^{10} + 1)$.

For numerical solution, the interval $[0,10]$ is subdivided into n equal subintervals of length $h = 10/n$. At each of the resulting abscissas $x_k = kh$ the differential equation is written down, whereby in place of $y''(x)$ one substitutes the approximate expression

$$y''(x) \approx \frac{y(x + h) - 2y(x) + y(x - h)}{h^2}. \tag{20}$$

There results the equation

$$-\frac{1}{h^2} y_{k+1} + \left[\frac{2}{h^2} + 1 \right] y_k - \frac{1}{h^2} y_{k-1} - 1 = 0, \tag{21}$$

which can be written down for $k = 1,2, \ldots, n - 1$, yielding, on account of the boundary conditions $y(0) = y(10) = 0$, a system of $n - 1$ linear equations in the same number of unknowns.

Examples. For $n = 5$, $h = 2$, the system of equations (21) reads:

y_1	y_2	y_3	y_4	
1.5	−.25	0	0	= 1
−.25	1.5	−.25	0	= 1
0	−.25	1.5	−.25	= 1
0	0	−.25	1.5	= 1
.827586207	.965517241	.965517241	.827586207	

It has the solution shown at the bottom of the tableau. With $n = 10$, $h = 1$, one obtains the system

y_1	y_2	y_3	y_4	y_5	$\cdots$	
3	−1				$\cdots$	= 1
−1	3	−1			$\cdots$	= 1
	−1	3	−1		$\cdots$	= 1
		−1	3	−1	$\cdots$	= 1
			−1	3	$\cdots$	= 1
					$\cdots$	.
					$\cdots$	.

.617886179 .853658537 .943089431 .975609756 .983739837

For reasons of symmetry one has here $y_9 = y_1$, $y_8 = y_2$, $y_7 = y_3$, $y_6 = y_4$. With $n = 1000$, $h = .01$, one finally would have:

y_1	y_2	y_3			
20001	−10000				= 1
−10000	20001	−10000			= 1
	−10000	20001	.		= 1
		.	.	.	
		.	.	.	.

Such systems of equations are rather easily solved, since the matrix is occupied only around the diagonal. One can apply the Gauss algorithm; the computational work for $n = 1000$ is roughly equivalent to the work involved in solving 18 linear equations in 18 unknowns with a dense matrix.

But how about *accuracy*? After all, the relation (20) does not hold exactly. Rather,

$$\frac{y(x + h) - 2y(x) + y(x - h)}{h^2} = y''(x)$$
$$+ \frac{h^2}{12} y^{(4)}(x) + \frac{h^4}{360} y^{(6)}(x) + \cdots , \tag{22}$$

so that in place of the equation

$$y'' + f(x)y = g(x) \tag{23}$$

we integrate in first approximation the differential equation

$$y'' + f(x)y = g(x) - \frac{h^2}{12} y^{(4)}(x).$$

Since the forcing term here is perturbed only by $O(h^2)$, the solution, by the superposition principle, is off by an amount of $O(h^2)$; the method therefore has order 2.

While the error for $h = 1$ is still about .1, it will thus for $h = .1$ be only about 10^{-3}, and for $h = .01$ one could already expect 5-digit accuracy.

There are now limits set, however, to refinements of the subdivision h; such a refinement, namely, increases not only the computational work, but also the sensitivity to rounding errors. In fact, for $n = 1000$ ($h = .01$), where the computational work is still modest, the difference between the differential equations $y'' - y + 1$ and $y'' + 1 = 0$ consists only in the diagonal elements of the coefficient matrix being 20001 in the first case, and 20000 in the second. But if such a small difference between the coefficient matrices can change the solution that much, this solution necessarily must be sensitive to rounding errors. (The exact solution of $y'' + 1 = 0$, $y(0) = y(10) = 0$, is $Y(x) = x(10 - x)/2$, so that, e.g., $Y(5) = 12.5$, as opposed to $Y(5) = .986524718 \ldots$ for the given problem.)

An enhancement of the accuracy, therefore, cannot be forced by an excessive increase of n, but only by a *refinement of the method*. One can observe, namely, on the basis of the expansion (22) that the error of the numerical solution of the differential equation (23) is expressible in terms of even powers $h^2, h^4, \ldots$ only, that is, as

$$y(x) = Y(x) + c_1(x)h^2 + c_2(x)h^4 + \cdots ,$$

where $c_1, c_2, \ldots$ are certain functions of x not further specified.

Whenever the error of a numerical process exhibits this behavior for decreasing h, one can proceed according to Romberg (cf. §6.10): Denoting by $y_0(x)$ the numerical solution obtained with $h = h_0$, by $y_1(x)$ the one with $h = h_1 = h_0/2$, in general by $y_k(x)$ the one obtained with $h = h_k = h_0 2^{-k}$, then one forms with them the additional functions

$$y_{0,1}(x) = \frac{4y_1(x) - y_0(x)}{3} \, ,$$

$$y_{1,1}(x) = \frac{4y_2(x) - y_1(x)}{3} \, ,$$

etc., then

$$y_{0,2}(x) = \frac{16y_{1,1}(x) - y_{0,1}(x)}{15} \, ,$$

etc., in general

$$y_{v,k}(x) = \frac{4^k y_{v+1,k-1} - y_{v,k-1}}{4^k - 1} \, . \tag{24}$$

Of course, $y_{0,1}(x)$ is defined only at the common abscissas of $y_1(x)$ and $y_0(x)$, likewise $y_{1,1}(x)$ only at the common abscissas of $y_2(x)$ and $y_1(x)$, etc. In general, $y_{v,k}$ is defined only at the same abscissas as $y_v(x)$, that is, at the multiples of $h_v = h_0 2^{-v}$.

In our example, one first obtains with $h_0 = 2$ (subdivision of the interval in 5 equal parts) the "basic solution" $y_0(x)$:

$$\left\{ \begin{array}{l} y_0(2) = .827586207 \\ y_0(4) = .965517241 \\ y_0(6) = .965517241 \\ y_0(8) = .827586207 \end{array} \right\} .$$

Then, with $h_1 = 1$:

$$\left\{ \begin{array}{l} y_1(1) = .617886179 \\ y_1(2) = .853658537 \\ y_1(3) = .943089431 \\ y_1(4) = .975609756 \\ y_1(5) = .983739837 \\ \qquad \cdot \qquad \qquad \cdot \\ \qquad \cdot \qquad \qquad \cdot \\ \qquad \cdot \qquad \qquad \cdot \\ \qquad \qquad \text{symmetric} \end{array} \right\} \, ,$$

etc. Considering only the function values at the points $x = 2$ and $x = 4$, one obtains the following Romberg schemes:

$y_v(2)$	$y_{v,1}(2)$	$y_{v,2}(2)$	$y_{v,3}(2)$
.827586207			
	.862349313		
.853658537		.864283902	
	.864162990		.864334966
.861536877		.864334168	
	.864323469		
.863626821			

$y_v(4)$	$y_{v,1}(4)$	$y_{v,2}(4)$	$y_{v,3}(4)$
.965517241			
	.978973928		
.975609756		.979202493	
	.979188208		.979206519
.978293595		.979206457	
	.979205316		
.978977386			

One so gets approximately

$$y(2) = .864334966, \quad y(4) = .979206519,$$

which is in good agreement with the exact solution

$$Y(2) = .864335413 \ldots , \quad Y(4) = .979206553 \ldots .$$

The problem is illustrated schematically in Fig. 9.6.

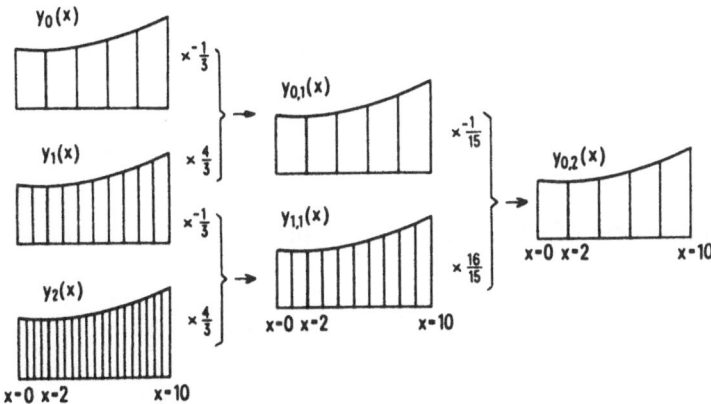

Figure 9.6. *To Romberg's convergence acceleration*

For the boundary value problem (19), the coefficient matrix turned out to be symmetric. This, to be sure, corresponds to the self-adjoint character of the boundary value problem, but is nevertheless kind of accidental. Consider as a further example the boundary value problem

$$y'' - xy + 1 = 0, \quad y(0) = y'(5) = 0. \tag{25}$$

Here the discretized differential equation is

$$\frac{y_{k+1} - 2y_k + y_{k-1}}{h^2} - x_k y_k + 1 = 0 \quad (k = 1, 2, \ldots, n-1),$$

or

$$-\frac{1}{h^2} y_{k-1} + \left[\frac{2}{h^2} + kh \right] y_k - \frac{1}{h^2} y_{k+1} - 1 = 0 \quad (k = 1, \ldots, n-1). \tag{26}$$

In addition, we have the equation

$$y_n' = \frac{y_{n+1} - y_{n-1}}{2h} = 0$$

for the boundary condition at $x=5$; from this, one gets $y_{n+1} = y_{n-1}$. One therefore writes down the differential equation in addition for $k = n$ and substitutes therein y_{n+1} by y_{n-1}:

$$-\frac{2}{h^2} y_{n-1} + \left[\frac{2}{h^2} + nh\right] y_n - 1 = 0. \tag{27}$$

Together with (26), this yields n equations for the n unknowns $y_1, \ldots, y_n$.

This system of equations for $n=5$, hence $h=1$, for example reads:

y_1	y_2	y_3	y_4	y_5	
3	−1				= 1
−1	4	−1			= 1
	−1	5	−1		= 1
		−1	6	−1	= 1
			−2	7	= 1

The coefficient matrix is not symmetric here, but one can restore symmetry by dividing the last equation by 2; it then becomes

$$-y_4 + 3.5y_5 = .5.$$

This, however, is not a generally applicable method to make a matrix symmetric, although it always helps for tridiagonal matrices. In the following §9.5 we discuss a method which − to the extent that one can expect it from the nature of the problem − produces symmetry quite generally.

§9.5. The energy method for discretizing continuous problems

The solution of the boundary value problem (25) discussed in §9.4,

$$y'' - xy + 1 = 0, \quad y(0) = y'(5) = 0,$$

is at the same time solution of the extremal problem

$$\frac{1}{2} \int_0^5 y'^2 dx + \frac{1}{2} \int_0^5 xy^2 dx - \int_0^5 y\, dx = \text{extremum} \qquad (28)$$

subject to the side condition $y(0) = 0$. Indeed, by treating this extremal problem with the methods of the calculus of variations, one immediately obtains again the original boundary value problem. However, it is better here to forgo the calculus of variations and to discretize the extremal problem directly.

Having subdivided the interval $0 \le x \le 5$ into n intervals of equal length h, one first approximates

$$\int_0^5 (y'(x))^2 dx \quad \text{by} \quad h \sum_{k=1}^n \left[y'\left(kh - \frac{h}{2}\right) \right]^2 ;$$

$y'(kh - h/2)$ in turn can be represented approximately as

$$\frac{y(kh) - y(kh - h)}{h},$$

so that one discretizes

$$\frac{1}{2} \int_0^5 (y'(x))^2 dx \quad \text{by} \quad \frac{1}{2h} \sum_{k=1}^n (y_k - y_{k-1})^2 .$$

The integrals

$$\frac{1}{2} \int_0^5 xy^2 dx \quad \text{and} \quad \int_0^5 y\, dx$$

are treated by the trapezoidal rule, so that altogether one obtains

$$\frac{1}{2h} \sum_{k=1}^{n} (y_k - y_{k-1})^2 + \frac{h}{2} \left[\frac{x_0 y_0^2}{2} + \sum_{k=1}^{n-1} x_k y_k^2 + \frac{x_n y_n^2}{2} \right]$$

$$- h \left[\frac{y_0}{2} + \sum_{k=1}^{n-1} y_k + \frac{y_n}{2} \right] = \text{extremum} .$$

After taking account of $y_0 = 0$, and division by h, there follows

$$\frac{1}{2} \sum_{k=1}^{n-1} y_k^2 \left[\frac{2}{h^2} + x_k \right] + \frac{1}{2} y_n^2 \left[\frac{1}{h^2} + \frac{x_n}{2} \right]$$

$$- \sum_{k=2}^{n} \frac{y_k y_{k-1}}{h^2} - \sum_{k=1}^{n-1} y_k - \frac{1}{2} y_n = \text{extremum} .$$

Now, as is well known, the extremal problem for a quadratic function $\frac{1}{2} \sum \sum a_{ik} x_i x_k + \sum b_i x_i$ is equivalent to the solution of a linear system of equations $\sum_{k=1}^{n} a_{ik} x_k + b_i = 0$ (cf. §3.7). In the present case, this system is given by

y_1	y_2	$\cdots$	y_n	1	
$\frac{2}{h^2} + x_1$	$-\frac{1}{h^2}$			-1	$= 0$
$-\frac{1}{h^2}$	$\frac{2}{h^2} + x_2$	$-\frac{1}{h^2}$		-1	$= 0$
$\cdot$	$\cdot$	$\cdot$		$\cdot$	$\cdot$
$\cdot$	$\cdot$	$\cdot$		$\cdot$	$\cdot$
	$\cdot$ $\frac{2}{h^2} + x_{n-1}$	$-\frac{1}{h^2}$		-1	$= 0$
	$-\frac{1}{h^2}$	$\frac{1}{h^2} + \frac{x_n}{2}$		$-\frac{1}{2}$	$= 0$

(29)

Note, in particular, that the coefficient matrix – as matrix of a quadratic form – has now automatically become symmetric.

Example. A beam with bending stiffness JE is laid on an elastic foundation (with spring constant k per unit length) and loaded with the weight $p(x)$ per unit length (see Fig. 9.7). What is desired is the deflection $y(x)$.

x-0 x-1

Figure 9.7. *Loaded beam on elastic ground*

We consider the energies:

1) Bending energy of the beam: $\frac{1}{2} \int_0^1 JE\, y''^2\, dx.$

2) Displacement energy of the ground: $\frac{1}{2} \int_0^1 ky^2\, dx.$

3) Virtual work of the exterior forces: $\int_0^1 p(x)y\, dx.$

Consequently, the deflection must satisfy

$$\frac{1}{2} \int_0^1 (JE\, y''^2 + ky^2)dx + \int_0^1 py\, dx = \text{extremum} ,$$

from which, with the help of the calculus of variations, one derives the differential equation

$$(JE\, y'')'' + ky + p = 0$$

with the boundary conditions $y''(0) = y'''(0) = y''(1) = y'''(1) = 0$. If one tries to solve this differential equation with difference methods, one arrives at a linear system of equations with a nonsymmetric coefficient matrix. In order to avoid this, one applies difference methods directly to the energy, approximating it by a quadratic function whose minimum is then determined.

Let, in particular, $JE \equiv 1$, $k \equiv 1$. (The problem, of course, could then be solved exactly). The beam is subdivided into 5 subintervals of length $h = .2$ and the deflections at the subdivision points are denoted by $y_0, y_1, y_2, y_3, y_4, y_5$. To begin with, one has

$$y_k'' \approx \frac{y_{k+1} - 2y_k + y_{k-1}}{h^2} = 25y_{k-1} - 50y_k + 25y_{k+1} ,$$

but for $k=0$ and $k=5$ one needs other expressions which avoid y_{-1} and y_6:

$$y_0'' \approx \frac{2y_0 - 5y_1 + 4y_2 - y_3}{h^2} = 50y_0 - 125y_1 + 100y_2 - 25y_3 ,$$

$$y_5'' \approx \frac{2y_5 - 5y_4 + 4y_3 - y_2}{h^2} = 50y_5 - 125y_4 + 100y_3 - 25y_2.$$

Computing the integrals by the trapezoidal rule, one obtains

$$\frac{1}{2} \int_0^1 y''^2 dx \approx \frac{h}{2} \sum_{k=1}^4 (25y_{k+1} - 50y_k + 25y_{k-1})^2$$

$$+ \frac{h}{4} (50y_0 - 125y_1 + 100y_2 - 25y_3)^2 + \frac{h}{4} (50y_5 - 125y_4 + 100y_3 - 25y_2)^2,$$

$$\frac{1}{2} \int_0^1 y^2 dx \approx \frac{h}{2} \left[\frac{1}{2} y_0^2 + y_1^2 + y_2^2 + y_3^2 + y_4^2 + \frac{1}{2} y_5^2 \right] ,$$

$$\int_0^1 py \, dx \approx h \left[\frac{1}{2} p_0 y_0 + \sum_{k=1}^4 p_k y_k + \frac{1}{2} p_5 y_5 \right] .$$

After deletion of the common factor h, the sum of these expressions is the quadratic function $F(y)$ belonging to the system of equations

y_0	y_1	y_2	y_3	y_4	y_5	1	
1875.5	−4375	3125	−625			$p_0/2$	= 0
−4375	10938.5	−8750	2187.5			p_1	= 0
3125	−8750	9063.5	−5000	2187.5	−625	p_2	= 0
−625	2187.5	−5000	9063.5	−8750	3125	p_3	= 0
		2187.5	−8750	10938.5	−4375	p_4	= 0
		−625	3125	−4375	1875.5	$p_5/2$	= 0

The solution of this system, therefore, yields the minimum of F and with it (approximately) the desired deflections.

Notes to Chapter 9

§9.1 Introductions to the numerical treatment of two-point boundary value problems for ordinary differential equations may be found in Fox [1957], Collatz [1960], Keller [1968, 1975, 1976], Bailey et al. [1968], Daniel & Moore [1970], Fried [1979], and Fox & Mayers [1987]. In addition, compare the survey articles in Hall & Watt [1976, Chapters 15–19] and in Gladwell & Sayers [1980, Chapters 9–11], as well as the contribution by Daniel in Childs et al. [1979, pp. 1–18]. These proceedings of a working conference contain some 30 contributions dealing with various aspects of computer codes for two-point boundary value problems, as well as an extensive bibliography.

Boundary value problems arising in applications are frequently not in the "standard" form required by many software packages. The paper by Ascher & Russell [1981] gives a survey of numerous relevant conversion devices.

Shooting methods are discussed in Keller [1968, Chapter 2], Keller [1976, Chapter 1], Hall & Watt [1976, Chapter 16], Stoer & Bulirsch [1980, Chapter 7], Gladwell & Sayers [1980, Chapter 10], and Fox & Mayers [1987, Chapter 5]. The general idea underlying the basic shooting method, say for the boundary value problem $y''(x) = f(x, y(x))$, $y(a) = \alpha$, $y(b) = \beta$, consists in determining a value for γ such that the solution $y = y(x; \gamma)$ of the initial value problem $y''(x) = f(x, y(x))$, $y(a) = \alpha$, $y'(a) = \gamma$, satisfies the given boundary condition at $x=b$. The resulting nonlinear equation, $y(b; \gamma) - \beta = 0$, has in general to be solved by iteration, usually Newton's method or one of its variants (see the paper by Deuflhard in Childs et al. [1979, pp. 40–66] for a detailed analysis of nonlinear equation solvers in boundary value problems). In practice, it is often difficult to obtain good initial estimates for γ; moreover, the initial value problem to be solved may be very sensitive to perturbations in the initial conditions. These difficulties motivated the search for more robust algorithms such as *invariant imbedding* methods (cf. Meyer [1973]) and *multiple* (or *parallel*) *shooting* methods. In the latter, the given interval $[a,b]$ is subdivided into subintervals $[x_i, x_{i+1}]$, with $a = x_0 < x_1 < \cdots < x_N = b$; the differential equation is then considered independently over each subinterval, choosing appropriate initial conditions at $x = x_i$ and integrating to $x = x_{i+1}$. The values of the solution at $x = x_i$ ($i = 0, \ldots, N$) are then simultaneously adjusted so as to satisfy the boundary conditions at $x = a$ and $x = b$ and the continuity conditions at x_i ($i = 1, \ldots, N - 1$). The choice of the "shooting points" x_i will depend on the behavior of the solution (especially in the case of discontinuous or singular solutions). Details on multiple shooting methods may be found, e.g., in Osborne [1969], Keller [1976], Stoer & Bulirsch [1980], Gladwell & Sayers [1980], and Fox & Mayers [1987].

Computer codes based on shooting techniques form part of standard software libraries such as IMSL or NAG. Compare also the contributions by Watts and Gladwell in Childs et al. [1979].

§9.2 Although any algorithm for solving nonlinear boundary value problems can in principle be used to solve linear problems, one cannot expect it to perform very efficiently in the linear case. Thus, there exist methods and computer codes tailored specifically to linear boundary value problems (and focusing on the efficient solution of the resulting systems of linear algebraic equations). Various aspects of this topic are discussed in Chapter 17 of Hall & Watt [1976]. An important source of "difficult" linear boundary value problems are the so-called *singular perturbation problems*, e.g.,

$$\varepsilon y'' + p(x)y' + q(x)y = r(x), \ \ a < x < b,$$

$$y(a) = \alpha, \ \ y(b) = \beta,$$

where the parameter ε is small: $0 < \varepsilon << 1$. (Solutions of such problems usually exhibit a very rapidly changing behavior in small boundary layer regions, or they possess internal turning points where there are sharp spikes.) The interested reader is referred to Hemker & Miller [1979] or to the recent survey paper by Kadalbajoo & Reddy [1989] for details on theoretical and numerical aspects of such problems, as well as for additional references.

§9.4 Theoretical and computational aspects of finite difference methods for two-point boundary value problems are discussed in, e.g., Fox [1957], Collatz [1960], Keller [1968, 1975, 1976], Hall & Watt [1976, Chapter 15], and Fox & Mayers [1987]. Russell [1977] compares finite difference methods with certain spline collocation methods (see the notes to §9.5). Approximations obtained by finite difference methods can often be improved iteratively (on the same mesh) by so-called (Richardson) *deferred correction* or *difference correction* methods. Details of these techniques (introduced originally by Fox in 1947) may be found in Fox [1957], Keller [1968, 1975, 1976], in the article by Fox in Gladwell & Sayers [1980], and in Fox & Mayers [1987]. Iterated deferred correction techniques form the basis of a series of computer codes developed by Pereyra and others for boundary value problems for first-order differential equations. The underlying discretization is the trapezoidal rule over a possibly nonuniform mesh (which is chosen adaptively). Descriptions of different versions of this method are given in the article by Pereyra in Childs et al. [1979] (see also the contribution by Fox in Gladwell & Sayers [1980]). A routine in the IMSL collection of codes is based on a version due to Lentini and Pereyra.

§9.5 A survey of some of the theoretical developments on projection methods (such as collocation, Ritz-Galerkin, and least squares) for two-point boundary value problems may be found in Reddien [1980]. This paper also contains an extensive bibliography. The monographs by Strang & Fix [1973], Prenter [1975] and Fried [1979] may be consulted for elementary introductions to projection methods based on spline functions.

Collocation methods for two-point boundary value problems are analyzed in Russell & Shampine [1972]; the paper also features a number of illuminating examples. As mentioned before, the paper by Russell [1977] investigates the relative merits of spline collocation and finite difference methods.

Spline collocation (employing Gauss points and B-spline bases together with automatic mesh selection) are the main ingredients of a powerful code due to Ascher et al. [1981]; compare also the contribution by Ascher, Christiansen & Russell in Childs et al. [1979] for a detailed description of this code.

References

Ascher, U., Christiansen, J. and Russell, R.D. [1981]: Collocation software for boundary value ordinary differential equations, *ACM Trans. Math. Software* **7**, 209–222.

Ascher, U. and Russell, R.D. [1981]: Reformulation of boundary value problems in "standard" form, *SIAM Rev.* **23**, 238–254.

Bailey, P.B., Shampine, L.F. and Waltman, P.E. [1968]: *Nonlinear Two Point Boundary Value Problems*, Academic Press, New York.

Childs, B. et al. (eds.) [1979]: *Codes for Boundary-Value Problems in Ordinary Differential Equations*, Lecture Notes in Computer Science **76**, Springer, New York.

Collatz, L. [1960]: *The Numerical Treatment of Differential Equations* (3rd ed.), Springer, New York.

Daniel, J.W. and Moore, R.E. [1970]: *Computation and Theory in Ordinary Differential Equations*, W.H. Freeman, San Francisco.

Fox, L. [1957]: *The Numerical Solution of Two-Point Boundary Problems in Ordinary Differential Equations*, Clarendon Press, Oxford.

Fox, L. and Mayers, D.F. [1987]: *Numerical Solution of Ordinary Differential Equations*, Chapman and Hall, London.

Fried, I. [1979]: *Numerical Solution of Differential Equations*, Academic Press, New York.

Gladwell, I. and Sayers, D.K. (eds.) [1980]: *Computational Techniques for Ordinary Differential Equations*, Academic Press, London.

Hall, G. and Watt, J.M. (eds.) [1976]: *Modern Numerical Methods for Ordinary Differential Equations*, Clarendon Press, Oxford.

Hemker, P.W. and Miller, J.J.H. (eds.) [1979]: *Numerical Analysis of Singular Perturbation Problems*, Academic Press, New York.

Kadalbajoo, M.K. and Reddy, Y.N. [1989]: Asymptotic and numerical analysis of singular perturbation problems: a survey, *Appl. Math. Comput.* **30**, 223–259.

Keller, H.B. [1968]: *Numerical Methods for Two-Point Boundary Value Problems*, Blaisdell, Waltham, Mass.

Keller, H.B. [1975]: Numerical solution of boundary value problems for ordinary differential equations: survey and some recent results in difference methods, in *Numerical Solution of Boundary Value Problems for Ordinary Differential Equations* (A.K. Aziz, ed.), pp. 27–88, Academic Press, New York.

Keller, H.B. [1976]: *Numerical Solution of Two Point Boundary Value Problems*, CBMS-NSF Regional Conference Series in Applied Math. **24**, *SIAM*, Philadelphia.

Meyer, G.H. [1973]: *Initial Value Methods for Boundary Value Problems. Theory and Application of Invariant Imbedding*, Mathematics in Science and Engineering **100**, Academic Press, New York.

Osborne, M.R. [1969]: On shooting methods for boundary value problems, *J. Math. Anal. Appl.* **27**, 417–433.

Prenter, P.M. [1975]: *Splines and Variational Methods*, Wiley, New York.

Reddien, G.W. [1980]: Projection methods for two-point boundary value problems, *SIAM Rev.* **22**, 156–171.

Russell, R.D. [1977]: A comparison of collocation and finite differences for two-point boundary value problems, *SIAM J. Numer. Anal.* **14**, 19–39.

Russell, R.D. and Shampine, L.F. [1972]: A collocation method for boundary value problems, *Numer. Math.* **19**, 1–28.

Stoer, J. and Bulirsch, R. [1980]: *Introduction to Numerical Analysis*, Springer, New York.

Strang, G. and Fix, G.J. [1973]: *An Analysis of the Finite Element Method*, Prentice-Hall, Englewood Cliffs, N.J.

CHAPTER 10

Elliptic Partial Differential Equations, Relaxation Methods

The classical model examples of partial differential equations are:

a) *Dirichlet problem* (elliptic case):

$$\frac{\partial^2 u}{\partial x^2} + \frac{\partial^2 u}{\partial y^2} = f(x,y) \quad \text{in the domain } B \text{ of the } (x,y)\text{-plane,} \qquad (1)$$

u (or $\partial u/\partial n$ in the so-called Neumann problem) given on the boundary of B.

b) *Heat equation* (parabolic case):

$$\frac{\partial u}{\partial t} = \frac{\partial^2 u}{\partial x^2} \quad \text{for } a \le x \le b, \quad t > 0, \qquad (2)$$

$u(x,t)$ given at $t = 0$ for *all* x,

u or $\partial u/\partial x$ given at $x = a$, $x = b$ for *all* t.

c) *Wave equation* (hyperbolic case):

$$\frac{\partial^2 u}{\partial t^2} = \frac{\partial^2 u}{\partial x^2} \quad \text{for } a \le x \le b, \quad t > 0, \qquad (3)$$

u and ∂u/∂t given at t = 0 for all x,

u or ∂u/∂x given at x = a, x = b for all t.

Problem a), which we now propose to solve (at least numerically), includes also the potential problem.

§10.1. Discretization of the Dirichlet problem

If a square grid with meshsize h is laid over the domain B, then at each interior point P of B (cf. Fig. 10.1)

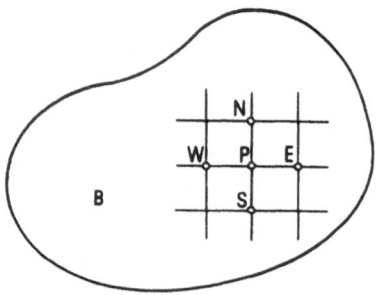

Figure 10.1. *Discretization of the Dirichlet problem*

one can approximate the second partial derivatives by difference quotients:

$$\frac{\partial^2 u}{\partial x^2} = \frac{u(x+h,y) - 2u(x,y) + u(x-h,y)}{h^2} + O(h^2),$$

$$\frac{\partial^2 u}{\partial y^2} = \frac{u(x,y+h) - 2u(x,y) + u(x,y-h)}{h^2} + O(h^2).$$

The differential equation at the point (x,y) is then approximated by

$$\frac{u(x+h,y) + u(x-h,y) + u(x,y+h) + u(x,y-h) - 4u(x,y)}{h^2} - f(x,y) = 0. \quad (4)$$

By formulating this equation at each grid point, one obtains a system of linear equations whose unknowns are the ordinates $u(x,y)$ for all grid points.

A certain difficulty, though, arises near the boundary of B. To avoid it, we assume for the time being that the boundary of B is composed of grid segments and that u is given on the boundary, as for example in Fig. 10.2.

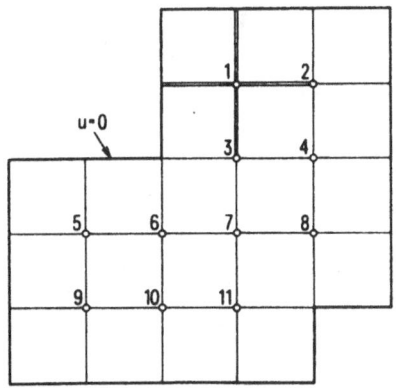

Figure 10.2. *Dirichlet problem with a boundary consisting of grid segments*

Thus, u is unknown in each of the 11 interior grid points. In each of these 11 points one now sets up the equation (4). If $u = 0$ is prescribed on the boundary, and $u_1, u_2, u_3, \ldots$ denote the unknown function values at the points $1, 2, 3, \ldots,$ one first has, say for the point 1:

$$
\begin{aligned}
u(x,y) &= u_1 \\
u(x + h, y) &= u_2 \\
u(x - h, y) &= 0 \qquad \text{(given boundary value),} \\
u(x, y + h) &= 0 \qquad \text{(given boundary value),} \\
u(x, y - h) &= u_3,
\end{aligned}
$$

hence the equation

$$u_2 + u_3 - 4u_1 = h^2 f_1$$

(where $f_k = f(x,y)$ at the point with number k). Altogether one obtains (after a change of sign):

u_1	u_2	u_3	u_4	u_5	u_6	u_7	u_8	u_9	u_{10}	u_{11}	1	
4	−1	−1									$h^2 f_1$	= 0
−1	4		−1								$h^2 f_2$	= 0
−1		4	−1		−1						$h^2 f_3$	= 0
	−1	−1	4			−1					$h^2 f_4$	= 0
				4	−1			−1			$h^2 f_5$	= 0
				−1	4	−1			−1		$h^2 f_6$	= 0
		−1			−1	4	−1			−1	$h^2 f_7$	= 0
			−1			−1	4				$h^2 f_8$	= 0
				−1				4	−1		$h^2 f_9$	= 0
					−1			−1	4	−1	$h^2 f_{10}$	= 0
						−1			−1	4	$h^2 f_{11}$	= 0

(5)

Since the coefficient matrix is symmetric and, as is easily checked, positive definite, the system of equations can be solved by Cholesky, whereby the band structure permits further simplifications. What is, and remains, decisive, however, is the fact that the number of equations and unknowns equals the number of grid points in the interior of the domain; one thus has to deal with extensive systems of equations, which require large computing times. If in the above example one would halve the meshsize h of the grid, one would already obtain $n = 61$ interior grid points, generally after division by p, $n = 20\ p^2 - 10\ p + 1$. This rapid increase of n will still concern us later.

First, however, we propose to treat

a) more general boundaries,

b) more general boundary conditions.

We find the following:

a) For a boundary which does not pass through the grid points, there are interior points (x,y), some of whose neighboring points $(x \pm h, y)$, $(x, y \pm h)$ are on the other side of the boundary. One must use, then, the points of intersection of the grid lines with the boundary of the domain (see Fig. 10.3).

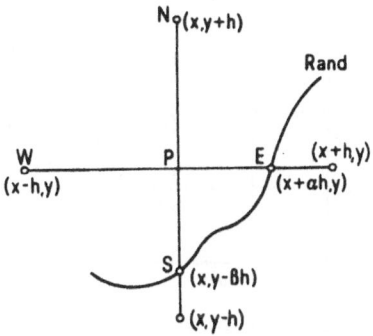

Figure 10.3. *Discretization for curvilinear boundary*

One approximates

$$\frac{1}{2}\left.\frac{\partial^2 u}{\partial x^2}\right|_P \approx \frac{1}{h^2}\left\{\frac{u(x-h,y)}{1+\alpha} - \frac{u(x,y)}{\alpha} + \frac{u(x+\alpha h,y)}{\alpha(1+\alpha)}\right\},$$

$$\frac{1}{2}\left.\frac{\partial^2 u}{\partial y^2}\right|_P \approx \frac{1}{h^2}\left\{\frac{u(x,y+h)}{1+\beta} - \frac{u(x,y)}{\beta} + \frac{u(x,y-\beta h)}{\beta(1+\beta)}\right\},$$

so that the equation for the point P becomes

$$\frac{2}{h^2}\left\{\frac{u_W}{1+\alpha} + \frac{u_N}{1+\beta} + \frac{u_E}{\alpha(1+\alpha)} + \frac{u_S}{\beta(1+\beta)} - \left[\frac{1}{\alpha} + \frac{1}{\beta}\right]u_P\right\} - f_P = 0. \tag{6}$$

b) In case of a boundary condition $\dfrac{\partial u}{\partial n} = 0$, one must introduce also the boundary values as unknowns and for each of them set up an additional equation.

In order, for example, to solve the problem described in Figure 10.4,

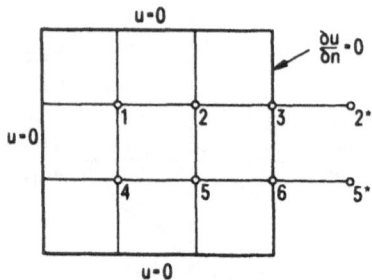

Figure 10.4. *Introduction of virtual grid points for boundary segments with prescribed normal derivative*

one can introduce two virtual grid points 2* and 5*; then

$$\frac{\partial^2 u}{\partial x^2} + \frac{\partial^2 u}{\partial y^2}\bigg|_3 = -\frac{4u_3 - u_2 - u_6 - u_{2*}}{h^2},$$

and on the other hand $u_{2*} = u_2$, because of $\partial u/\partial n = 0$ at the point 3; therefore, the equation for this point becomes:

$$4u_3 - 2u_2 - u_6 + h^2 f_3 = 0.$$

Altogether one obtains

u_1	u_2	u_3	u_4	u_5	u_6	1
4	−1		−1			$h^2 f_1$
−1	4	−1		−1		$h^2 f_2$
	−2	4			−1	$h^2 f_3$
−1			4	−1		$h^2 f_4$
	−1		−1	4	−1	$h^2 f_5$
		−1		−2	4	$h^2 f_6$

Now in case a) as well as in case b), however, symmetry is lost, which is a great disadvantage, since symmetric systems of equations can be solved much more easily.

It is possible, however, to force symmetry of the system also in case of more complicated boundaries and boundary conditions, by carrying out the discretization with the help of the *energy method*. For the example of Figure 10.4, this means the following:

$$\frac{\partial^2 u}{\partial x^2} + \frac{\partial^2 u}{\partial y^2} = f(x,y)$$

is the Euler equation for the variational problem

$$\delta \iint \left[\frac{1}{2} \left(\frac{\partial u}{\partial x} \right)^2 + \frac{1}{2} \left(\frac{\partial u}{\partial y} \right)^2 + uf \right] dx \, dy = 0 \tag{7}$$

with the boundary condition $u = 0$ on 3 of the 4 sides of the square. In a square, say the one with vertices 1, 2, 4, 5, one has approximately

$$\iint \left(\frac{\partial u}{\partial x} \right)^2 dx \, dy = h^2 \left\{ \frac{1}{2} \left(\frac{u_2 - u_1}{h} \right)^2 + \frac{1}{2} \left(\frac{u_5 - u_4}{h} \right)^2 \right\},$$

$$\iint uf \, dx \, dy = \frac{h^2}{4} \{ u_1 f_1 + u_2 f_2 + u_5 f_5 + u_4 f_4 \}.$$

Taking account of the boundary conditions, one thus has altogether

$$\iint \left[\frac{1}{2} \left(\frac{\partial u}{\partial x} \right)^2 + \frac{1}{2} \left(\frac{\partial u}{\partial y} \right)^2 + uf \right] dx \, dy$$

$$= \frac{1}{4} u_1^2 + \frac{1}{4} u_1^2 + \frac{1}{4} u_1^2 + \frac{1}{4} (u_1 - u_2)^2 + \frac{1}{4} u_2^2 + \frac{1}{4} u_2^2 + \frac{1}{4} (u_2 - u_3)^2 + \frac{1}{4} u_3^2 + \frac{1}{4} u_1^2 + \frac{1}{4} u_4^2$$

$$+ \frac{1}{4} (u_1 - u_4)^2 + \frac{1}{4} (u_1 - u_4)^2 + \frac{1}{4} (u_1 - u_2)^2 + \frac{1}{4} (u_4 - u_5)^2 + \frac{1}{4} (u_2 - u_5)^2 + \frac{1}{4} (u_2 - u_5)^2$$

$$+\frac{1}{4}(u_2-u_3)^2+\frac{1}{4}(u_5-u_6)^2+\frac{1}{4}(u_3-u_6)^2+\frac{1}{4}u_4^2+\frac{1}{4}u_4^2+\frac{1}{4}u_4^2+\frac{1}{4}(u_4-u_5)^2$$

$$+\frac{1}{4}u_5^2+\frac{1}{4}u_5^2+\frac{1}{4}(u_5-u_6)^2+\frac{1}{4}u_6^2+h^2[u_1f_1+u_2f_2+\frac{1}{2}u_3f_3+u_4f_4+u_5f_5+\frac{1}{2}u_6f_6]$$

$$=2u_1^2-u_1u_2-u_1u_4+2u_2^2-u_2u_3-u_2u_5+u_3^2-\frac{1}{2}u_3u_6+2u_4^2-u_4u_5+2u_5^2-u_5u_6$$

$$+u_6^2+h^2[u_1f_1+u_2f_2+\frac{1}{2}u_3f_3+u_4f_4+u_5f_5+\frac{1}{2}u_6f_6]$$

$$=\frac{1}{2}(\mathbf{u},\mathbf{Au})+(\mathbf{u},\mathbf{b})=Q(\mathbf{u}),$$

where

$$\mathbf{A}=\begin{bmatrix}4&-1&0&-1&0&0\\-1&4&-1&0&-1&0\\0&-1&2&0&0&-\frac{1}{2}\\-1&0&0&4&-1&0\\0&-1&0&-1&4&-1\\0&0&-\frac{1}{2}&0&-1&2\end{bmatrix},\quad \mathbf{b}=h^2\begin{bmatrix}f_1\\f_2\\f_3/2\\f_4\\f_5\\f_6/2\end{bmatrix}. \qquad (8)$$

The minimum of this function $Q(\mathbf{u})$ is achieved when $\mathbf{Au}+\mathbf{b}=0$; one thus has to solve this linear system. The matrix $\mathbf{A}$ is symmetric, even positive definite, the latter since $(\mathbf{u},\mathbf{Au})$ is a sum of pure squares, which can only be 0 if $\mathbf{u}=\mathbf{0}$([1]).

---

[1] A systematic further development of the idea described here leads to the method of finite elements, a method which today is widely used. (Editors' remark)

§10.2. The operator principle

If in the first example of §10.1, depicted in Fig. 10.2, one reduces the meshsize by a factor of 10 (the solution then becomes 100-times as accurate), the number of interior grid points becomes 1901. The coefficient matrix of the system of equations then contains 3613801 coefficients, which can no longer be stored in the usual high-speed memory. Since most of these coefficients are 0, however, this can also be avoided if one defines the matrix not by its n^2 coefficients, but by a *computational rule*, which for an arbitrary vector $\mathbf{v}$ shows how to compute the vector $\mathbf{Av}$.

Example of such a computational rule:

```
procedure op(n,x) res:(ax);
    value n;
    integer n; array x,ax;
    begin
        integer j;
        ax[1] := 2 × x[1] − x[2];
        ax[n] := 2 × x[n] − x[n − 1];
        for j := 2 step 1 until n − 1 do
            ax[j] := 2 × x[j] − x[j − 1] − x[j + 1];
    end op;
```

This procedure evidently embodies the matrix

$$
\mathbf{A} = \begin{bmatrix}
2 & -1 & & & & 0 \\
-1 & 2 & -1 & & & \\
 & -1 & 2 & -1 & & \\
 & & & \ddots & & \\
 & & & & \ddots & \\
 & & & & \ddots & -1 \\
0 & & & & -1 & 2
\end{bmatrix}.
$$

One question, though, still remains unanswered: How do I solve a linear system of equations $\mathbf{Ax} + \mathbf{b} = \mathbf{0}$, if I have at my disposal only the vector $\mathbf{b}$ and a computational rule such as the above procedure *op*? Neither Gauss elimination nor the Cholesky method can be applied in this case, since these require an **array** $a[1:n,1:n]$ explicitly. We therefore will have to develop new methods to deal with this situation.

For the Dirichlet problem, the matrix **A** is determined by the difference operator in (4), which one represents most conveniently as in Fig. 10.5,

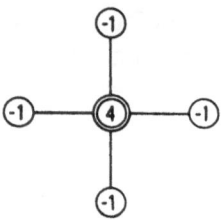

Figure 10.5. *The difference operator for the Dirichlet problem*

and by the shape of the domain. This operator means that when the components of a vector **x** are arranged as a *field* in the shape of the domain B:

$$
\begin{array}{cccc}
x_1 & x_2 & & \\
x_3 & x_4 & & \\
x_5 & x_6 & x_7 & x_8 \\
x_9 & x_{10} & x_{11} &
\end{array}
\qquad (9)
$$

then an arbitrary component of **Ax** is obtained by placing the difference operator on top of the field (the double circle on top of the component x_k which corresponds to the desired component $(\mathbf{Ax})_k$), multiplying the operator coefficients with the x-values lying underneath, and then adding everything up; thus, for example,

$$(\mathbf{Ax})_6 = 4x_6 - x_5 - x_{10} - x_7.$$

In some cases, different components of **Ax** are computed by different operators. For example, in case of the domain in Fig. 10.6, with the free boundary on the right, the operator valid for the interior points is the one in Fig. 10.5, while for the points on the right-hand boundary it is the one in Fig. 10.7.

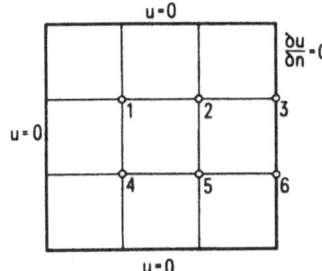

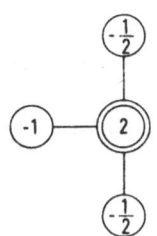

Figure 10.6. *Domain with free*   Figure 10.7. *Difference operator for*
boundary on the right   *points on the free boundary*

The same operators are valid also for half the meshsize (cf. Fig. 10.8).

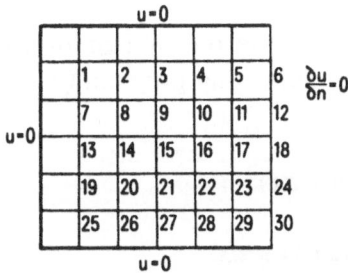

Figure 10.8. *Halving of the meshsize for the domain of Fig. 10.6*

Now it is not all that simple, though, to describe to a machine the shape of the domain and the type of difference operator. The domain can be described, for example, by specifying for each interior point the four neighboring points N, E, S, W, as well as the type of operator (these are numbered). The domain above in Fig. 10.6, with 6 grid points and 10 boundary points (with prescribed function values) is thus characterized by an **array** v[1:6, 0:4] whose meaning is as follows:

$v[k, 0]$ indicates which operator is valid at the point k:

operator 1:   4 -1 -1 -1 -1

operator 2:   2 $-\dfrac{1}{2}$ 0 $-\dfrac{1}{2}$ -1

(These numbers are stored, e.g., as **array** *dop* [1:2, 0:4].)

$v[k,j]$ $(j = 1,2,3,4)$ denotes the neighbors of the point k in the order N, E, S, W. (0 means nonexisting point.)

Therefore, v contains the following values:

k \\ j	0	1	2	3	4
1	1	0	2	4	0
2	1	0	3	5	1
3	2	0	0	6	2
4	1	1	5	0	0
5	1	2	6	0	4
6	2	3	0	0	5

These data uniquely determine the system of equations; assuming that $x[0] = 0$, one obtains the following program (which admittedly is not particularly efficient):

```
procedure op(n,x,ax);
    value n;
    integer n; array x,ax;
    begin
        integer j, k, vk0; real s;
        for k := 1 step 1 until n do
        begin
            vk0 := v[k,0];
            s := dop[vk0,0] × x[k];
            for j := 1,2,3,4 do
                s := s + x[v[k,j]] × dop[vk0,j];
            ax[k] := s
        end
    end op;
```

    If the shape of the domain is simple, as for example in the case of the above square with free right-hand boundary, which now, however, is assumed to be covered more generally by a grid of meshsize $h=1/n$, one can easily get by without the **array** v and *dop*, by introducing an auxiliary procedure *ap* which corresponds to the single application of a difference operator (we again assume $x[0] := 0$):

```
procedure op(n,x) res:(ax);
    value n;
    integer n;  array x, ax;
    begin
        procedure ap (k,optyp,n,e,s,w);
            value k, optyp,n,e,s,w;
            integer k, optyp,n,e,s,w;
            ax[k] := if optyp = 1 then
                    4 × x[k] − x[n] − x[e] − x[s] − x[w]
                else
                    2 × x[k] − (x[n] + x[s])/2 − x[w];
            comment end ap;
        integer j,k,ℓ;
        ap(1,1,0,2,n+1,0);
        for j := 2 step 1 until n−1 do ap(j,1,0,j+1,j+n,j−1);
        ap(n,2,0,0,2 × n,n−1);
        for ℓ := 2 × n step n until n × (n−2) do
        begin
            k := ℓ − n + 1;
            ap(k,1,k−n,k+1,k+n,0);
            for j := k+1 step 1 until ℓ − 1 do
                ap(j,1,j−n,j+1,j+n,j−1);
            ap(ℓ,2,ℓ − n,0,ℓ + n, ℓ − 1);
        end ℓ;
        k := (n−2) × n+1;
        ap(k, 1,k−n,k+1,0,0);
        for j := k+1 step 1 until (n−1) × n−1 do
            ap(j, 1,j−n,j+1,0,j−1);
        k := (n−1) × n;
        ap(k,2,k−n,0,0,k−1);
    end op;
```

§10.3. The general principle of relaxation

A linear system of equations $\mathbf{Ax} + \mathbf{b} = \mathbf{0}$ with symmetric and positive definite coefficient matrix, according to §3.7, is equivalent to the minimum problem for the quadratic function

$$F(\mathbf{x}) = \frac{1}{2}\,(\mathbf{x},\mathbf{Ax}) + (\mathbf{b},\mathbf{x}). \tag{10}$$

One can actually solve it by systematically searching for the minimum of the function F in the x-space. The general principle of such methods, called *relaxation methods*, can be described as follows:

One chooses in the x-space an initial point $\mathbf{x}_0$ and a relaxation direction $\mathbf{h}_0$, and then moves a certain distance from $\mathbf{x}_0$ in the direction of $\mathbf{h}_0$. One so arrives at a point $\mathbf{x}_1$, chooses here again a relaxation direction $\mathbf{h}_1$, etc. (cf. Fig. 10.9).

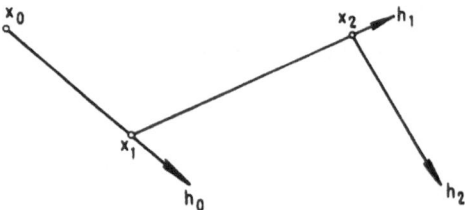

Figure 10.9. *Relaxation methods*

Of course, one chooses the relaxation directions, and also the distances by which one travels along these directions, in such a way that $F(\mathbf{x})$ continually decreases. One then hopes that the sequence $\mathbf{x}_0$, $\mathbf{x}_1$, $\mathbf{x}_2$, . . . converges to the minimum point, that is, to the solution of the system of linear equations.

For the practical implementation, one needs the gradient of the function F. One has

$$\frac{\partial F}{\partial x_j} = \frac{1}{2}\sum_{i=1}^{n} a_{ij}x_i + \frac{1}{2}\sum_{k=1}^{n} a_{jk}x_k + b_j = \sum_{k=1}^{n} a_{jk}x_k + b_j,$$

thus

$$\text{grad } F = \mathbf{A}\mathbf{x} + \mathbf{b} = \mathbf{r}. \tag{11}$$

The components of grad F are thus precisely the deficits (residuals) which are obtained when the respective $\mathbf{x}$ is substituted into the equations. The residual vector $\mathbf{r}$ therefore not only indicates how well $\mathbf{x}$ satisfies the equations, but also in which direction F decreases. What is important to note is that the rule for computing $\mathbf{A}\mathbf{x}$ from $\mathbf{x}$ is sufficient for computing this gradient, so that in a relaxation method the matrix $\mathbf{A}$ is not explicitly needed.

When moving from $\mathbf{x}$ in the relaxation direction $\mathbf{h}$, thus traveling through the points $\mathbf{x}_t = \mathbf{x} + t\mathbf{h}$, one has for the quadratic function $F(\mathbf{x}_t)$ in dependence of t:

$$F(\mathbf{x}_t) = F(\mathbf{x} + t\mathbf{h}) = \frac{1}{2}(\mathbf{x} + t\mathbf{h}, \mathbf{A}(\mathbf{x} + t\mathbf{h})) + (\mathbf{b}, \mathbf{x} + t\mathbf{h})$$

$$= F(\mathbf{x}) + \frac{t}{2}(\mathbf{h}, \mathbf{A}\mathbf{x}) + \frac{t}{2}(\mathbf{x}, \mathbf{A}\mathbf{h}) + \frac{t^2}{2}(\mathbf{h}, \mathbf{A}\mathbf{h}) + t(\mathbf{b}, \mathbf{h}),$$

hence, because of $(\mathbf{h}, \mathbf{A}\mathbf{x}) = (\mathbf{x}, \mathbf{A}\mathbf{h})$,

$$F(\mathbf{x}_t) = F(\mathbf{x}) + t(\mathbf{h}, \mathbf{r}) + \frac{t^2}{2}(\mathbf{h}, \mathbf{A}\mathbf{h}). \tag{12}$$

By virtue of $(\mathbf{h}, \mathbf{A}\mathbf{h}) > 0$, this is a quadratic polynomial in t whose only minimum lies at

$$t_M = -\frac{(\mathbf{h}, \mathbf{r})}{(\mathbf{h}, \mathbf{A}\mathbf{h})}, \tag{13}$$

where (with $\mathbf{x}_M = \mathbf{x}_{t_M}$)

$$F(\mathbf{x}_M) = F(\mathbf{x}) - \frac{1}{2}\frac{(\mathbf{h}, \mathbf{r})^2}{(\mathbf{h}, \mathbf{A}\mathbf{h})}. \tag{14}$$

Actually, $F(\mathbf{x}_t)$ lies below $F(\mathbf{x})$ in the whole interval $(0, 2t_M)$, so that it is only necessary to choose t somewhere in this interval in order to go down "lower" (see Fig. 10.10).

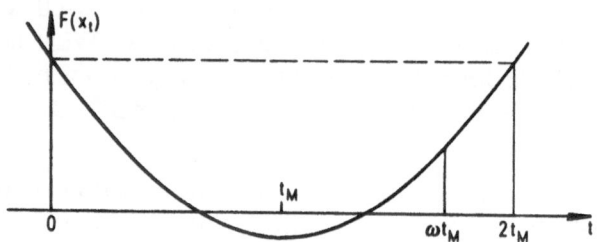

Figure 10.10. $F(\mathbf{x}_t)$ *as a function of* t

The residual $\mathbf{r}_t$ at the point $\mathbf{x}_t$ is likewise a function of t:

$$\mathbf{r}_t = A\mathbf{x}_t + \mathbf{b} = A(\mathbf{x} + t\mathbf{h}) + \mathbf{b} = \mathbf{r} + tA\mathbf{h}. \tag{15}$$

Because of

$$(\mathbf{h}, \mathbf{r}_t) = (\mathbf{h}, \mathbf{r}) + t(\mathbf{h}, A\mathbf{h}) = (t - t_M)(\mathbf{h}, A\mathbf{h}),$$

the minimum point $\mathbf{x}_M$ is characterized by the new residual $\mathbf{r}_M$ being perpendicular to the relaxation direction.

   The various relaxation methods (Gauss-Seidel, steepest descent, conjugate gradients, etc.) differ only in the choice of the relaxation direction $\mathbf{h}_k$ (in the kth step) and the choice of t in the formula $\mathbf{x}_{k+1} = \mathbf{x}_k + t\mathbf{h}_k$ for the new point.

   One certainly would expect that the best relaxation method is the one in which at each point $\mathbf{x}_k$ one chooses the negative gradient $-\mathbf{r}_k$ as relaxation direction (optimal direction) and travels in this direction to the minimum point $\mathbf{x}_M$ (optimal point), that is, uses the recursion formula

$$\mathbf{x}_{k+1} = \mathbf{x}_k - \frac{(\mathbf{r}_k, \mathbf{r}_k)}{(\mathbf{r}_k, A\mathbf{r}_k)} \, \mathbf{r}_k \quad \text{(where } \mathbf{r}_k = A\mathbf{x}_k + \mathbf{b}). \tag{16}$$

In reality, however, this *method of steepest descent* is rather poor.

§10.4. The method of Gauss-Seidel, overrelaxation

In the method of Gauss-Seidel, which is occasionally applied also to nonsymmetric systems of equations, one starts with a suitable approximation vector $\mathbf{x} = [x_1, x_2, \ldots, x_n]^T$. Then the n unknowns, one after another (for example in the order $x_1, x_2, \ldots, x_n, x_1, x_2, \ldots, x_n, x_1, \ldots$), are continually improved, whereby x_j *is changed in such a way that, with the remaining unknowns held fixed, the jth equation is fulfilled.*

Example. In solving the system of equations

	x_1	x_2	x_3	1
0 =	5	−2	−1	−3
0 =	2	5	−1	−2
0 =	1	1	5	−1

one so obtains for x_1, x_2, x_3 in turn (if one starts with $\mathbf{x}=0$):

x_1	x_2	x_3
0	0	0
.6000	.1600	.0480
.6736	.1402	.0372
.6635	.1420	.0389
.6646	.1419	.0387
.6645	.1419	.0387

This method, which for a positive definite coefficient matrix always converges, can be subsumed under general relaxation methods: if the unknown x_j is improved such that the jth equation is satisfied, one obtains indeed

$$x'_j = -\frac{b_j + \sum_{k \neq j} a_{jk} x_k}{a_{jj}}, \quad x'_k = x_k \text{ for } k \neq j. \tag{17}$$

One then has

$$x'_j - x_j = -\frac{b_j + \sum_{k=1}^{n} a_{jk} x_k}{a_{jj}} = -\frac{r_j}{a_{jj}},$$

hence

$$\mathbf{x'} - \mathbf{x} = t\mathbf{e}_j \ \text{ with } \ t = -\frac{r_j}{a_{jj}} = -\frac{(\mathbf{r},\mathbf{e}_j)}{(\mathbf{e}_j,\mathbf{A}\mathbf{e}_j)} . \tag{18}$$

One step of the Gauss-Seidel method thus corresponds to a relaxation step with $\mathbf{h} = \mathbf{e}_j$ and $t = t_M$. The method as a whole consists in choosing successively the relaxation directions $\mathbf{e}_1, \mathbf{e}_2, \ldots, \mathbf{e}_n, \mathbf{e}_1, \mathbf{e}_2, \ldots, \mathbf{e}_n, \ldots$ and always traveling in the respective direction to the minimum point.

Computational experience shows, however, that in many cases the method of Gauss-Seidel converges exceptionally slowly. But now, convergence can be improved by traveling in the respective relaxation direction not only to the minimum point $\mathbf{x}_M = \mathbf{x} + t_M \mathbf{h}$, but beyond it by a certain percentage. The extent of overshooting beyond the position t_M is determined by an *overrelaxation factor* ω, that is, one selects a fixed factor $\omega \ (> 1)$, and then in each relaxation step travels to the position $t = \omega t_M$ (cf. Fig. 10.10). The computational rule in the kth step therefore is, when $k \equiv j \pmod{n}$:

$$x'_j = x_j - \omega \frac{r_j}{a_{jj}}, \ \ x'_i = x_i \ \text{ for } \ i \neq j . \tag{19}$$

For this method, which contains with $\omega = 1$ the Gauss-Seidel method, the following is true:

Theorem 10.1. *If A is symmetric and positive definite, then the overrelaxation method for every fixed ω with $0 < \omega < 2$ converges to the solution of the system of equations $\mathbf{A}\mathbf{x} + \mathbf{b} = \mathbf{0}$.*

Proof. For the kth step on has, when $k \equiv j \pmod{n}$:

$$\mathbf{x}_{k+1} = \mathbf{x}_k - \omega \frac{r_j}{a_{jj}} \mathbf{e}_j. \tag{20}$$

Therefore,

$$F(\mathbf{x}_{k+1}) = \frac{1}{2}(\mathbf{x}_k,\mathbf{A}\mathbf{x}_k) - \frac{\omega r_j}{a_{jj}}(\mathbf{e}_j,\mathbf{A}\mathbf{x}_k) + \frac{\omega^2 r_j^2}{2a_{jj}^2}(\mathbf{e}_j,\mathbf{A}\mathbf{e}_j)$$

$$+ (\mathbf{b},\mathbf{x}_k) - \frac{\omega r_j}{a_{jj}}(\mathbf{b},\mathbf{e}_j).$$

In view of $A\mathbf{x}_k + \mathbf{b} = \mathbf{r}_k$, $(\mathbf{e}_j, A\mathbf{e}_j) = a_{jj}$, one thus obtains

$$F(\mathbf{x}_{k+1}) = F(\mathbf{x}_k) - \left[\omega - \frac{\omega^2}{2}\right]\frac{r_j^2}{a_{jj}}. \tag{21}$$

For $0 < \omega < 2$ and $a_{jj} > 0$, the numerical sequence $F_k = F(\mathbf{x}_k)$ decreases monotonically, and is therefore convergent, since F cannot fall below the minimum (guaranteed by Theorem 3.6). Therefore, $\lim(F_{k+1} - F_k) = 0$, hence also

$$\lim_{k\to\infty} r_j = 0.$$

Here, $j = j(k)$ denotes as before the index of the equation that is processed in the kth step; it is defined by $j \equiv k \pmod{n}$ and $1 \le j \le n$.

One cannot immediately conclude, however, that

$$\lim_{k\to\infty} \mathbf{r}_k = \mathbf{0};$$

all one knows at the moment is that

$$\lim_{k\to\infty} r_{k,j(k)} = 0.$$

(Here, $r_{k,i}$ denotes the ith component of the residual vector $\mathbf{r}_k$ of the kth step; in particular, $r_{k,j} = r_j$ for $j = j(k)$.) Thus,

$$|r_{k,j(k)}| < \varepsilon \quad \text{for} \quad k > M(\varepsilon).$$

Also, by virtue of (20),

$$\lim_{k\to\infty} (\mathbf{x}_{k+1} - \mathbf{x}_k) = \mathbf{0},$$

hence

$$\mathbf{r}_{k+1} - \mathbf{r}_k = \mathbf{A}(\mathbf{x}_{k+1} - \mathbf{x}_k) \to 0 \text{ as } k \to \infty.$$

Consequently, for each individual i:

$$|r_{k+p,i} - r_{k,i}| < p\varepsilon \; (p = 1, \ldots, n) \text{ if } k > N(\varepsilon).$$

If k is a multiple of n, $k > N(\varepsilon)$, $k > M(\varepsilon)$, we have in particular that

$$|r_{k+1,1} - r_{k,1}| < \varepsilon, \; |r_{k+1,1}| < \varepsilon \; => \; |r_{k,1}| < 2\varepsilon,$$

$$|r_{k+2,2} - r_{k,2}| < 2\varepsilon, \; |r_{k+2,2}| < \varepsilon \; => \; |r_{k,2}| < 3\varepsilon,$$

$$\vdots \qquad\qquad \vdots \qquad\qquad \vdots$$

$$|r_{k+n,n} - r_{k,n}| < n\varepsilon, \; |r_{k+n,n}| < \varepsilon \; => \; |r_{k,n}| < (n+1)\varepsilon.$$

Therefore, $||\mathbf{r}_k|| < n^2\varepsilon$, and this is true for arbitrarily small ε, if only k is made large enough. From $\mathbf{r}_k \to 0$ there finally follows that $\mathbf{x}_k$ converges to the solution of the system of equations, q.e.d.

The proof reveals nothing at all about the *speed of convergence*; what happens, actually, is that convergence is optimal for a certain ω between 1 and 2. For ill-conditioned matrices (cf. §10.7) the optimum lies very close to 2. Unfortunately, the optimal ω is generally not known a priori, but must be determined experimentally.

For the example

$$137\,x - 100\,y = 11$$
$$-100\,x + 73\,y = -8$$

(exact solution: $x=3$, $y=4$), the following number of cycles, in dependence of ω, are required in order to achieve 6-digit accuracy (1 cycle = n individual steps, i.e., each variable is corrected *once*):

ω	cycles	ω	cycles
1	150000	1.98	1000
1.5	50000	1.9802	800
1.8	16000	1.99	1600
1.9	8000	1.995	3500
1.95	4000	1.999	14000
1.97	1500	1.9999	150000

The optimal ω equals 1.9802, for which 800 cycles are required.

Programming technique. Since in each step one needs only one component of **r**, the procedure *op* must be modified somewhat. Let

> **real procedure** *axj*(*n*,*j*,*x*);
>   **value** *n*,*j*;
>   **integer** *n*,*j*; **array** *x*;
>   **begin**
>   .
>   .
>   .
>   **end**;

be a procedure which computes the *j*th component of **Ax**, and $d[j]$ be the *j*th diagonal element of **A**. Then one can program as follows:

```
        for j := 1 step 1 until n do x[j] := 0;
reen:
        s := 0;
        for j := 1 step 1 until n do
        begin
            rj := b[j] + axj(n,j,x);
            s := s + rj × rj;
            x[j] := x[j] − omega × rj/d[j];
        end;
        if s > eps then goto reen;
```

§10.5. The method of conjugate gradients

The method of steepest descent, already written off as inadequate, can surprisingly be improved as follows: After one has arrived at the minimum point x_k (coming from x_{k-1} along the straight line $x_{k-1} + th_{k-1}$), one does not simply seek the minimum along the gradient r_k emanating from x_k, but rather the minimum in the whole plane E_k spanned by the vectors $-r_k$ and h_{k-1} which pass through the point x_k (see Fig. 10.11).

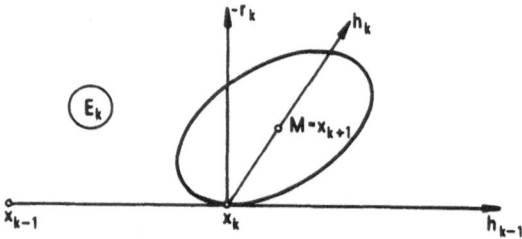

Figure 10.11. *Choice of the new relaxation direction h_k conjugate to h_{k-1}*

If we consider the curves in which this plane E_k intersects the level surfaces $F = $ const, we find that these are concentric ellipses, one of which touches the vector h_{k-1} at x_k, and whose common center M is the desired minimum of F on E_k. In order to reach this center, one chooses at x_k a new relaxation direction in the plane E_k conjugate to h_{k-1} (because this conjugate direction passes through M):

$$h_k = -r_k + \varepsilon_{k-1}h_{k-1} \quad (k \neq 0), \tag{22}$$

$$(h_k, Ah_{k-1}) = 0. \tag{23}$$

Through substitution of (22) in (23) one obtains

$$\varepsilon_{k-1} = \frac{(r_k, Ah_{k-1})}{(h_{k-1}, Ah_{k-1})}, \tag{24}$$

which determines the relaxation direction (22). In this direction $\mathbf{h}_k$ one travels to the minimum point, which necessarily is the center of the ellipse, and thus puts

$$\mathbf{x}_{k+1} = \mathbf{x}_k + \lambda_k \mathbf{h}_k, \tag{25}$$

where, by (13),

$$\lambda_k = - \frac{(\mathbf{h}_k, \mathbf{r}_k)}{(\mathbf{h}_k, A\mathbf{h}_k)}. \tag{26}$$

With this, a step would now be completed, except that in the new point $\mathbf{x}_{k+1}$ one still has to determine the residual $\mathbf{r}_{k+1}$. For this, one obtains from (25):

$$A\mathbf{x}_{k+1} + \mathbf{b} = A\mathbf{x}_k + \mathbf{b} + \lambda_k A\mathbf{h}_k,$$

or

$$\mathbf{r}_{r+1} = \mathbf{r}_k + \lambda_k A\mathbf{h}_k. \tag{27}$$

Since the point $\mathbf{x}_{k+1}$ furnishes the minimum of the function $F(\mathbf{x})$ in the plane E_k, the gradient at this point, that is $\mathbf{r}_{k+1}$, must be perpendicular to E_k:

$$(\mathbf{r}_{k+1}, \mathbf{r}_k) = (\mathbf{r}_{k+1}, \mathbf{h}_k) = (\mathbf{r}_{k+1}, \mathbf{h}_{k-1}) = 0. \tag{28}$$

On the basis of this orthogonality, the formulae (24) and (26) can still be simplified somewhat. With (22) one first gets

$$(\mathbf{r}_k, \mathbf{h}_k) = - ||\mathbf{r}_k||^2 + \varepsilon_{k-1}(\mathbf{r}_k, \mathbf{h}_{k-1}) = - ||\mathbf{r}_k||^2. \tag{29}$$

We thus obtain in place of (26):

$$\lambda_k = \frac{||\mathbf{r}_k||^2}{(\mathbf{h}_k, A\mathbf{h}_k)}. \tag{30}$$

Then, from (27) and (28), there follows

$$\lambda_{k-1}(\mathbf{r}_k, \mathbf{A}\mathbf{h}_{k-1}) = (\mathbf{r}_k, \lambda_{k-1}\mathbf{A}\mathbf{h}_{k-1}) = (\mathbf{r}_k, \mathbf{r}_k - \mathbf{r}_{k-1}) = ||\mathbf{r}_k||^2 \ ;$$

by (24), (26) and (29), however,

$$\varepsilon_{k-1} = \frac{\lambda_{k-1}(\mathbf{r}_k, \mathbf{A}\mathbf{h}_{k-1})}{||\mathbf{r}_{k-1}||^2} \ ,$$

so that finally

$$\varepsilon_{k-1} = \frac{||\mathbf{r}_k||^2}{||\mathbf{r}_{k-1}||^2} \ . \tag{31}$$

As is shown by the new formulae (30), (31), all coefficients λ_k, ε_k are necessarily *positive*.

The whole computing process is started with an arbitrary vector $\mathbf{x}_0$, for which one computes $\mathbf{r}_0 = \mathbf{A}\mathbf{x}_0 + \mathbf{b}$ and chooses $\mathbf{h}_0 = -\mathbf{r}_0$ as first relaxation direction. Thereafter, this *method of conjugate gradients* proceeds according to the following instructions:

for $k := 0$ step 1 until m do
begin
    if $k \neq 0$ then begin comment *evaluate here formulae* (31), (22) end;
    comment *compute* Ah_k;
    comment *evaluate in turn formulae* (30), (25), (27);
end;

As is seen, only the products $\mathbf{A}\mathbf{x}_0$, $\mathbf{A}\mathbf{h}_k$ ($k = 0, 1, \ldots$) are needed here; one can thus apply the operator principle.

Special properties of the method of conjugate gradients. The most important property of this computing process, discovered by E. Stiefel and M.R. Hestenes[1], is

[1] Hestenes M.R., Stiefel E.: Methods of conjugate gradients for solving linear systems, *J. Res. Nat. Bur. Standards* **49**, 409–436 (1952). Cf. also Engeli M., Ginsburg Th., Rutishauser H., Stiefel E.: *Refined Iterative Methods for Computation of the Solution and the Eigenvalues of Self-Adjoint Boundary Value Problems*, Mitt. Inst. f. angew. Math. ETH Zürich, Nr. 8, Birkhäuser, Basel 1959.

$$(\mathbf{r}_i, \mathbf{r}_j) = 0, \quad (\mathbf{h}_i, \mathbf{A}\mathbf{h}_j) = 0 \quad \text{for} \quad i \neq j, \tag{32}$$

that is, the residuals are orthogonal and the relaxation directions conjugate.

Proof by mathematical induction: One assumes that (32) is valid for $i, j \leq k$. It is assumed, further, that $\mathbf{r}_0, \ldots, \mathbf{r}_k \neq 0$.

For $i, j \leq k = 1$, (32) is true, because $(\mathbf{r}_0, \mathbf{r}_1) = 0$ by (28) and $(\mathbf{h}_1, \mathbf{A}\mathbf{h}_0) = 0$ by (23). What needs to be proved, therefore, is that the induction hypothesis implies

$$(\mathbf{r}_{k+1}, \mathbf{r}_j) = 0 \quad (j = 0, \ldots, k), \tag{33}$$

$$(\mathbf{h}_{k+1}, \mathbf{A}\mathbf{h}_j) = 0 \quad (j = 0, \ldots, k). \tag{34}$$

For $j = k$, (33) follows directly from (28), and (34) from (23). In the case $j < k$ we make use of (27), (22) and the induction hypothesis:

$$\begin{aligned}
(\mathbf{r}_{k+1}, \mathbf{r}_j) &= (\mathbf{r}_k, \mathbf{r}_j) + \lambda_k (\mathbf{A}\mathbf{h}_k, \mathbf{r}_j) \\
&= 0 + \lambda_k (\mathbf{A}\mathbf{h}_k, -\mathbf{h}_j + \varepsilon_{j-1} \mathbf{h}_{j-1}) \\
&= -\lambda_k (\mathbf{A}\mathbf{h}_k, \mathbf{h}_j) + \lambda_k \varepsilon_{j-1} (\mathbf{A}\mathbf{h}_k, \mathbf{h}_{j-1}) = 0
\end{aligned}$$

(for $j = 0$ the term with $\mathbf{h}_{j-1}$ is absent). This proves (33). Furthermore, from (22) and the assumption, there follows

$$(\mathbf{h}_{k+1}, \mathbf{A}\mathbf{h}_j) = -(\mathbf{r}_{k+1}, \mathbf{A}\mathbf{h}_j) + \varepsilon_k (\mathbf{h}_k, \mathbf{A}\mathbf{h}_j) = -(\mathbf{r}_{k+1}, \mathbf{A}\mathbf{h}_j).$$

Because of $\mathbf{r}_j \neq 0$, one has $\lambda_j \neq 0$, thus by (27) and (33),

$$\mathbf{A}\mathbf{h}_j = \frac{1}{\lambda_j} (\mathbf{r}_{j+1} - \mathbf{r}_j),$$

$$(\mathbf{h}_{k+1}, \mathbf{A}\mathbf{h}_j) = -(\mathbf{r}_{k+1}, \mathbf{r}_{j+1} - \mathbf{r}_j) \frac{1}{\lambda_j} = 0; \qquad \text{q.e.d.}$$

What is happening, therefore, is the following: r_0, r_1, ..., r_k are mutually orthogonal and either $r_{k+1} = 0$ or r_{k+1} is also perpendicular to r_0, ..., r_k. For $k + 1 = n$ at the latest, however, the first case must occur, that is, one has $r_n = 0$ at the latest, and thus in x_n the desired solution.

The method of conjugate gradients thus yields (theoretically) the solution of the system of equations after at most n steps as x_n. It is therefore, on the one hand, a relaxation method which in each step reduces the function F, but on the other hand, the solution, as in elimination methods, is obtained in a finite number of steps (without, however, the matrix A as such being needed). This means that the so-called iterative methods and the direct methods are not mutually exclusive entities.

Now it is true that this interesting property is considerably disturbed by rounding errors. The inner products (r_i, r_j) in practice do not become 0 exactly, especially not if i and j lie far apart. As a consequence, r_n will not vanish, and in fact may not even be very small. If this happens, one simply goes on computing, without worrying too much.

Example. We again solve the simple system of equations

$$137\, x - 100\, y - 11 = 0$$
$$-100\, x + 73\, y + 8 = 0$$

and start at the point $x_0 = [0,0]^T$, for which $r_0 = [-11,8]^T$, $||r_0||^2 = 185$. One further obtains:

$$h_0 = [11,-8]^T, \quad Ah_0 = [2307,-1684]^T,$$

$$(h_0, Ah_0) = 38849, \quad \lambda_0 = .004762027,$$

$$x_1 = [.05238230, -.03809622]^T,$$

$$r_1 = [-.01400400, -.01925300]^T,$$

$$||r_1||^2 = .0005667900, \quad \varepsilon_0 = .000003063730,$$

$$h_1 = [.01403770, .01922849]^T,$$

$$Ah_1 = [.0003160, -.000090]^T,$$

$$(h_1, Ah_1) = .000002705349, \quad \lambda_1 = 209.5072,$$

$$x_2 = [2.993382, 3.990411]^T,$$

$$r_2 = [.05220028, -.03810865]^T.$$

The computation of r_1 and Ah_1 was subject to severe cancellation. r_2 should be equal to 0, but we even have $||r_2|| > ||r_1||$. We therefore continue:

$$||r_2||^2 = .004177138, \quad \varepsilon_1 = 7.369816,$$

$$h_2 = [.05125500, .1798191]^T,$$

$$Ah_2 = [-10.95998, 8.001295]^T,$$

$$(h_2, Ah_2) = .8770320, \quad \lambda_2 = .004762811,$$

$$x_3 = [2.993626, 3.991267]^T,$$

$$r_3 = [-3_{10}-8, 10-8]^T.$$

At first sight, x_3 does not appear to be better than x_2. But in reality, x_3 is at least at the bottom of the valley which the function $F(x)$ forms in three-dimensional space. $F(x_3)$ lies only about $2.8_{10}-7$ above the minimum value, in contrast to $F(x_2)$, which exceeds it by $2_{10}-5$. Also, r_3 is noticeably shorter than r_2.

§10.6. Application to a more complicated problem

To be computed is the deflection $u(x,y)$ of a square plate, clamped at all four sides. If $p(x,y)$ denotes the load on the plate at the point (x,y), this deflection satisfies:

$$\Delta^2 u = p(x,y) \quad \text{in the interior,} \tag{35}$$

$$u = 0, \quad \frac{\partial u}{\partial n} = 0 \quad \text{on the boundary.} \tag{36}$$

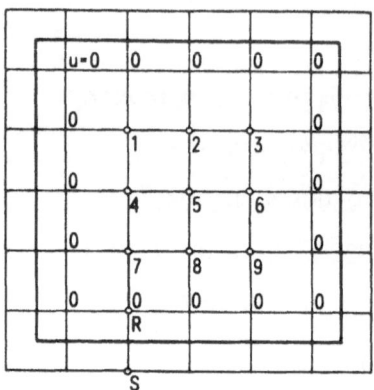

Figure 10.12. *Discretization for the problem of the clamped plate*

Exploiting the boundary conditions, we lay a dual net, that is a net whereby the boundary of the square does not run along mesh lines, but bisects them (see Fig. 10.12). Then $u=0$ on the boundary means $u(R) + u(S) = 0$, while $\partial u/\partial n = 0$ means $u(R) - u(S) = 0$, so that we have to put u equal to zero at all grid points in the neighborhood of the boundary. There so remain only 9 interior points in which the differential equation (35) has to be applied. The Laplace operator

$$\Delta = \frac{\partial^2}{\partial x^2} + \frac{\partial^2}{\partial y^2} \; ,$$

as we know, is approximated on the grid by the operator of Fig. 10.5, multiplied by $1/h^2$. For the biharmonic operator

$$\Delta^2 = \frac{\partial^4}{\partial x^4} + 2\,\frac{\partial^4}{\partial x^2 \partial y^2} + \frac{\partial^4}{\partial y^4} \; ,$$

we have to apply it twice. To that end, we choose in the u-field around the value u_P the following notations:

$$u_{NN}$$

$$u_{NW} \quad u_N \quad u_{NE}$$

$$u_{WW} \quad u_W \quad u_P \quad u_E \quad u_{EE}$$

$$u_{SW} \quad u_S \quad u_{SE}$$

$$u_{SS}$$

In the Δu-field one then has at the points N, W, P, E, S the values (apart from the factor $1/h^2$):

$$4u_N - u_P - u_{NW} - u_{NN} - u_{NE}$$

$$4u_W - u_P - u_{SW} - u_{WW} - u_{NW} \quad 4u_P - u_W - u_N - u_E - u_S \quad 4u_E - u_P - u_{NE} - u_{EE} - u_{SE}$$

$$4u_S - u_P - u_{SE} - u_{SS} - u_{SW}$$

Consequently, one has in P approximately

$$\Delta^2 u \approx \frac{1}{h^4}\,(20u_P - 8u_N - 8u_W - 8u_E - 8u_S + 2u_{NW} + 2u_{NE} + 2u_{SW} + 2u_{SE} + u_{NN} + u_{WW} + u_{EE} + u_{SS}),$$

that is, $h^4 \Delta^2$ is to be replaced on the grid by the operator of Fig. 10.13.

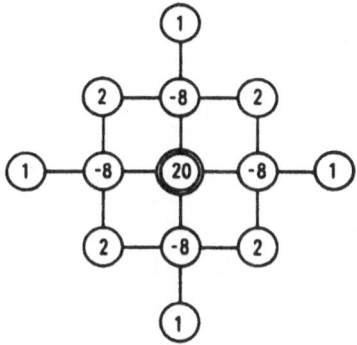

Figure 10.13. *Difference operator for the plate problem*

(One could have obtained this result also in a simpler way, namely by application of the operator of Fig. 10.5 onto itself.)

In this way one obtains for our problem the system of equations (p_k is the load at the point k):

u_1	u_2	u_3	u_4	u_5	u_6	u_7	u_8	u_9	1	
20	−8	1	−8	2	0	1	0	0	$-h^4 p_1$	= 0
−8	20	−8	2	−8	2	0	1	0	$-h^4 p_2$	= 0
1	−8	20	0	2	−8	0	0	1	$-h^4 p_3$	= 0
−8	2	0	20	−8	1	−8	2	0	$-h^4 p_4$	= 0
2	−8	2	−8	20	−8	2	−8	2	$-h^4 p_5$	= 0
0	2	−8	1	−8	20	0	2	−8	$-h^4 p_6$	= 0
1	0	0	−8	2	0	20	−8	1	$-h^4 p_7$	= 0
0	1	0	2	−8	2	−8	20	−8	$-h^4 p_8$	= 0
0	0	1	0	2	−8	1	−8	20	$-h^4 p_9$	= 0

(37)

The matrix here is fairly dense, but if the grid is refined, the zeros begin to predominate. In the general case, the matrix has the form

$$
\begin{bmatrix}
A & B & I & & & 0 \\
B & A & B & I & & \\
I & B & A & B & I & \\
 & I & B & A & B & I \\
 & & \cdot & \cdot & \cdot & \cdot & \cdot \\
0 & & & \cdot & \cdot & \cdot & \cdot
\end{bmatrix}
\tag{38}
$$

with

$$
A =
\begin{bmatrix}
20 & -8 & 1 & & & 0 \\
-8 & 20 & -8 & 1 & & \\
1 & -8 & 20 & -8 & 1 & \\
 & 1 & -8 & 20 & -8 & 1 \\
 & & \cdot & \cdot & \cdot & \cdot & \cdot \\
0 & & \cdot & \cdot & \cdot & \cdot & \cdot
\end{bmatrix},
\tag{39}
$$

$$\mathbf{B} = \begin{bmatrix} -8 & 2 & & & & 0 \\ 2 & -8 & 2 & & & \\ & 2 & -8 & 2 & & \\ & & 2 & -8 & 2 & \\ & & & \ddots & \ddots & \ddots \\ & & & & \ddots & \ddots & \ddots \\ 0 & & & & \ddots & \ddots & \ddots \end{bmatrix}.$$ (40)

But here too, it is better not to write down the equations, but instead indicate for each point the operator type and the neighbors:

	Op	N	E	S	W	NN	NE	EE	SE	SS	SW	WW	NW
point 1	1	0	2	4	0	0	0	3	5	7	0	0	0
point 2	1	0	3	5	1	0	0	0	6	8	4	0	0
.	.	.											
.	.	.	.										
.	.	.	.										
point 9	1	6	0	0	8	3	0	0	0	0	0	7	5
oper. 1	20	-8	-8	-8	-8	1	2	1	2	1	2	1	2

(The last row of the tableau contains the operator definition.) For large grids, this involves a significant reduction in data. Of course, it would suffice to indicate the immediate neighbors N, E, S, W, but during computation one would then lose too much time with searching this list.

If the plate is not clamped everywhere, matters become significantly more complicated. One finds the desired deflection u generally as solution of the following minimum problem:

$$\frac{1}{2} \int_{\Omega} (\Delta u)^2 \, dx dy + (1 - \mu) \int_{\Omega} \left[\left(\frac{\partial^2}{\partial x \partial y} \right)^2 - \frac{\partial^2 u}{\partial x^2} \frac{\partial^2 u}{\partial y^2} \right] dx dy$$

$$- 12 \frac{1 - \mu^2}{Ed^3} \int_{\Omega} up \, dx dy = \text{minimum}.$$ (41)

The integral

$$\int_{\Omega} (u_{xy}^2 - u_{xx} u_{yy}) dx dy$$

can here be transformed into the line integral

$$I_2 = \oint_{\partial Q} u_y\, du_x = \oint_{\partial Q} u_y(u_{xx}\, dx + u_{xy}\, dy) \tag{42}$$

(extended over the boundary ∂Q of the domain). When the whole boundary is clamped, this term drops out, since then $u_x = u_y = 0$ along the boundary.

We now treat in particular a plate which is clamped at left, simply supported at right, and is free on top and bottom (see Fig. 10.14).

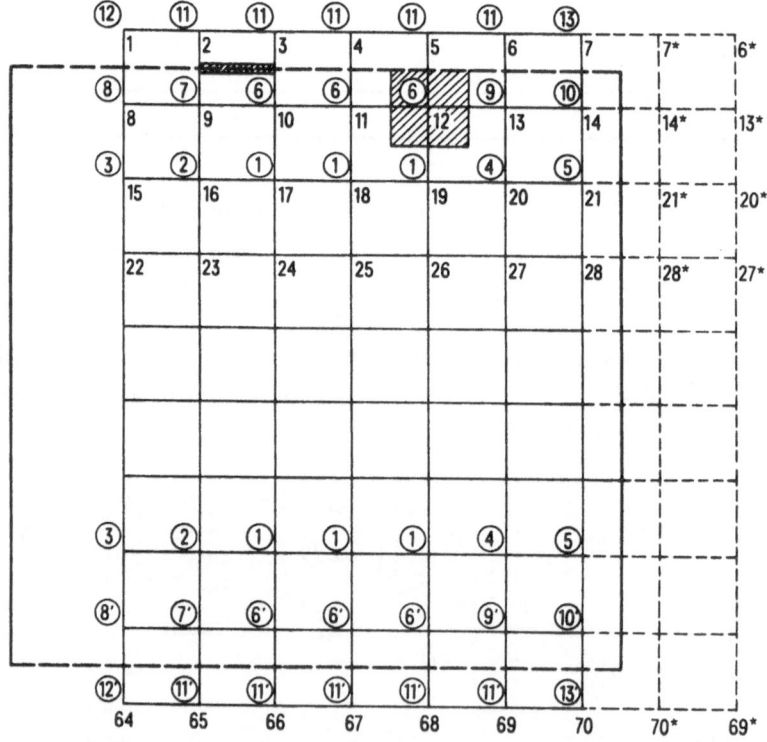

Figure 10.14. *Discretization for a special plate problem*

To the integral

$$I_1 = \frac{1}{2} \int_{Q} (\Delta u)^2 \, dx dy, \tag{43}$$

the square with center at 12 and the neighboring squares yield the following contributions:

$$\frac{1}{2} h^2 (\Delta u)^2 \bigg|_{12} \approx \frac{1}{2h^2} (4u_{12} - u_5 - u_{11} - u_{13} - u_{19})^2,$$

$$\frac{1}{2} h^2 (\Delta u)^2 \bigg|_{11} \approx \frac{1}{2h^2} (4u_{11} - u_4 - u_{10} - u_{12} - u_{18})^2,$$

$$\frac{1}{2} h^2 (\Delta u)^2 \bigg|_{13} \approx \frac{1}{2h^2} (4u_{13} - u_6 - u_{12} - u_{14} - u_{20})^2,$$

$$\frac{1}{2} h^2 (\Delta u)^2 \bigg|_{19} \approx \frac{1}{2h^2} (4u_{19} - u_{12} - u_{18} - u_{20} - u_{26})^2.$$

These are the only squares in which the value u_{12} occurs. (Around the point 5 there is no square, since this point lies outside the plate.) One therefore has

$$\frac{\partial I_1}{\partial u_{12}} \approx \frac{1}{h^2} (19u_{12} - 4u_5 - 8u_{11} - 8u_{13} - 8u_{19} + u_4 + u_6 + u_{10} + u_{14}$$
$$+ 2u_{18} + 2u_{20} + u_{26}). \tag{44}$$

Analogously for u_{19} (in this case the squares with the centers 12, 18, 19, 20, 26 contribute):

$$\frac{\partial I_1}{\partial u_{19}} \approx \frac{1}{h^2} (20u_{19} - 8u_{12} - 8u_{18} - 8u_{20} - 8u_{26} + u_5 + 2u_{11} + 2u_{13}$$
$$+ u_{17} + u_{21} + 2u_{25} + 2u_{27} + u_{33}). \tag{45}$$

u_5, however, occurs only in the contribution of the square with center 12:

$$\frac{\partial I_1}{\partial u_5} \approx \frac{1}{h^2} \, (-4u_{12} + u_5 + u_{11} + u_{13} + u_{19}). \qquad (46)$$

Thus, in the uppermost sequence of grid points (with the exception of the points 1 and 7), the operator on the left in Fig. 10.15 is valid, while in the second sequence from the top, according to (44), it is the one on the right in that figure (exceptions: 8, 9, 13, 14). In the interior, according to (45), one uses the normal discrete biharmonic operator of Fig. 10.13.

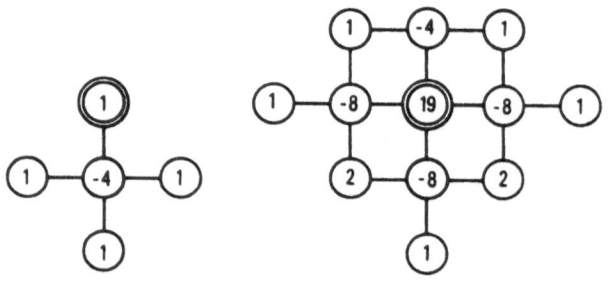

Figure 10.15. *Operators for the two top grid sequences*

The same operators would be valid also at the left and right boundary. One must only note that $u=0$ to the left of the sequence 1–64, and that the u-values in the sequence 7*–70* are the negatives of those in the sequence 7–70. (An analogous statement holds also for the sequence 6*–69*.) In this way one finds for the points 15, 16, 20, 21 the operators depicted in Fig. 10.16 (as always up to the factor $1/h^2$).

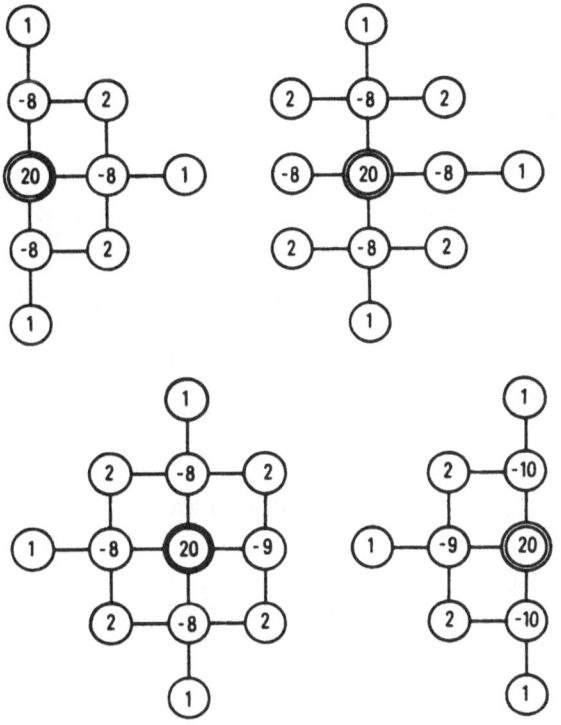

Figure 10.16. *Operators for the left and right boundary*

Now a corrective treatment is still required to take account of the term $(1 - \mu)I_2$. Since on the vertical boundaries, $u \equiv 0$, hence $u_y = 0$, there remain in I_2 only the integrals over the upper boundary Γ_1 and the lower boundary Γ_2:

$$
\begin{aligned}
I_2 &= - \int_{\Gamma_1} u_y u_{xx} dx + \int_{\Gamma_2} u_y u_{xx} dx \\
&= - u_y u_x \bigg|_{\partial \Gamma_1} + \int_{\Gamma_1} u_{xy} u_x dx + u_x u_y \bigg|_{\partial \Gamma_2} - \int_{\Gamma_2} u_{xy} u_x dx \\
&= \frac{1}{2} \frac{\partial}{\partial y} \int_{\Gamma_1} u_x^2 dx - \frac{1}{2} \frac{\partial}{\partial y} \int_{\Gamma_2} u_x^2 dx.
\end{aligned}
$$

Now for the hatched boundary piece in Fig. 10.14, for example, one has

$$\frac{\partial}{\partial y} \int u_x^2 dx \approx \frac{1}{h}\left[\int_2^3 u_x^2 dx - \int_9^{10} u_x^2 dx\right] \approx \frac{1}{h^2}\left[(u_3 - u_2)^2 - (u_{10} - u_9)^2\right].$$

Altogether, therefore:

$$I_2 \approx \frac{1}{2h^2}\left[u_1^2 + (u_2 - u_1)^2 + (u_3 - u_2)^2 + (u_4 - u_3)^2 + (u_5 - u_4)^2 + (u_6 - u_5)^2\right.$$

$$+ (u_7 - u_6)^2 + \frac{1}{2}(u_{7*} - u_7)^2 - u_8^2 - (u_9 - u_8)^2 - (u_{10} - u_9)^2$$

$$- (u_{11} - u_{10})^2 - (u_{12} - u_{11})^2 - (u_{13} - u_{12})^2 - (u_{14} - u_{13})^2$$

$$\left. - \frac{1}{2}(u_{14*} - u_{14})^2\right] + \text{analogous contributions from } \Gamma_2,$$

where $u_{7*} = - u_7$, $u_{14*} = - u_{14}$. Furthermore:

$$h^2 \frac{\partial I_2}{\partial u_1} \approx 2u_1 - u_2, \quad h^2 \frac{\partial I_2}{\partial u_2} \approx - u_1 + 2u_2 - u_3, \dots,$$

$$h^2 \frac{\partial I_2}{\partial u_6} \approx - u_5 + 2u_6 - u_7, \quad h^2 \frac{\partial I_2}{\partial u_7} \approx 3u_7 - u_6,$$

$$h^2 \frac{\partial I_2}{\partial u_8} \approx - 2u_8 + u_9, \quad h^2 \frac{\partial I_2}{\partial u_9} \approx u_8 - 2u_9 + u_{10}, \dots,$$

$$h^2 \frac{\partial I_2}{\partial u_{13}} \approx u_{12} - 2u_{13} + u_{14}, \quad h^2 \frac{\partial I_2}{\partial u_{14}} \approx - 3u_{14} + u_{13}.$$

If we assume, say $\mu = .167$, we must thus take account of the term $.833 I_2$ in (41). This requires that (46) and (44) be supplemented by

$$.833 \frac{\partial I_2}{\partial u_5} \approx - .833u_4 + 1.666u_5 - .833u_6,$$

$$.833\ \frac{\partial I_2}{\partial u_{12}} \approx .833u_{11} - 1.666u_{12} + .833u_{13}.$$

It is by these amounts that the operators at the upper boundary are to be corrected. Their new form can be seen from Fig. 10.17: above on the left is depicted the operator for the point 1, to the right the one for the points 2–6, below on the left the operator for the point 7, and to the right the one for 10, 11, 12.

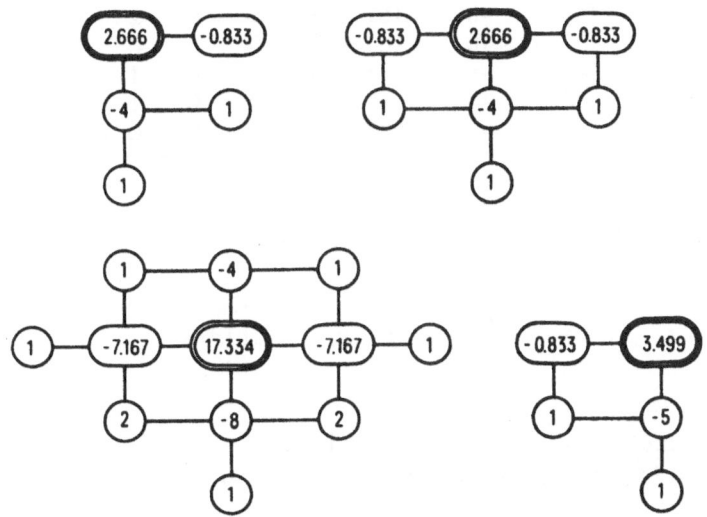

Figure 10.17. *Corrected operators for the upper boundary*

The operators belonging to the points 8, 9, 13, 14 can easily be determined analogously. The operators for the points at the lower boundary are of course obtained immediately by reflection. As a summary, Table 10.1 contains a complete list of the 13 essentially different operators (up to the factor $1/h^2$) occurring in the problem. In Fig. 10.14 the encircled numbers then indicate which operator belongs to what point, while the numbers $6', 7', \ldots, 13'$ denote the operators $6, 7, \ldots, 13$ reflected in the north-south direction.

Table 10.1. *Complete list of the operators of a special plate problem*

Op.	P	N	E	S	W	NN	NE	EE	SE	SS	SW	WW	NW
1	20	-8	-8	-8	-8	1	2	1	2	1	2	1	2
2	20	-8	-8	-8	-8	1	2	1	2	1	2	0	2
3	20	-8	-8	-8	0	1	2	1	2	1	0	0	0
4	20	-8	-9	-8	-8	1	2	0	2	1	2	1	2
5	28	-10	0	-10	-9	1	0	0	0	1	2	1	2
6	17.334	-4	-7.167	-8	-7.167	0	1	1	2	1	2	1	1
7	17.334	-4	-7.167	-8	-7.167	0	1	1	2	1	2	0	1
8	17.334	-4	-7.167	-8	0	0	1	1	2	1	0	0	0
9	17.334	-4	-8.167	-8	-7.167	0	1	0	2	1	2	1	1
10	24.501	-5	0	-10	-8.167	0	0	0	0	1	2	1	1
11	2.666	0	-.833	-4	-.833	0	0	0	1	1	1	0	0
12	2.666	0	-.833	-4	0	0	0	0	1	1	0	0	0
13	3.499	0	0	-5	-.833	0	0	0	0	1	1	0	0

In the total energy (41) there is still the term

$$I_3 = -12 \frac{1-\mu^2}{Ed^3} \int_\varrho p(x,y)u(x,y)\,dx\,dy, \qquad (47)$$

where $p(x,y)$ means the load. Since we have multiplied the other terms (arbitrarily) by h^2, we must do the same here and find

$$h^2 I_3 \approx -\gamma \sum_{k=1}^{70} p_k u_k \quad \text{with} \quad \gamma = \frac{12(1-\mu^2)}{Ed^3} h^4.$$

Differentiation yields

$$h^2 \frac{\partial I_3}{\partial u_k} \approx -\gamma p_k.$$

These values $-\gamma p_k$ ($k = 1, \ldots, 70$) form the vector **b** of the system of equations, while the operators define the matrix **A**. Note, however, that $p_1, \ldots, p_7, p_{64}, \ldots, p_{70}$ must all be equal to 0.

§10.7. Remarks on norms and the condition of a matrix

Norms generalize the notion of absolute value to vectors and matrices. They serve, among other things, to estimate vectors and matrices in iterative processes.

We consider here only *vector norms* of the form

$$||x||_p = \left[\sum_{i=1}^{n} |x_i|^p \right]^{1/p} \tag{48}$$

with $p \geq 1$, so-called *Hölder norms*. Specifically, this norm is called in the case

$$p = 1: \; L_1\text{-norm}$$
$$p = 2: \; \text{Euclidean norm,}$$
$$p = \infty: \; \text{maximum norm.}$$

$p = \infty$ is a limit case, for which (48) becomes

$$||x||_\infty = \max_{1 \leq i \leq n} |x_i|. \tag{49}$$

The following is valid in all these cases:

$$||x|| > 0 \text{ for } x \neq 0,$$
$$||kx|| = |k| \; ||x|| \; (k \text{ arbitrary scalar}), \tag{50}$$
$$||x + y|| \leq ||x|| + ||y||.$$

The sets $\{x| \; ||x||_p \leq 1\}$ in $\mathbf{R}^n$ have the following geometric meaning:

$$p = 1: \; \text{hyperoctahedron,}$$
$$p = 2: \; \text{hypersphere,}$$
$$p = \infty: \; \text{hypercube.}$$

Matrix norms could be defined independently. We restrict ourselves, however, to subordinate matrix norms: one calls

$$||A||_p = \max_{x \neq 0} \frac{||Ax||_p}{||x||_p} \tag{51}$$

the matrix norm subordinate to the vector norm $||\cdot||_p$.

The characteristic properties of vector norms are transmitted to
matrix norms:

$$
\begin{aligned}
&||A|| > 0, \quad ||A|| = 0 \text{ only for } A = 0, \\
&||kA|| = |k| \ ||A|| \quad (k \text{ arbitrary scalar}), \\
&||A+B|| \leq ||A|| + ||B||,
\end{aligned}
\tag{52}
$$

but in addition, one has

$$||AB|| \leq ||A|| \ ||B||, \tag{53}$$

since

$$\frac{||ABx||}{||x||} = \frac{|A(Bx)|}{||Bx||} \ \frac{||Bx||}{||x||} \leq \left[\max_{y \neq 0} \frac{||Ay||}{||y||} \right] \left[\max_{x \neq 0} \frac{||Bx||}{||x||} \right].$$

Normally, there holds strict inequality, since the two maxima are not
attained simultaneously as a rule.

In the three cases $p = 1, 2, \infty$ one can compute $||A||_p$ directly:

Case $p = 2$:

$$||Ax||_2^2 = (Ax, Ax) = (x, A^T Ax),$$

thus,

$$\frac{||Ax||_2^2}{||x||_2^2} = \max_{x \neq 0} \frac{(x, A^T Ax)}{(x, x)} , \tag{54}$$

which is the maximum Rayleigh quotient of $A^T A$. Now it is known, however, that the maximum Raleigh quotient of a symmetric matrix is equal to the largest eigenvalue of this matrix[1]. Since $A^T A$ is positive definite, one thus has

$$|| A ||_2 = \sqrt{\lambda_{max}(A^T A)} \ . \tag{55}$$

Because of

$$\lambda_{max}(A^T A) \le \text{trace } (A^T A) = \sum_{i=1}^{n} \sum_{j=1}^{n} |a_{ij}|^2,$$

there further follows the estimate

$$|| A ||_2 \le \left[\sum_{i=1}^{n} \sum_{j=1}^{n} |a_{ij}|^2 \right]^{1/2} . \tag{56}$$

The bound on the right is the so-called *Schur norm* of A[2].

Case p = 1:

$$|| Ax ||_1 = \sum_{i=1}^{n} | \sum_{j=1}^{n} a_{ij} x_j | \le \sum_i \sum_j |a_{ij}| \, |x_j| = \sum_j |x_j| \sum_i |a_{ij}|$$
$$\le (\max_j \sum_i |a_{ij}|) \sum_j |x_j| = || x ||_1 \max_j \sum_i |a_{ij}|.$$

The bound is in fact attained, namely for

$$x_i = 0 \ (i \ne j), \ x_j = 1,$$

if the *j*th column yields the largest sum. We thus have

---

[1] See, for example, Schwarz H.R., Rutishauser H., Stiefel E.: *Numerical Analysis of Symmetric Matrices*, Prentice-Hall, Englewood Cliffs, N.J., 1973, Theorem 4.3. (Editors' remark)

[2] Also called Frobenius norm. (Translator's remark)

$$||\mathbf{A}||_1 = \max_{1 \le j \le n} \sum_{i=1}^{n} |a_{ij}|, \tag{57}$$

that is, $||\mathbf{A}||_1$ is equal to the largest "column sum".

Case $p = \infty$:

$$||\mathbf{Ax}||_\infty = \max_i |\sum_j a_{ij}x_j| \le \max_i \sum_j |a_{ij}| \, |x_j|$$

$$\le (\max_j |x_j|)(\max_i \sum_j |a_{ij}|) = ||\mathbf{x}||_\infty \max_i \sum_j |a_{ij}|.$$

Since here, too, the bound is attained when all x_j have modulus 1 and appropriate sign, one finds:

$$||\mathbf{A}||_\infty = \max_{1 \le i \le n} \sum_{j=1}^{n} |a_{ij}|, \tag{58}$$

that is, $||\mathbf{A}||_\infty$ is equal to the largest "row sum".

Example. For the matrix

$$\mathbf{A} = \begin{bmatrix} 1 & 10 \\ 0 & 1 \end{bmatrix}$$

one obtains $||\mathbf{A}||_1 = ||\mathbf{A}||_\infty = 11$. Furthermore,

$$\mathbf{A}^T\mathbf{A} = \begin{bmatrix} 1 & 10 \\ 10 & 101 \end{bmatrix}, \quad \lambda_{\max}(\mathbf{A}^T\mathbf{A}) = 51 + 10\sqrt{26} = 101.990195,$$

$$||\mathbf{A}||_2 = \sqrt{\lambda_{\max}} = 5 + \sqrt{26} = 10.0990195.$$

Already the estimate (56) with the Schur norm gives a good bound here, namely $\sqrt{102} = 10.099505$.

The *condition number* plays an important role in the solution of linear systems of equations. We are guided by the following idea: The solution x of the system $Ax + b = 0$ evidently cannot be determined more accurately than is permitted by the inaccuracy in the computation of Ax near the solution.

If the elements of the matrix A in each row have about the same order of magnitude, one can assume, as a rough approximation, that the computation of Ax, and hence also of $r = Ax + b$, is falsified by a δr of the order of magnitude

$$||\delta r||_2 \approx \theta ||A||_2 ||x||_2,$$

where θ is a unit in the last position of the mantissa of the computer (cf. also Appendix §A3.4). But if r cannot be determined more accurately than with such an error δr, then also $x = A^{-1}(r - b)$ cannot be computed more accurately than with an error $\delta x = A^{-1}\delta r$, that is, one has, roughly,

$$||\delta x||_2 \approx ||A^{-1}||_2 ||\delta r||_2 \approx \theta ||A^{-1}||_2 ||A||_2 ||x||_2. \tag{59}$$

If we introduce

$$K = ||A||_2 ||A^{-1}||_2 \tag{60}$$

as condition number of A, then

$$||\delta x||_2 \approx \theta K ||x||_2. \tag{61}$$

Consequently, under these circumstances, one must expect a relative error θK, thus an inaccuracy of about K units in the last position of the mantissa for the vector x.

If A is symmetric and positive definite,

$$\lambda_{\max}(A^T A) = \lambda_{\max}(A^2) = (\lambda_{\max}(A))^2,$$

thus $||A||_2 = \lambda_{\max}(A)$; likewise, $||A^{-1}||_2 = 1/\lambda_{\min}(A)$, and hence

$$K = \frac{\lambda_{max}(\mathbf{A})}{\lambda_{min}(\mathbf{A})} . \tag{62}$$

Note: *All iterative methods for the solution of* $\mathbf{Ax} + \mathbf{b} = \mathbf{0}$ *converge more slowly the larger K.*

Examples. 1) The matrix

$$\mathbf{A} = \begin{bmatrix} 137 & -100 \\ -100 & 73 \end{bmatrix}$$

has the eigenvalues $\lambda_1 \approx 210$, $\lambda_2 \approx 1/210$, thus $K \approx 44100$. In §10.5, when we obtained with the method of conjugate gradients and 7-digit computation the result (2.993626, 3.991267) in place of (3,4), we can be satisfied; one could in no way expect anything better.

2) For the matrix of spline interpolation

$$\begin{bmatrix} 2 & 1 & & & & & & 0 \\ 1 & 4 & 1 & & & & & \\ & 1 & 4 & 1 & & & & \\ & & \cdot & \cdot & \cdot & & & \\ & & & \cdot & \cdot & \cdot & & \\ & & & & \cdot & \cdot & \cdot & \\ & & & & & 1 & 4 & 1 \\ 0 & & & & & & 1 & 2 \end{bmatrix},$$

as we have seen in §6.8, one has $\lambda_{min} > 1$, $\lambda_{max} < 6$, hence $K < 6$, that is, the condition is always good (independent of the order).

3) For the unit matrix $\mathbf{I}$ one has $\lambda_{min} = \lambda_{max} = 1$, $K = 1$, which is the best possible condition.

4) For the matrix

$$
\begin{bmatrix}
2 & -1 & & & & & & 0 \\
-1 & 2 & -1 & & & & & \\
 & -1 & 2 & -1 & & & & \\
 & & \cdot & \cdot & \cdot & & & \\
 & & & \cdot & \cdot & \cdot & & \\
 & & & & \cdot & \cdot & \cdot & \\
 & & & & & -1 & 2 & -1 \\
0 & & & & & & -1 & 2
\end{bmatrix}
$$

of order n one has ([3])

$$
\lambda_{max} = 4 \cos^2 \frac{\pi}{2n+2}, \quad \lambda_{min} = 4 \sin^2 \frac{\pi}{2n+2}, \quad K = \cot^2 \frac{\pi}{2n+2} \ .
$$

For large n, one gets approximately $K \approx 4n^2/\pi^2$. This is a moderately bad condition. More or less the same holds true for the matrix of the Dirichlet problem for a domain that is covered and discretized by an $n \times n$-grid.

5) For the matrix

$$
\begin{bmatrix}
37 & 5 & 12 & 2 \\
 & 62 & 58 & -1 \\
\text{sym.} & & 66 & 17 \\
 & & & 30
\end{bmatrix}
$$

one has $\lambda_{max} \approx 125$, $\lambda_{min} \approx 6.59_{10}-6$, $K \approx 19_{10}6$; it is thus extremely ill-conditioned.

6) The matrix of the plate problem of Fig. 10.14 has the eigenvalues $\lambda_{max} \approx 62$, $\lambda_{min} \approx .04$, that is, $K \approx 1550$, which leads to a loss of accuracy of 3–4 digits.

---

[3] See Zurmühl R.: *Matrizen*, 4th ed., Springer, Berlin 1964, pp. 229f. (Editors' remark)

7) For the nonsymmetric matrix

$$
A = \begin{bmatrix}
1 & 10 & & & & & & 0 \\
 & 1 & 10 & & & & & \\
 & & 1 & 10 & & & & \\
 & & & \cdot & \cdot & \cdot & & \\
 & & & & \cdot & \cdot & & \\
 & & & & & \cdot & \cdot & \\
 & & & & & & 1 & 10 \\
0 & & & & & & & 1
\end{bmatrix}
$$

of order n, one first obtains the rough estimate[4] $\lambda_{max}(A^T A) \approx 121$. Moreover,

$$
A^{-1} = \begin{bmatrix}
1 & -10 & 100 & -1000 & 10000 & \cdots & (-10)^{n-1} \\
 & 1 & -10 & 100 & -1000 & \cdots & (-10)^{n-2} \\
 & & 1 & -10 & 100 & \cdots & (-10)^{n-3} \\
 & & & \cdot & & & \\
 & & & & \cdot & & \\
 & & & & & \cdot & \\
0 & & & & & & 1
\end{bmatrix},
$$

hence $\lambda_{max}(A^{-T} A^{-1}) \approx 1.01 \times 10^{2n-2}$. Therefore, $K \approx 1.1 \times 10^n$, even though all eigenvalues of A are here equal to 1.

---

[4] Here the fact is used that every matrix norm, thus in particular (58), is an upper bound for the absolute values of the eigenvalues. This follows at once from (51), (52) and the definition of eigenvalues. (Editors' remark)

Notes to Chapter 10

A modern introduction which treats the main classes of methods available for approximately solving elliptic problems, including a description of the software package ELLPACK, is in Birkhoff & Lynch [1984].

§10.1 High-order approximation at curved boundaries is cumbersome in finite difference methods, cf. (6). A general polynomial extrapolation type method is described and analyzed in Pereyra, Proskurowski & Widlund [1977].

The "Energy Method" involves ideas similar to the so-called *Galerkin* (or *Ritz-Galerkin*) methods in which the energy functional in (7) is minimized over a finite-dimensional space of functions. The method is frequently used in its finite element setting, in which one first triangulates the domain into small pieces (finite elements) and then constructs the finite-dimensional function space to consist of (continuous) piecewise polynomials of a certain degree. (On a uniform triangulation, and with piecewise linear functions, this leads for the Laplace operator to the five-point formula in (4), when the functions are expressed in terms of their nodal values.) To a certain extent, boundary conditions at curved boundaries are easier to implement to high accuracy in finite element methods than in standard finite difference methods on a rectangular mesh. This is particularly so for natural boundary conditions. Also, in many problems, minimization of energy is a basic principle, and the resulting partial differential equation is derived from the Euler equation for that minimization problem. In such situations, it may be argued that Galerkin methods are closer to the physics of the problem than finite difference methods.

Much research in approximation of elliptic problems focuses on the selection of a suitable (nonuniform) mesh. This can be done either a priori, e.g. if the location of a singularity is known, or adaptively during the computation, with information drawn from one approximation being used to alter the mesh, after which a new, presumably better, approximation is computed. Reference is made to Schatz & Wahlbin [1979] and Eriksson & Johnson [1988] for analyses of representative examples of the two approaches – a priori and adaptive – in the finite element context.

§10.2 The "Operator Principle", which at first glance may appear to be a rather trivial concept, is actually quite a useful idea. Typically, in iterative methods for solving a discrete equation $Ax = y$, it is enough that one knows how to evaluate the action of the operator A on any vector; one does not need a matrix representation of the operator A. In certain applications, evaluating the operator A in itself involves the solution of one or more elliptic boundary value problems, cf. Bramble [1981] for a simple example. The corresponding matrix for A would not only be dense, but also extremely hard to compute.

§§10.3, 10.4, 10.5 and 10.7 When using an iterative method for solving an equation $Ax = b$ with the operator (!) ill-conditioned, and hence the convergence of the iterative method slow, current opinion favors *preconditioning*. Here one iterates on the equivalent equation $M^{-1}Ax = M^{-1}b$, where the operator M^{-1}, the *preconditioner*, should have the following two properties: (i) the equation $My = d$ is "easy" to solve, i.e., the action of

M^{-1} is "easy" to compute; (ii) the operator $M^{-1}A$ is "well"-conditioned, so that only a "few" iterations are necessary to solve the preconditioned problem. Preconditioners, and indeed solution algorithms in general, are often tailored to take advantage of a particular computer architecture, see Ortega & Voigt [1985] for a survey and a comprehensive list of references. Sometimes preconditioners are constructed from considerations of the matrix representation for the operator A; an example is the incomplete Cholesky factorization, Meijerink & van der Vorst [1977]. Other investigations adopt the operator principle and construct a preconditioning problem $My = d$ having some "natural" relationship with the original problem, see e.g. Bramble, Pasciak & Schatz [1986], where parallel computer architectures are taken advantage of via substructuring of the physical domain.

Another popular iterative method for solving the equations coming from discretization of an elliptic problem is the *multigrid method*, cf. Hackbusch [1985].

For further details on iterative methods for solving large sparse systems of linear algebraic equations, the reader is referred to the texts by Varga [1962], Wachspress [1966], Young [1971] and Hageman & Young [1981]. The software package ITPACK (Kincaid, Respess & Young [1982]) contains subroutines implementing adaptive accelerated iterative algorithms.

§10.6 It should be remarked that the "Energy Method" used by Rutishauser in the plate bending problem, (41) et seq., is not the basis for most commonly used finite element methods for that problem (in contrast to the situation for Poisson's problem in §10.1). We refer to the survey by Glowinski & Pironneau [1979] for details.

References

Birkhoff, G. and Lynch, R.E. [1984]: *Numerical Solution of Elliptic Problems*, Studies in Applied Mathematics 6, SIAM, Philadelphia.

Bramble, J.H. [1981]: The Lagrange multiplier method for Dirichlet's problem, *Math. Comp.* **37**, 1–11.

Bramble, J.H., Pasciak, J.E. and Schatz, A.H. [1986]: The construction of preconditioners for elliptic problems by substructuring. I, *Math. Comp.* **47**, 103–134.

Eriksson, K. and Johnson, C. [1988]: An adaptive finite element method for linear elliptic problems, *Math. Comp.* **50**, 361–383.

Glowinski, R. and Pironneau, O. [1979]: Numerical methods for the first biharmonic equation and for the two-dimensional Stokes problem, *SIAM Rev.* **21**, 167–212.

Hackbusch, W. [1985]: *Multigrid Methods and Applications*, Springer Series in Computational Mathematics 4, Springer, New York.

Hageman, L.A. and Young, D.M. [1981]: *Applied Iterative Methods*, Academic Press, New York.

Kincaid, D.R., Respess, J.R. and Young, D.M. [1982]: Algorithm 586 – ITPACK 2C: A FORTRAN package for solving large sparse linear systems by adaptive accelerated iterative methods, *ACM Trans. Math. Software* **8**, 302–322.

Meijerink, J.A. and van der Vorst, H.A. [1977]: An iterative solution method for linear systems of which the coefficient matrix is a symmetric M-matrix, *Math. Comp.* **31**, 148–162.

Ortega, J.M. and Voigt, R.G. [1985]: Solution of partial differential equations on vector and parallel computers, *SIAM Rev.* **27**, 149–240.

Pereyra, V., Proskurowski, W. and Widlund, O. [1977]: High order fast Laplace solvers for the Dirichlet problem on general regions, *Math. Comp.* **31**, 1–16.

Schatz, A.H. and Wahlbin, L.B. [1979]: Maximum norm estimates in the finite element method on plane polygonal domains. II: Refinements, *Math. Comp.* **33**, 465–292.

Varga, R.S. [1962]: *Matrix Iterative Analysis*, Prentice-Hall, Englewood Cliffs, N.J.

Wachspress, E.L. [1966]: *Iterative Solution of Elliptic Systems, and Applications to the Neutron Diffusion Equations of Reactor Physics*, Prentice-Hall, Englewood Cliffs, N.J.

Young, D.M. [1971]: *Iterative Solution of Large Linear Systems*, Academic Press, New York.

CHAPTER 11

Parabolic and Hyperbolic Partial Differential Equations

§11.1. One-dimensional heat conduction problems

We consider the temperature distribution $y(x,t)$ along a homogeneous rod of length L, which at one end $(x=L)$ is held at temperature 0, while at the other end $(x=0)$ the temperature is prescribed as a function $b(t)$ of time. Let the thermal conductivity of the rod be $f(x)$, the initial temperature be given as $a(x)$, and let there be interior heat generation $g(x,t)$ (cf. Fig. 11.1).

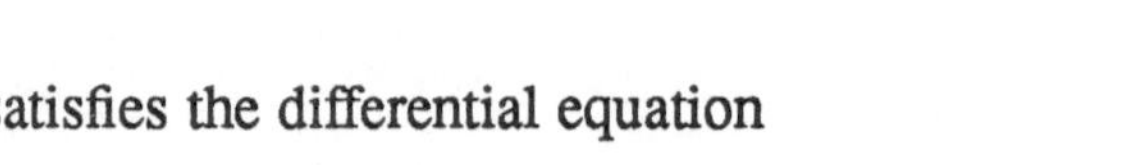

$$y = b(t) \boxed{\qquad y(x, 0) = a(x) \qquad} y = 0$$

$$x = 0 \qquad\qquad\qquad x = L$$

Figure 11.1. *Initial and boundary conditions for a heat conduction problem in a rod*

Then $y(x,t)$ satisfies the differential equation

$$\frac{\partial y}{\partial t} = f(x)\,\frac{\partial^2 y}{\partial x^2} + g(x,t) \quad (0 \le x \le L,\ t \ge 0) \tag{1}$$

with initial and boundary conditions

$$y(x, 0) = a(x), \quad y(0,t) = b(t), \quad y(L,t) \equiv 0. \tag{2}$$

For the solution of the problem one first divides the interval $0 \le x \le L$ into n equal parts of length h and introduces the following notations:

$$x_k \;=\; kh \qquad \text{(abscissas)},$$

$$y_k(t) \;=\; y(x_k,t) \qquad \begin{array}{l}\text{(temperature at the point } x_k \\ \text{as a function of time)},\end{array}$$

$$a_k \;=\; a(x_k) \qquad \text{(initial temperature at the point } x_k),$$

$$f_k \;=\; f(x_k) \qquad \text{(thermal conductivity at the point } x_k),$$

$$g_k(t) \;=\; g(x_k,t) \qquad \begin{array}{l}\text{(heat generation at the point } x_k \\ \text{in function of time)}.\end{array}$$

Now we know that in first approximation,

$$\frac{\partial^2 y}{\partial x^2} \approx \frac{y(x+h,t) - 2y(x,t) + y(x-h,t)}{h^2},$$

so that through substitution in the differential equation one obtains the discretization in the space coordinate,

$$\frac{dy_k}{dt} = f_k \frac{y_{k+1}(t) - 2y_k(t) + y_{k-1}(t)}{h^2} + g_k(t) \quad (k = 1, \ldots, n-1). \quad (3)$$

Since $y_0(t) = b(t)$ and $y_n(t) \equiv 0$ are given functions, (3) represents a system of $n-1$ ordinary differential equations of the first order for the unknown functions $y_1(t), \ldots, y_{n-1}(t)$, which describes approximately the temperature variations at the abscissas of the rod.

This system is in fact linear and has a constant coefficient matrix

$$\mathbf{A} = -\frac{1}{h^2}
\begin{bmatrix}
2f_1 & -f_1 & & & & & 0 \\
-f_2 & 2f_2 & -f_2 & & & & \\
& -f_3 & 2f_3 & -f_3 & & & \\
& & \cdot & \cdot & \cdot & & \\
& & & \cdot & \cdot & \cdot & \\
& & & & \cdot & \cdot & \cdot \\
& & & & -f_{n-2} & 2f_{n-2} & -f_{n-2} \\
0 & & & & & -f_{n-1} & 2f_{n-1}
\end{bmatrix}. \quad (4)$$

As forcing terms one has the functions $g_k(t)$, and the initial conditions are $y_k(0) = a_k$ $(k = 1, \ldots, n - 1)$. In vector notation:

$$\frac{d\mathbf{y}}{dt} = \mathbf{A}\mathbf{y} + \mathbf{g}(t) \quad \text{with} \quad \mathbf{y}(0) = \mathbf{a}, \tag{5}$$

where the given boundary values $y_0(t)$, $y_n(t)$ have been included in $\mathbf{g}(t)$.

If the rod, instead, is thermally isolated at one end (e.g. at $x=L$), one has the boundary condition $\partial y/\partial x = 0$ for $x=L$ and all t. The way this condition is realized in the discretization is by first putting $y_{n+1} = y_{n-1}$, which at the point $x = x_n$ then yields for the 2nd derivative the approximation

$$\left. \frac{d^2y}{dx^2} \right|_{x = x_n} \approx \frac{2y_{n-1} - 2y_n}{h^2}$$

and thus the following differential equation for the function $y_n(t)$, which is now unknown:

$$\frac{dy_n}{dt} = f_n \frac{2y_{n-1}(t) - 2y_n(t)}{h^2} + g_n(t). \tag{6}$$

The system of differential equations also for this problem thus has the form (5), only $\mathbf{A}$ is now the $n \times n$-matrix

$$\mathbf{A} = -\frac{1}{h^2} \begin{bmatrix} 2f_1 & -f_1 & & & & \\ -f_2 & -2f_2 & -f_2 & & & \\ & \cdot & \cdot & \cdot & & \\ & & \cdot & \cdot & \cdot & \\ & & & \cdot & \cdot & \cdot \\ & & & -f_{n-1} & 2f_{n-1} & -f_{n-1} \\ & & & & -2f_n & 2f_n \end{bmatrix}. \tag{7}$$

Such a system of differential equations, however, can (in both cases) easily be integrated numerically with methods already discussed (Euler, Runge-Kutta, trapezoidal rule, etc.).

a) *Numerical integration by Euler.* The integration step from $t = t_\ell$ to $t + \tau = t_{\ell + 1}$ for the system (3) reads:

$$\mathbf{y}(t_{\ell + 1}) = \mathbf{y}(t_\ell) + \tau \mathbf{A} \mathbf{y}(t_\ell) + \tau \mathbf{g}(t_\ell), \tag{8}$$

or, if $y_{\ell,k}$ denotes the kth component of the vector $\mathbf{y}(t_\ell)$,

$$y_{\ell + 1,k} = y_{\ell,k} - \frac{\tau f_k}{h^2} (-y_{\ell,k-1} + 2y_{\ell,k} - y_{\ell,k+1}) + \tau g_{\ell,k}$$

($k = 1,2, \ldots, n-1$, resp. n), where $y_{\ell,0}, y_{\ell,n}$ are to be replaced by the given boundary values. With this explicit recursion formula one can compute all quantities $y_{\ell + 1,k}$ directly from the $y_{\ell,k}$ and in this way carry out one *time step* (integration step with respect to the variable t).

b) *Integration with the trapezoidal rule.* It is amazing how long people held on to Euler's method for the numerical integration of the heat equation, while all along no efforts were spared to improve the numerical integration of ordinary differential equations. It was only after 1950 that also the trapezoidal rule, under the name "implicit recursion formula", was introduced. Applied to the system (5), it reads

$$\mathbf{y}(t_{\ell+1}) - \mathbf{y}(t_\ell) = \frac{\tau}{2} (\mathbf{A}\mathbf{y}(t_{\ell+1}) + \mathbf{A}\mathbf{y}(t_\ell) + \mathbf{g}(t_{\ell+1}) + \mathbf{g}(t_\ell)),$$

which yields for the components of the unknown vector $\mathbf{y}(t_{\ell + 1})$ the system of equations

$$(\mathbf{I} - \frac{\tau}{2} \mathbf{A})\mathbf{y}(t_{\ell+1}) = (\mathbf{I} + \frac{\tau}{2} \mathbf{A})\mathbf{y}(t_\ell) + \frac{\tau}{2} \mathbf{g}(t_{\ell+1}) + \frac{\tau}{2} \mathbf{g}(t_\ell). \tag{9}$$

These equations (written out componentwise) are called the "implicit recursion formulas", since they no longer permit to compute the $y_{\ell + 1,k}$ directly, but rather necessitate in each time step the solution of a linear system of equations. The disadvantage which thus accrues, however, is not serious, at least not for the one-dimensional heat conduction problem, since:

1) The matrix $\mathbf{A}$ is tridiagonal([1]).

2) The eigenvalues of the matrix $\mathbf{A}$ are real and negative, so that $\mathbf{I} - \frac{1}{2}\tau\mathbf{A}$ is never singular.

Note: *One should not compute* $\left[\mathbf{I} - \frac{1}{2}\tau\mathbf{A}\right]^{-1}$ *in order to obtain an explicit formula. This would only increase the expenditure, both in computing and storage.*

§11.2. Stability of the numerical solution

We first observe that for the discussion of stability we can restrict ourselves to the homogeneous equation

$$\mathbf{z}' = \mathbf{A}\mathbf{z}, \quad \mathbf{z}(0) = \mathbf{a}. \tag{10}$$

Assuming, indeed, that $\mathbf{g}(t)$ does not vary too rapidly, we can account for this by replacing $\mathbf{g}(t)$ in (5) by $\mathbf{g}_0 + t\mathbf{g}_1$:

$$\mathbf{y}' = \mathbf{A}\mathbf{y} + \mathbf{g}_0 + t\mathbf{g}_1 \quad \text{with} \quad \mathbf{y}(0) = \mathbf{a}.$$

If we now integrate the system

$$\mathbf{u}' = \mathbf{A}\mathbf{u} + \mathbf{g}_0 + t\mathbf{g}_1, \quad \mathbf{u}(0) = -\mathbf{A}^{-1}\mathbf{g}_0 - \mathbf{A}^{-2}\mathbf{g}_1,$$
$$\mathbf{v}' = \mathbf{A}\mathbf{v}, \quad\quad\quad\quad\quad \mathbf{v}(0) = \mathbf{a} + \mathbf{A}^{-1}\mathbf{g}_0 + \mathbf{A}^{-2}\mathbf{g}_1,$$

and note that $\mathbf{u}(0) + \mathbf{v}(0) = \mathbf{a}$, one gets precisely $\mathbf{u} + \mathbf{v} = \mathbf{y}$. As is easily verified, however,

$$\mathbf{u} = -\mathbf{A}^{-1}\mathbf{g}_0 - \mathbf{A}^{-2}\mathbf{g}_1 - t\mathbf{A}^{-1}\mathbf{g}_1,$$

---

[1] In addition, the coefficient matrix for fixed τ is constant. It suffices to compute its triangular decomposition once; then in each step, only forward and backward substitution is required. (Editors' remark)

that is, $\mathbf{u}$ is linear in t and hence is exactly integrated by the trapezoidal rule. The whole error of the discretized solution therefore is borne by $\mathbf{v}$, a solution of the homogeneous equation.

As earlier in §8.5, a solution of the homogeneous system (10) is analyzed componentwise: To each eigenvalue λ of the matrix $\mathbf{A}$ there belongs a component $\mathbf{v}_\lambda(t)$ which theoretically ought to behave like $e^{\lambda t} \cdot \mathbf{v}_\lambda(0)$, where $\mathbf{v}_\lambda(0)$ depends on the initial value $\mathbf{z}(0) = \mathbf{a}$. The solution of the system is obtained by superposition of all these components:

$$\mathbf{z}(t) = \sum_\lambda \mathbf{v}_\lambda(t) = \sum_\lambda e^{\lambda t}\mathbf{v}_\lambda(0). \tag{11}$$

Now for the one-dimensional heat conduction problem, where, depending on the boundary conditions, $\mathbf{A}$ may have the form (4) or (7), all eigenvalues are real, negative and simple ([1]); more precisely:

$$0 > \lambda > -4M/h^2, \text{ where } M = \max_{0 \le k \le n} f_k. \tag{12}$$

All components of the exact solution of (5) are therefore damped; one must thus see to it that also the numerical solution is damped.

a) For the *Euler method* (8), the component of the solution belonging to the eigenvalue λ, when integrating with the step τ, behaves like

$$\mathbf{v}(t_{\ell+1}) = (1 + \tau\lambda)\mathbf{v}(t_\ell),$$

that is

$$\mathbf{v}(t_\ell) = (1 + \tau\lambda)^\ell \mathbf{v}(0), \tag{13}$$

and this is damped precisely when $|1 + \tau\lambda| < 1$. This implies $1 + \tau\lambda > -1$ for all λ, so that τ, according to (12), must satisfy

---

[1] A tridiagonal matrix in which all elements of the two side diagonals are positive has real and simple eigenvalues. In fact, such a matrix can first be symmetrized by a similarity transformation with a diagonal matrix. The fact that a symmetric matrix of this type has simple eigenvalues then follows from Theorem 4.9 in Schwarz H.R., Rutishauser H., Stiefel E.: *Numerical Analysis of Symmetric Matrices*, Prentice-Hall, Englewood Cliffs, N.J. 1973. (Editors' remark)

$1 - 4\tau M/h^2 \geq -1$, or

$$\tau \leq \frac{h^2}{2M} \, . \tag{14}$$

It is to be noted, however, that this is merely the maximum stepsize admissible on the basis of the stability requirement, which does not yet produce great accuracy.

We demonstrate the necessity of this restriction with a simple example:

$$\frac{\partial y}{\partial t} = \frac{\partial^2 y}{\partial x^2} \quad (0 \leq x \leq 1, \ t \geq 0),$$

$$y(x, 0) \equiv 0, \quad y(0,t) = 10^6 t, \quad \frac{\partial y}{\partial x} (1,t) \equiv 0. \tag{15}$$

(Here, $f(x) \equiv 1$, hence $M = 1$.) We first integrate with $h = .2$, $\tau = .01$, in which case $h^2/2M = .02$ and condition (14) is thus satisfied. Then τ is held fixed, but h is halved, causing (14) to be violated. The results (of fixed-point computation) are summarized in Tables 11.1 and 11.2.

Table 11.1. *Integration of (15) by Euler; case of stability;*
$\tau = .01, \ h^2/2M = .02$

t $\quad$ x	0	.2	.4	.6	.8	1.0
0	0	0	0	0	0	0
.01	10000	0	0	0	0	0
.02	20000	2500	0	0	0	0
.03	30000	6250	625	0	0	0
.04	40000	10781	1875	156	0	0
.05	50000	15859	3672	547	39	0
.06	60000	21347	5937	1201	156	19
.07	70000	27158	8605	2124	383	87
.08	80000	33230	11623	3309	744	235
.	.					
.	.					
.	.					

Table 11.2. *Integration of (15) by Euler; case of instability;*
$$\tau = .01, \; h^2/2M = .005$$

t \ x	0	.1	.2	.3	.4	.5	.6	$\cdots$
0	0	0	0		0	0	0	$\cdots$
.01	10000	0	0	0	0	0	0	
.02	20000	10000	0	0	0	0	0	
.03	30000	10000	10000	0	0	0	0	
.04	40000	30000	0	10000	0	0	0	
.05	50000	10000	40000	−10000	10000	0	0	
.06	60000	80000	−40000	60000	−20000	10000	0	
.07	70000	−60000	180000	−120000	90000	−30000	10000	
.08	80000	310000	−360000	390000	−240000	130000	−40000	
.	.							
.	.							
.	.							

One recognizes immediately that in the second case the solution is completely unstable. The refinement of the subdivision in the space coordinate, rather than giving the expected improvement, causes the accuracy to deteriorate catastrophically. Yet, such an improvement would be highly desirable, because the solution obtained with $\tau = .01$, $h = .2$ is still very inaccurate, as is shown by a comparison with Table 11.3, which gives the exact (rounded) solution.

Table 11.3. *Exact solution of the heat conduction problem* (15)

t \ x	0	.2	.4	.6	.8	1.0
0	0	0	0	0	0	0
.01	10000	568	8	0	0	0
.02	20000	3014	231	8	0	0
.03	30000	6707	968	85	4	0
.04	40000	11194	2272	321	31	4
.05	50000	16239	4093	777	109	22
.06	60000	21700	6369	1482	272	74
.07	70000	27488	9039	2444	547	186
.08	80000	33542	12055	3660	956	384
.	.					
.	.					
.	.					

b) In the *trapezoidal rule* the component of the numerical solution belonging to the eigenvalue λ behaves like

$$v(t_{\ell+1}) = \frac{1 + \dfrac{\tau\lambda}{2}}{1 - \dfrac{\tau\lambda}{2}}\, v(t_\ell),$$

thus like

$$v(t_\ell) = \left[\frac{1 + \dfrac{\tau\lambda}{2}}{1 - \dfrac{\tau\lambda}{2}} \right]^{\ell} v(t_0). \tag{16}$$

This component, therefore, is integrated as accurately as

$$\frac{1 + \dfrac{\tau\lambda}{2}}{1 - \dfrac{\tau\lambda}{2}} \text{ agrees with } e^{\tau\lambda}.$$

The deviation between these two quantities, and hence the approximate error per step for the component with eigenvalue λ, is in first approximation given by $\lambda^3\tau^3/12$. Therefore, if the error for no component is to exceed ε in absolute value, one must have, by (12),

$$\tau \le \frac{h^2}{4M} \sqrt[3]{12\varepsilon} . \tag{17}$$

Thus, for example, when $h = .2$, $M = 1$, $\varepsilon = 10^{-5}$, one obtains the condition $\tau \le .01 \sqrt[3]{.00012} \approx 5_{10}-4$.

Now, however, this choice is overly cautious because of two reasons:

1) The components corresponding to the dangerous (strongly negative) eigenvalues yield only a small contribution to the complete solution.

2) As time increases, the components of the solution belonging to large negative eigenvalues of $\mathbf{A}$ are so much damped that eventually one has to deal only with the eigenvalues near 0; these, however, allow a larger τ.

In order to be able to account for the various components according to their strengths, we assume that the contributions of the eigenvalues at time $t = 0$ are uniformly distributed, so that the eigenvalues between λ and $\lambda + d\lambda$ contribute by $d\lambda$. The solution at time t then is

$$\int_0^\Lambda e^{-\lambda t} d\lambda \quad \text{with } \Lambda = \frac{4M}{h^2} .$$

The total error in the step from t to $t + \tau$ would then amount to

$$\int_0^\Lambda e^{-\lambda t} \frac{\lambda^3 \tau^3}{12} d\lambda.$$

Now, however, the contributions of the various eigenvalues add up according to the law of Pythagoras, since the eigenvectors are mutually perpendicular. (The matrix **A** is easily symmetrized.) For the relative error $\varphi(t)$ (associated with the step from t to $t + \tau$) one thus has:

$$\varphi^2(t) \approx \frac{\int_0^\Lambda e^{-2\lambda t} \dfrac{\lambda^6 \tau^6}{144} \, d\lambda}{\int_0^\Lambda e^{-2\lambda t} \, d\lambda} .$$

With the substitution $2\lambda t = \kappa$, $2\Lambda t = K$ one finally gets

$$\varphi^2(t) \approx \frac{\dfrac{\tau^6}{144 \times 128 t^7} \int_0^K e^{-\kappa} \kappa^6 d\kappa}{\dfrac{1}{2t} \int_0^K e^{-\kappa} d\kappa} < \frac{6! \tau^6}{144 \times 64 t^6} = \frac{5}{64} \left[\frac{\tau}{t} \right]^6 ,$$

from which one concludes that the condition

$$\tau \le t \sqrt[6]{\frac{64}{5} \varepsilon^2} \tag{18}$$

guarantees that always $\varphi(t) \le \varepsilon$. The step τ must thus be chosen proportional to the elapsed time, which, while allowing a large τ for large t, is nevertheless too severe a restriction when t is very small. The condition (17), after all, is sufficient in any case, and the bound occurring there is larger than the right-hand side of (18) when

$$t < \frac{h^2}{4M} \sqrt[6]{\frac{45}{4}} \approx \frac{h^2}{4M} \sqrt[3]{\frac{10}{3}} .$$

This leads to the following recipe:

1) Determine an integer ν in the vicinity of

$$\frac{1}{\sqrt[3]{3.6\varepsilon}} . \tag{19}$$

2) Integrate over v steps, each time with

$$\tau_0 = \frac{h^2}{4M}\sqrt[3]{12\varepsilon} \ , \tau_0, 2\tau_0, 4\tau_0, 8\tau_0, 16\tau_0 \ , \ \ldots \ . \tag{20}$$

(After the first v steps, one reaches approximately the t-value $\sqrt[3]{10/3}\ h^2/4M$, for which the two bounds derived for τ agree approximately.)

In a practical example, the effect of this may be as follows:

Let us solve the heat conduction problem (15), assuming the interval $0 \le x \le 1$ subdivided into 100 subintervals (i.e., $h = .01$) and an accuracy request of $\varepsilon = 10^{-4}$. One obtains

$$1/\sqrt[3]{.00036} \approx 15 = v,$$

$$\tau_0 = .000025 \sqrt[3]{.0012} \approx 2.66_{10}-6.$$

Therefore, one integrates over 30 steps with τ_0, then over 15 steps each time with $2\tau_0$, $4\tau_0$, $8\tau_0$, etc., and thus arrives with

$$
\begin{aligned}
&30 \text{ steps at } t = .00008,\\
&45 \text{ steps at } t = .00016,\\
&60 \text{ steps at } t = .00032,\\
&75 \text{ steps at } t = .00064,\\
&90 \text{ steps at } t = .00128,\\
&105 \text{ steps at } t = .00256,\\
&120 \text{ steps at } t = .00512,\\
&135 \text{ steps at } t = .01024,\\
&150 \text{ steps at } t = .02048,\\
&165 \text{ steps at } t = .04096,\\
&180 \text{ steps at } t = .08192.
\end{aligned}
$$

Euler's method on the same problem, for reasons of stability alone, would have required $\tau = .00005$, hence 1600 steps till $t = .08$. Here, however, beyond mere stability, we still managed to achieve a certain accuracy.

§11.3. The one-dimensional wave equation

When a wave equation

$$\frac{\partial^2 y}{\partial t^2} = f(x) \frac{\partial^2 y}{\partial x^2} + g(x,t) \tag{21}$$

(with suitable initial and boundary conditions) has to be integrated, one proceeds completely analogously: The x-domain $a \le x \le b$ is first subdivided into n subintervals of equal length, and then the differential equation formulated at each subdivision point. This yields the system

$$\frac{d^2 y_k}{dt^2} = f_k \frac{y_{k+1}(t) - 2y_k(t) + y_{k-1}(t)}{h^2} + g_k(t).$$

As initial conditions, $y_k(0)$ and $y_k'(0)$ are prescribed. As to the functions $y_0(t) = y(a,t)$ and $y_n(t) = y(b,t)$, they are either given, or additional equations can be set up for them. (This too is completely analogous to the heat conduction problem.)

In each case, there results a system of differential equations of the second order,

$$\mathbf{y}'' = \mathbf{A}\mathbf{y} + \mathbf{b}, \tag{22}$$

where the coefficient matrix $\mathbf{A}$ is tridiagonal and has usually negative real eigenvalues λ satisfying (12). This being the case, however, one knows that the solution of the homogeneous equation

$$\mathbf{y}'' = \mathbf{A}\mathbf{y} \tag{23}$$

is composed of particular solutions of the form

$$\mathbf{v}(t) = \mathbf{v}(0)e^{\pm i\sqrt{|\lambda|}\, t} . \tag{24}$$

The wave character of the solution, therefore, is not destroyed by the discretization in the space variable. We have to see to it that also the discretization in the time variable, that is, the numerical integration of (23), does not destroy the oscillatory character.

To this end, one first introduces a new variable $z = y'$, with which the system (23) goes over into

$$\frac{d}{dt}\begin{bmatrix} y \\ z \end{bmatrix} = \begin{bmatrix} O & I \\ A & O \end{bmatrix} \begin{bmatrix} y \\ z \end{bmatrix} . \tag{25}$$

If $-\omega^2$ is an eigenvalue of A, then $\pm i\omega$ are eigenvalues of the combined matrix

$$\begin{bmatrix} O & I \\ A & O \end{bmatrix}, \tag{26}$$

since

$$\det\begin{bmatrix} i\omega I & I \\ A & i\omega I \end{bmatrix} = \det\begin{bmatrix} i\omega I & I \\ A + \omega^2 I & O \end{bmatrix} = (-1)^n \det(I) \det(A + \omega^2 I).$$

If the eigenvalues of A satisfy the relation (12), then those of (26) lie on the imaginary axis between $\pm i 2\sqrt{M}/h$.

Now for the integration of a linear system whose coefficient matrix has exclusively purely imaginary eigenvalues, the trapezoidal rule is predestined by virtue of its amplitude fidelity. In view of the form (26) of the coefficient matrix, one obtains the following equation for a time step (with $y_\ell = y(t_\ell)$, $z_\ell = z(t_\ell)$):

$$\left\{\begin{bmatrix} I & O \\ O & I \end{bmatrix} - \frac{\tau}{2}\begin{bmatrix} O & I \\ A & O \end{bmatrix}\right\}\begin{bmatrix} y_{\ell+1} \\ z_{\ell+1} \end{bmatrix} = \left\{\begin{bmatrix} I & O \\ O & I \end{bmatrix} + \frac{\tau}{2}\begin{bmatrix} O & I \\ A & O \end{bmatrix}\right\}\begin{bmatrix} y_\ell \\ z_\ell \end{bmatrix},$$

that is,

$$y_{\ell+1} - \frac{\tau}{2} z_{\ell+1} = y_\ell + \frac{\tau}{2} z_\ell ,$$

$$z_{t+1} - \frac{\tau}{2} Ay_{t+1} = z_t + \frac{\tau}{2} Ay_t \ .$$

Multiplication of the first of these two equations by $\frac{1}{2} \tau A$ and addition to the second yields

$$\left[I - \frac{\tau^2}{4} A \right] z_{t+1} = \left[I + \frac{\tau^2}{4} A \right] z_t + \tau Ay_t \ ,$$

or

$$\left[I - \frac{\tau^2}{4} A \right] \Delta z_t = \frac{\tau^2}{2} Az_t + \tau Ay_t , \text{ where } \Delta z_t = z_{t+1} - z_t \ . \quad (27)$$

We thus arrive at the following computational process for a time step:

1) $r = y_t + \frac{\tau}{2} z_t ,$

2) $w = \tau Ar,$

3) $\left[I - \frac{\tau^2}{2} A \right] v = w$ solve for $v,$

4) $z_{t+1} = z_t + v,$

5) $y_{t+1} = y_t + \frac{\tau}{2} (z_t + z_{t+1}).$

As to the choice of the time step τ, the controlling factor is how well

$$\frac{1 + \dfrac{i\omega\tau}{2}}{1 - \dfrac{i\omega\tau}{2}} \text{ agrees with } e^{i\omega\tau} , \quad (28)$$

because these are the quantities by which the component of the solution belonging to the eigenvalue $i\omega$ of the matrix (26) is multiplied in one step, the first for computation with the trapezoidal rule, the second for exact integration. The difference of these two quantities can be estimated in first approximation by

$$\frac{1}{12}\left[\frac{2\sqrt{M}}{h}\,\tau\right]^3 .$$

If this is not to exceed ε, the condition

$$\tau \leq \frac{h}{2\sqrt{M}}\,\sqrt[3]{12\varepsilon} \tag{29}$$

must be satisfied. The decisive difference in comparison with the heat conduction problem is the factor

$$\frac{h}{2\sqrt{M}} \quad \text{as opposed to} \quad \frac{h^2}{4M} .$$

This means that during the reduction of the spacial stepsize h, the time step τ must be reduced only proportionally to, not quadratically in, h.

On the other hand, there is no damping here, and therefore no progressive disappearance of those components of the solution which require a small time step. The integration step, therefore, cannot be continually doubled.

Furthermore, the trapezoidal rule, while faithfully reproducing amplitudes, distorts the phase. The first factor in (28) can namely be written as

$$\frac{1 + \dfrac{i\omega\tau}{2}}{1 - \dfrac{i\omega\tau}{2}} = e^{2i\,\tan^{-1}\frac{\omega\tau}{2}} .$$

Since

$$2 \tan^{-1} \frac{\omega \tau}{2} < \omega \tau ,$$

the phase thus increases more slowly in the numerical solution than in the exact one.   This difference is more pronounced in the high-frequency components and therefore causes those to lag behind.

Numerical example. We consider a rope stretched between $x = - 10$ and $x = 10$. Suppose at time $t=0$ there exists a triangular deflection around the point $x=0$, caused by a blow to the rope. For $t>0$ this deflection divides into two triangles which move away from each other in opposite directions, are reflected at $x=10$ and $x = - 10$, respectively, and then again return towards $x=0$, where at time $t=20$ they ought to form again the original wave form with opposite sign.

To be solved, therefore, is the problem

$$\frac{\partial^2}{\partial t^2} y(x,t) = \frac{\partial^2}{\partial x^2} y(x,t), \quad -10 \le x \le 10, \quad t \ge 0,$$

$$y(-10,t) = y(10,t) \equiv 0,$$

$$y(x, 0) = 10000 \max \{1 - |x|, 0\},$$

$$\frac{\partial y}{\partial t} (x, 0) \equiv 0.$$

Because of symmetry we can restrict ourselves to the interval $0 \le x \le 10$, which we subdivide into 100 parts of length .1. The resulting system of 100 differential equations of second order is then integrated with the method described above, whereby $\tau = .025$ is chosen as time step.

Table 11.4. *Numerical solution of a wave equation*

x \ t =	0	.5	1.0	1.5	2.0	2.5	3.0	3.5	4.0 …	16.0	16.5	17.0	17.5	18.0	18.5	19.0	19.5	20.0
0	10000	4889	352	-14	-41	14	-11	20	-34	-13	9	-4	-1	3	-66	-747	-4239	-8793
	9000	5127	327	92	-4	-6	11	-22	36	13	-9	4	1	-8	-58	-827	-4266	-8628
	8000	4852	980	-88	97	-23	-8	25	-38	-12	8	-3	-2	1	-106	-980	-4418	-8075
	7000	5066	1468	-69	-124	60	-6	-22	37	11	-7	2	2	-15	-142	-1289	-4565	-7284
	6000	5267	2065	-162	9	-63	23	10	-28	-10	5	-0	-5	-7	-252	-1660	-4734	-6266
	5000	4903	2553	133	96	-6	-19	3	13	7	-3	-2	5	-37	-366	-2163	-4779	-5189
	4000	4340	2869	646	-4	101	-26	1	-0	-5	-0	5	-10	-39	-583	-2679	-4735	-4064
	3000	3963	3465	908	6	-97	85	-29	1	1	3	-8	7	-98	-827	-3253	-4503	-3004
	2000	3574	4194	1520	-245	-1	-88	60	-17	3	-7	10	-20	-132	-1201	-3742	-4157	-2007
	1000	2937	4779	2093	-96	2	6	-51	29	-7	11	-14	0	-250	-1610	-4169	-3663	-1148
1	0	2555	4694	2479	237	70	68	-17	-10	12	-15	16	-43	-358	-2141	-4406	-3100	-467
	0	1937	4475	2795	523	89	-37	89	-46	-17	19	-21	-31	-582	-2676	-4484	-2471	-5
	0	1574	3938	3596	936	-158	-2	-85	99	22	-22	17	-101	-825	-3254	-4351	-1809	188
	0	968	3531	4291	1637	-233	-67	14	-92	-26	25	-29	-126	-1199	-3758	-4045	-1167	211
	0	376	3000	4717	2050	3	40	11	21	29	-26	6	-246	-1620	-4166	-3612	-541	80
	0	97	2447	4714	2324	183	153	24	38	-30	24	-43	-360	-2139	-4424	-3046	-85	34
	0	18	2066	4355	2911	442	-13	10	-20	29	-23	-34	-568	-2703	-4464	-2465	260	-2
	0	3	1535	4013	3628	1116	-246	-67	-20	-26	13	-88	-839	-3249	-4364	-1770	309	124
	0	0	880	3501	4378	1610	-93	-72	4	19	-15	-141	-1178	-3794	-4013	-1167	293	126
	0	0	384	2935	4702	1952	-49	178	-1	-10	-14	-218	-1651	-4151	-3627	-502	99	118
2	0	0	130	2528	4612	2325	83	91	79	-2	-13	-387	-2117	-4454	-3018	-86	70	-96
	0	0	36	2085	4406	2912	580	-148	-48	11	-68	-531	-2745	-4437	-2476	269	32	-219
	0	0	8	1468	3991	3766	1133	-109	-152	-28	-47	-872	-3235	-4376	-1748	287	156	-268
	0	0	2	843	3429	4397	1596	-136	114	29	-177	-1148	-3829	-3992	-1153	257	156	-98
	0	0	0	395	2972	4623	1885	-127	142	-55	-182	-1678	-4148	-3621	-491	93	80	138
	0	0	0	153	2580	4638	2271	132	-27	21	-405	-2116	-4457	-3023	-55	45	-101	240
	0	0	0	50	2060	4405	3035	621	-66	-85	-521	-2753	-4449	-2445	234	90	-278	212
	0	0	0	14	1417	3919	3836	1193	-120	-44	-859	-3270	-4339	-1773	304	142	-216	-52
	0	0	0	4	823	3419	4352	1537	-183	-156	-1177	-3806	-4026	-1090	181	204	-93	-139
	0	0	0	1	405	3024	4633	1778	-150	-215	-1639	-4206	-3563	-522	120	13	209	-218
3	0	0	0	0	171	2597	4656	2334	147	-351	-2178	-4405	-3071	5	8	-106	216	-1

Table 11.4 contains the results, namely for x-values in the interval $0 \le x \le 3$ (with step .1) and t-values in the intervals $0 \le t \le 4$ and $16 \le t \le 20$ (with step .5). At time $t=20$ the original wave form is actually no longer achieved exactly, but it is at least reproduced qualitatively. Owing to the lagging of the high-frequency components, the vertices of the triangle indeed appear rounded at $x=0$ and $x=1$.

§11.4. Remarks on two-dimensional heat conduction problems

In a two-dimensional medium which fills a domain G, the heat equation reads

$$\frac{\partial u}{\partial t} = \frac{\partial^2 u}{\partial x^2} + \frac{\partial^2 u}{\partial y^2} , \tag{30}$$

and the boundary conditions are in general of the form

$$\alpha \frac{\partial u}{\partial n} + \beta u + \gamma = 0 \text{ on the boundary of } G. \tag{31}$$

If now, as in §10.1 for the Dirichlet problem, one lays a grid over the (x,y)-plane and approximates $\partial^2 u/\partial x^2 + \partial^2 y/\partial y^2$ at the point P by

$$\frac{1}{h^2} \left(-4u_P + u_N + u_E + u_S + u_W \right) ,$$

one obtains a system of differential equations

$$\frac{du_P}{dt} = - \frac{1}{h^2} \left(4u_P - u_N - u_E - u_S - u_W \right) \tag{32}$$

(namely one such for each grid point P), from which the approximate temperature behavior $u_P(t)$ can be computed for each grid point P.

For points near the boundary, (32) must be modified: in the case of Fig. 11.2, for example, where the values of u are prescribed on the boundary, one gets

$$\frac{du_P}{dt} = -\frac{1}{h^2}(4u_P - u_E - u_S) + \frac{1}{h^2},$$

which contains also a contribution (namely the term $1/h^2$) to the forcing term. If, however, as in Fig. 11.3, $\partial u/\partial n = 0$ is prescribed,

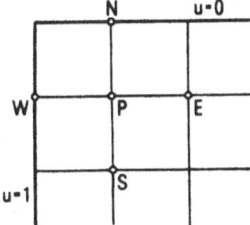

Figure 11.2. *Point near the boundary*

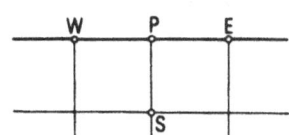

Figure 11.3. *Point on the boundary*

one obtains, following the model of elliptic differential equations,

$$\frac{du_P}{dt} = -\frac{1}{h^2}(4u_P - u_E - u_W - 2u_S).$$

In each case, there results also here a system of differential equations of the form

$$\frac{du}{dt} = Au + g,$$

wherein the forcing term g incorporates boundary effects.

Unfortunately, the coefficient matrix **A** is not automatically symmetric. This would be unimportant for the method of Euler, but could be a disadvantage for the integration by means of the trapezoidal rule. It is therefore worthwhile trying to enforce symmetry of **A**, which indeed is possible in the following way:

One subdivides the domain into elementary squares (one square each around every grid point) and sets up the heat balance, keeping the temperature constant over each square (cf. Fig. 11.4). The heat flow into the neighboring

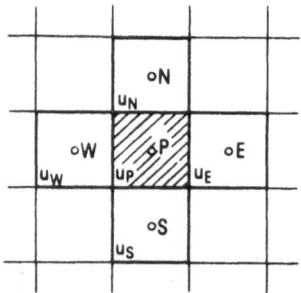

Figure 11.4. *Subdivision into elementary squares with constant temperature*

square is always proportional to the length h of the common side and to the temperature drop, which equals, for example, $(u_P - u_E)/h$. The following therefore holds for a time interval τ:

$$h^2 \delta u_P = -\tau \left[h\, \frac{u_P - u_N}{h} + h\, \frac{u_P - u_E}{h} + h\, \frac{u_P - u_S}{h} + h\, \frac{u_P - u_W}{h} \right],$$

which leads again to the differential equation (32) for the interior points P. The same holds true for points near the boundary with prescribed temperature.

The matter is different for a true boundary point P, as in Fig. 11.3 with boundary condition $\partial u/\partial n = 0$. The corresponding elementary square then indeed is a "half square" which also has only half the heat content (cf. Fig. 11.5). The balance, therefore, is given here by

$$\frac{1}{2}\,h^2\delta u_P = -\tau\left[\frac{h}{2}\,\frac{u_P - u_E}{h} + \frac{h}{2}\,\frac{u_P - u_W}{h} + h\,\frac{u_P - u_S}{h}\right],$$

which yields for u_P the differential equation

$$\frac{1}{2}\,\frac{du_P}{dt} = -\frac{1}{h^2}\left(2u_P - \frac{1}{2}\,u_E - \frac{1}{2}\,u_W - u_S\right).$$

Analogously, in the case of Fig. 11.6, one has at the vertex P

$$\frac{1}{4}\,h^2\delta u_P = -\tau\left[\frac{h}{2}\,\frac{u_P - u_W}{h} + \frac{h}{2}\,\frac{u_P - u_S}{h}\right],$$

and thus

$$\frac{1}{4}\,\frac{du_P}{dt} = -\frac{1}{h^2}\left(u_P - \frac{1}{2}\,u_W - \frac{1}{2}\,u_S\right).$$

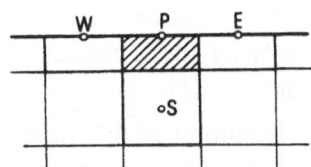

Figure 11.5. *Boundary point P with "half square"*

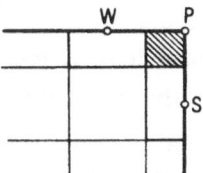

Figure 11.6. *Boundary point P at a vertex*

In this manner one obtains a system of differential equations of the form

$$\mathbf{D} \frac{d\mathbf{u}}{dt} = \mathbf{A}\mathbf{u} + \mathbf{g}, \tag{33}$$

where $\mathbf{D}$ is a diagonal matrix with positive diagonal elements, and $\mathbf{A}$ a symmetric negative definite matrix, which, by the way, is the same as the one obtained if the boundary value problem $\Delta u = 0$ for the same domain with the same boundary conditions is discretized by means of the energy method.

The eigenvalues of the matrix $\mathbf{D}^{-1}\mathbf{A}$ lie in the interval $0 > \lambda > -8/h^2$. If one integrates by Euler (where it is convenient to work with the matrix $\mathbf{D}^{-1}\mathbf{A}$), the time step therefore is subject to the condition

$$\tau \le \frac{h^2}{4} . \tag{34}$$

For the trapezoidal rule, however, one writes (with $\mathbf{u}_\ell = \mathbf{u}(t_\ell)$):

$$\mathbf{D}(\mathbf{u}_{\ell+1} - \mathbf{u}_\ell) = \tfrac{1}{2}\,\tau\mathbf{D}(\mathbf{u}'_{\ell+1} + \mathbf{u}_\ell{}') = \frac{\tau}{2}\,(\mathbf{A}\mathbf{u}_{\ell+1} + \mathbf{A}\mathbf{u}_\ell) + \frac{\tau}{2}\,(\mathbf{g}_{\ell+1} + \mathbf{g}_\ell),$$

$$(\mathbf{D} - \frac{\tau}{2}\,\mathbf{A})\mathbf{u}_{\ell+1} = (\mathbf{D} + \frac{\tau}{2}\,\mathbf{A})\mathbf{u}_\ell + \frac{\tau}{2}\,(\mathbf{g}_{\ell+1} + \mathbf{g}_\ell). \tag{35}$$

One now has to solve *in each time step* this linear system of equations for $\mathbf{u}_{\ell+1}$, which is as voluminous as the one for solving the Dirichlet problem for the same domain. Nevertheless, the matrix $\mathbf{D} - \frac{1}{2}\,\tau\mathbf{A}$ is symmetric and positive definite[1]. In addition, the condition of this matrix is significantly better than the one of $\mathbf{A}$, at least as long as τ is small. For the domain $0 \le x \le 1, 0 \le y \le 1$, for example, if $u(x,y,t)$ is prescribed on the boundary, and this square is subdivided into m^2 small squares with sides $h = 1/m$, one has,

$$\mathbf{D} = \mathbf{I}_{(m-1)^2} = \text{unit matrix of order } (m-1)^2,$$

---

[1] Cf. footnote[1] in §11.1.

$$\max \; |\lambda(\mathbf{A})| = \frac{8}{h^2} \cos^2 \frac{\pi}{2m}, \quad \min \; |\lambda(\mathbf{A})| = \frac{8}{h^2} \sin^2 \frac{\pi}{2m} \;,$$

so that the condition of **A** equals

$$\cot^2 \frac{\pi}{2m} \approx \frac{4m^2}{\pi^2} \;,$$

while the one for $\mathbf{D} - \frac{1}{2} \tau \mathbf{A}$ is (if $m > 2$):

$$K = \frac{1 + 4 \dfrac{\tau}{h^2} \cos^2 \dfrac{\pi}{2m}}{1 + 4 \dfrac{\tau}{h^2} \sin^2 \dfrac{\pi}{2m}} << \cot^2 \frac{\pi}{2m} \;. \tag{36}$$

Actually, $K \approx 2$ if τ is chosen as the maximum admissible time step $h^2/4$ for Euler's method; for larger τ, in any case,

$$K < 1 + \frac{4\tau}{h^2} \;.$$

This favorable condition has the consequence that the overrelaxation method (see §10.4) applied to the system of equations (35) converges very rapidly. If τ is continuously doubled, then so is, approximately, the condition of the coefficient matrix $\mathbf{D} - \frac{1}{2} \tau \mathbf{A}$, which slows down the convergence; but then, on the other hand, one advances more quickly.

Example. To be solved is a heat conduction problem (30) for the domain of Fig. 11.7, where $u \equiv 0$ is prescribed on the boundary and $u(x,y,0) \equiv 1000$ as initial value in the interior.

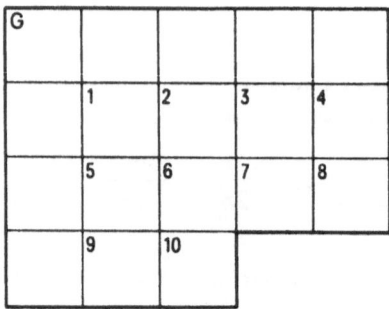

Figure 11.7. *The domain G and its discretization*

We work with the grid shown ($h = .2$) and first carry out two steps of Euler's method with $\tau = .005$. (One thus has $\tau = h^2/8$, that is, by (34), half of the admissible maximum.) The results are arranged in accordance with the geometric position of the points, and are rounded to integers:

$t = 0:$	1000	1000	1000	1000
	1000	1000	1000	1000
	1000	1000		

$t = .005:$	750	875	875	750
	875	1000	875	750
	750	750		

$t = .01:$	594	766	750	578
	750	922	766	578
	578	594		

We now add one step of the trapezoidal rule with $\tau = .02$. The system of equations (35) is solved by overrelaxation ($\omega = 1.143$), whereby as starting vector the approximation $u(.01)$ found by Euler's method is used, which must also be substituted on the right-hand side of (35). After 1, 2 and 3 cycles, respectively, there results:

$t = .03$, after one cycle:

$$
\begin{array}{cccc}
348 & 503 & 458 & 255 \\
456 & 610 & 447 & 210 \\
259 & 254 & &
\end{array}
$$

$t = .03$, after two cycles:

$$
\begin{array}{cccc}
304 & 448 & 400 & 240 \\
402 & 545 & 423 & 257 \\
249 & 291 & &
\end{array}
$$

$t = .03$, after three cycles:

$$
\begin{array}{cccc}
295 & 437 & 401 & 249 \\
397 & 554 & 434 & 254 \\
255 & 288 & &
\end{array}
$$

A two-dimensional heat conduction problem with circular symmetry. In the following example we wish to show how a two-dimensional problem is treated that could be reduced analytically to a one-dimensional one.

To be solved is

$$
\frac{\partial u}{\partial t} = \frac{\partial^2 u}{\partial x^2} + \frac{\partial^2 u}{\partial y^2}
$$

on the disc $x^2 + y^2 \le 1$, where for $t=0$

$$
u(x,y,0) = \begin{cases} 0 & \text{if} \quad x^2 + y^2 \le .01, \\ 1 & \text{if} \quad x^2 + y^2 > .01, \end{cases}
$$

and for $x^2 + y^2 = 1$, thus on the boundary, $\partial u/\partial n = 0$. Physically, we are dealing here with the hot shrinking of a disc of radius 1 onto a shaft of radius .1 made of the same material (cf. Fig. 11.8);

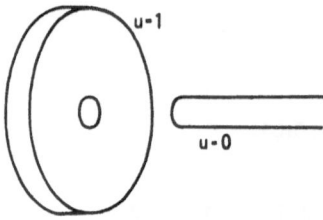

Figure 11.8. *Shrinking of a disc onto a shaft*

the edge of the disc is isolated. The question arises as to how the temperature is equalizing.

Since the disc has radial symmetry, one could transform the differential equation in the usual way to

$$\frac{\partial u}{\partial t} = \frac{1}{r} \frac{\partial}{\partial r} \left[r \frac{\partial u}{\partial r} \right] \tag{37}$$

with $\partial u / \partial r = 0$ on the boundary. This is a one-dimensional problem whose solution is given by

$$u(r,t) = \sum_{k=1}^{\infty} c_k J_0(n_k r) \exp(-n_k^2 t) + c_0, \tag{38}$$

where the n_k $(k = 1, 2, \dots)$ are the zeros of the Bessel function J_1, and the c_k must be chosen so as to satisfy the initial conditions. The series (38), however, converges slowly for small t and we do not want to make use of it.

We rather think of the disc $0 \leq r \leq 1$ as being subdivided into circular rings K_p of thickness $h = 1/n$, whereby the innermost and outermost, to be sure, are to have only thickness $h/2$:

$$K_0: \qquad 0 \le r \le h/2,$$

$$K_1: \qquad h/2 \le r \le 3h/2,$$

$$\cdot$$
$$\cdot$$
$$\cdot$$

$$K_{n-1}: \qquad (n-1)h - \frac{1}{2} h \le r \le (n-1)h + \frac{1}{2} h,$$

$$K_n: \qquad 1 - \frac{1}{2} h \le r \le 1 .$$

The temperature on the ring K_p is assumed to be (spacially) constant and is denoted by $u_p(t)$. One then draws up the heat balance, where it is to be noted that K_0 has the area $\pi h^2/4$, K_p ($p = 1, \ldots, n-1$) the area $2\pi ph^2$, and K_n the area $\pi(n - 1/4)h^2$. The heat flow between K_p and K_{p+1} is proportional to the length $\pi(2p + 1)h$ of the borderline and to the temperature gradient $\frac{1}{h} (u_{p+1} - u_p)$, so that the following equations hold:

$$\frac{\pi}{4} h^2 \delta u_0 = \tau \pi(u_1 - u_0),$$

$$2\pi ph^2 \delta u_p = \tau \pi((2p + 1)(u_{p+1} - u_p) + (2p - 1)(u_{p-1} - u_p))$$
$$(p = 1, \ldots, n-1),$$
$$\pi h^2 \left[n - \frac{1}{4} \right] \delta u_n = \tau \pi(2n - 1)(u_{n-1} - u_n) .$$

From this, one obtains the following $n+1$ differential equations of first order for the $n+1$ unknown functions $u_0(t), \ldots, u_n(t)$:

$$\frac{1}{4} \frac{du_0}{dt} = \frac{1}{h^2} (u_1 - u_0),$$

$$2p \frac{du_p}{dt} = \frac{1}{h^2} ((2p - 1)u_{p-1} - 4pu_p + (2p + 1)u_{p+1})$$
$$(p = 1, \ldots, n-1), \tag{39}$$

$$\left[n - \frac{1}{4} \right] \frac{du_n}{dt} = \frac{1}{h^2} ((2n - 1)u_{n-1} - (2n - 1)u_n) .$$

This system has the form

$$D \frac{d\mathbf{u}}{dt} = A\mathbf{u}, \tag{40}$$

where the matrix A is symmetric and negative definite and D is a positive diagonal matrix:

$$D = \begin{bmatrix} \frac{1}{4} & & & & & & 0 \\ & 2 & & & & & \\ & & 4 & & & & \\ & & & \cdot & & & \\ & & & & \cdot & & \\ & & & & & 2n-2 & \\ 0 & & & & & & n-\frac{1}{4} \end{bmatrix}, \tag{41}$$

$$A = \frac{1}{h^2} \begin{bmatrix} -1 & 1 & & & & & \\ 1 & -4 & 3 & & & & \\ & 3 & -8 & 5 & & & \\ & & 5 & -12 & 7 & & \\ & & \cdot & & \cdot & \cdot & \\ & & & \cdot & & \cdot & \cdot \\ & & & & 2n-3 & -4n+4 & 2n-1 \\ & & & & & 2n-1 & -2n+1 \end{bmatrix}. \tag{42}$$

The systems of equations (35) to be solved here, describing the use of the trapezoidal rule, are

$$\left[D - \frac{\tau}{2} A \right] \mathbf{u}_{k+1} = \left[D + \frac{\tau}{2} A \right] \mathbf{u}_k.$$

For an analysis of the stability and accuracy, the controlling factor would be the eigenvalues of the generalized eigenvalue problem (cf. §12.1) $\lambda D\mathbf{x} = A\mathbf{x}$.

Numerical example. For the spacial discretization we choose $h = .02$. The system (39) then consists of 51 differential equations. We solve it by the trapezoidal rule, choosing at the beginning as time step $\tau = \tau_0 = 5_{10}-6$. In analogy with the procedure in §11.2 (cf. Eq. (20)), τ is doubled for the first time after 40 steps, and then again after every 20 additional steps. Table 11.5 shows the result on the interval $0 \le r \le .3$, where the temperature, though, is tabulated only every 20th integration step.

Table 11.5. *Hot shrinking of a disc of radius 1 onto a cold shaft of radius .1*

t \ r	.00	.02	.04	.06	.08	.10	.12	.14	.16	.18	.20	.22	.24	.26	.28	.30
0	0	0	0	0	0	.50000	1.0000	1.0000	1.0000	1.0000	1.0000	1.0000	1.0000	1.0000	1.0000	1.0000
.0001	.00003	.00014	.00151	.01518	.11429	.52009	.91109	.99047	.99928	.99996	1.0000	1.0000	1.0000	1.0000	1.0000	1.0000
.0002	.00051	.00155	.00874	.04637	.19156	.53386	.85681	.97175	.99595	.99955	.99996	1.0000	1.0000	1.0000	1.0000	1.0000
.0003	.00250	.00563	.02192	.08183	.24692	.54434	.82179	.95154	.99010	.99839	.99978	1.0000	1.0000	1.0000	1.0000	1.0000
.0004	.00704	.01307	.03949	.11680	.28867	.55294	.79805	.93265	.98259	.99636	.99936	.99990	.99999	1.0000	1.0000	1.0000
.0006	.02539	.03759	.08173	.17971	.34849	.56707	.76910	.90137	.96564	.99001	.99753	.99947	.99990	.99998	1.0000	1.0000
.0008	.05479	.07164	.12698	.23251	.39074	.57892	.75305	.87805	.94901	.98173	.99432	.99845	.99962	.99992	.99998	1.0000
.0012	.13145	.15185	.21349	.31566	.45036	.59916	.73796	.84754	.92119	.96372	.98506	.99446	.99814	.99943	.99984	.99996
.0016	.21221	.23178	.28927	.37987	.49390	.61698	.73317	.82976	.90079	.94721	.97432	.98855	.99530	.99822	.99938	.99980
.0024	.35097	.36598	.40950	.47654	.55943	.64896	.73594	.81293	.87540	.92206	.95423	.97476	.98692	.99362	.99707	.99873
.0032	.45455	.46564	.49783	.54756	.60954	.67761	.74561	.80839	.86236	.90578	.93857	.96185	.97744	.98729	.99317	.99650
.0048	.59051	.59697	.61584	.64540	.68309	.72584	.77048	.81410	.85437	.88969	.91922	.94282	.96088	.97413	.98346	.98978
.0064	.67353	.67768	.68988	.70920	.73427	.76338	.79470	.82645	.85707	.88533	.91039	.93179	.94943	.96348	.97432	.98241
.0096	.76840	.77051	.77673	.78674	.80001	.81586	.83354	.85227	.87127	.88987	.90751	.92373	.93825	.95092	.96170	.97066
.0128	.82075	.82201	.82577	.83185	.84002	.84993	.86122	.87347	.88627	.89921	.91194	.92415	.93559	.94607	.95548	.96377
.0192	.87665	.87725	.87904	.88197	.88594	.89086	.89658	.90295	.90982	.91701	.92438	.93176	.93902	.94603	.95271	.95896
.0256	.90600	.90635	.90739	.90910	.91145	.91437	.91781	.92170	.92596	.93050	.93526	.94013	.94505	.94994	.95473	.95936
.0384	.93633	.93649	.93697	.93776	.93885	.94022	.94186	.94374	.94583	.94810	.95053	.95309	.95573	.95844	.96118	.96392
.0512	.95186	.95195	.95223	.95268	.95331	.95410	.95506	.95616	.95740	.95876	.96022	.96179	.96343	.96513	.96688	.96866
.0768	.96765	.96769	.96781	.96802	.96830	.96867	.96911	.96962	.97019	.97084	.97154	.97229	.97310	.97395	.97483	.97575
.1024	.97562	.97564	.97571	.97583	.97599	.97620	.97645	.97674	.97707	.97744	.97784	.97828	.97875	.97925	.97978	.98033
.1536	.98343	.98344	.98347	.98352	.98359	.98367	.98378	.98390	.98404	.98419	.98437	.98455	.98475	.98497	.98519	.98543
.2048	.98687	.98688	.98689	.98691	.98694	.98698	.98703	.98709	.98715	.98722	.98730	.98739	.98748	.98758	.98768	.98779
.3072	.98923	.98923	.98923	.98924	.98924	.98925	.98926	.98928	.98929	.98931	.98932	.98934	.98936	.98938	.98941	.98943
.4096	.98975	.98975	.98975	.98975	.98975	.98976	.98976	.98976	.98976	.98977	.98977	.98978	.98978	.98979	.98979	.98980
.6144	.98989	.98989	.98989	.98989	.98989	.98989	.98989	.98989	.98989	.98989	.98989	.98989	.98989	.98989	.98989	.98989

Notes to Chapter 11

For basic finite difference theory in time-dependent problems, the reader is referred to Richtmyer & Morton [1967] (which has aged well), and for finite element methods in parabolic problems to Thomée [1984]. A representative investigation of the finite element method in a second-order hyperbolic problem is given in Bales [1984]. Surveys of additional methods such as special finite difference methods in shock problems, spectral methods and particle (vortex) methods can be found in Brezzi [1985].

§§11.1, 11.2 and 11.4 Current wisdom concerning time-discretization of parabolic problems favors implicit methods. The reason for this is the very stringent timestep constraint (14) typical of explicit methods. Economical methods have been developed for solving the resulting systems of equations; see, e.g., the incomplete iteration idea of Douglas, Dupont & Ewing [1979].

Rutishauser's derivation of (18) is an example of an a priori attempt to find efficient timesteps. The problem of doing this, especially adaptively during the computation, has a long history in the context of ordinary differential equations. Such ideas are being extended to partial differential equations; see, e.g., Eriksson & Johnson [1987].

§11.3 For a comprehensive account of the questions of amplitude and phase errors considered by Rutishauser, see Vichnevetsky & Bowles [1982].

A final note to Chapters 10 and 11: the problem of assessing the accuracy of a numerical approximation to a partial differential equation can be rather tricky. For an amusing example, see Symonds & Yu [1985, Fig. 3].

References

Bales, L.A. [1984]: Semidiscrete and single step fully discrete approximations for second order hyperbolic equations with time-dependent coefficients, *Math. Comp.* **43**, 383–414.

Brezzi, F. (ed.) [1985]: *Numerical Methods in Fluid Dynamics*, Lecture Notes Math. **1127**, Springer, New York.

Douglas, J., Jr., Dupont, T. and Ewing, R.E. [1979]: Incomplete iteration for time-stepping a Galerkin method for a quasilinear parabolic problem, *SIAM J. Numer. Anal.* **16**, 503–522.

Eriksson, K. and Johnson, C. [1987]: Error estimates and automatic time step control for nonlinear parabolic problems. I, *SIAM J. Numer. Anal.* **24**, 12–23.

Richtmyer, R.D. and Morton, K.W. [1967]: *Difference Methods for Initial-Value Problems*, 2nd ed., Interscience Publishers, New York.

Symonds, P.S. and Yu, T.X. [1985]: Counterintuitive behavior in a problem of elastic-plastic beam dynamics, *J. Appl. Mech.* **52**, 517–522.

Thomée, V. [1984]: *Galerkin Finite Element Methods for Parabolic Problems*, Lecture Notes Math. **1054**, Springer, New York.

Vichnevetsky, R. and Bowles, J.B. [1982]: *Fourier Analysis of Numerical Approximations of Hyperbolic Equations*, SIAM Studies in Applied Mathematics **5**, SIAM, Philadelphia.

CHAPTER 12

The Eigenvalue Problem For Symmetric Matrices

§12.1. Introduction

Matrix eigenvalue problems arise, for example, from Hamilton's principle; the latter states: A mechanical system whose kinetic and potential energy are given by

$$T = \sum_{i=1}^{n} \sum_{j=1}^{n} P_{ij}(q_1, \ldots, q_n)\dot{q}_i\dot{q}_j, \quad U = U(q_1, \ldots, q_n), \tag{1}$$

evolves between the time instances t_0 and t_1 in such a way that the functions $q_i(t)$ describing the motion make the action integral

$$J = \int_{t_0}^{t_1} (T - U)dt$$

stationary, the values $q_i(t_0)$ and $q_i(t_1)$ being held fixed.

The Euler equations for this variational problem are

$$\frac{d}{dt}\frac{\partial T}{\partial \dot{q}_k} - \frac{\partial}{\partial q_k}(T - U) = 0 \quad (k = 1, \ldots, n),$$

thus, since P_{ij} is of course symmetric,

$$2\sum_{j} P_{kj}\ddot{q}_j + \sum_{i}\sum_{j}\frac{\partial P_{kj}}{\partial q_i}\dot{q}_i\dot{q}_j - \sum_{i}\sum_{j}\frac{\partial P_{ij}}{\partial q_k}\dot{q}_i\dot{q}_j + \frac{\partial U}{\partial q_k} = 0 \quad (k = 1, \ldots, n).$$

If the system has a stable equilibrium point, that is, if there is a point $\bar{q}_1, \ldots, \bar{q}_n$ at which

$$\frac{\partial U}{\partial q_k}(\bar{q}_1, \ldots, \bar{q}_n) = 0 \quad (k = 1, \ldots, n)$$

and

$$\frac{\partial^2 U}{\partial q_i \partial q_j}(\bar{q}_1, \ldots, \bar{q}_n) = 2a_{ij} \quad (i, j = 1, \ldots, n)$$

are the elements of a positive definite matrix, then, as long as the quantities $x_i = q_i - \bar{q}_i$, $\dot{x}_i = \dot{q}_i$ remain small, and letting $P_{kj}(\bar{q}_1, \ldots, \bar{q}_n) = b_{kj}$, there holds in first approximation:

$$\sum_j b_{kj} \ddot{x}_j + \sum_j a_{kj} x_j = 0 \quad (k = 1, \ldots, n).$$

With $\mathbf{A} = [a_{kj}]$, $\mathbf{B} = [b_{kj}]$, $\mathbf{x} = [x_1, \ldots, x_n]^T$, one so obtains

$$\mathbf{B}\ddot{\mathbf{x}} + \mathbf{A}\mathbf{x} = \mathbf{0}.$$

Here, $\mathbf{B}$ according to its meaning is positive definite, since a kinetic energy cannot be negative.

Seeking $\mathbf{x}$ in the form $\mathbf{x} = e^{i\omega t}\mathbf{z}$, one then obtains

$$-\omega^2 \mathbf{B}\mathbf{z} + \mathbf{A}\mathbf{z} = \mathbf{0},$$

that is, $\lambda = \omega^2$ must be a solution of the *generalized eigenvalue problem*

$$(\mathbf{A} - \lambda \mathbf{B})\mathbf{z} = \mathbf{0}, \tag{2}$$

where $\mathbf{A}$ and $\mathbf{B}$ are symmetric and positive definite matrices.

If $\mathbf{B}$ is the unit matrix, one has an *ordinary eigenvalue problem*

$$\mathbf{A}\mathbf{z} = \lambda \mathbf{z}, \tag{3}$$

and we now show that the generalized problem can be reduced to the ordinary one: Since $\mathbf{B}$ in (2) is positive definite, it can be decomposed by Cholesky: $\mathbf{B} = \mathbf{R}^T \mathbf{R}$. Letting $\mathbf{z} = \mathbf{R}^{-1}\mathbf{y}$, (2) becomes

$$(\mathbf{A} - \lambda \mathbf{R}^T \mathbf{R})\mathbf{R}^{-1}\mathbf{y} = 0,$$

$$(\mathbf{R}^{-1})^T(\mathbf{A} - \lambda \mathbf{R}^T \mathbf{R})\mathbf{R}^{-1}\mathbf{y} = 0,$$

hence, with $\mathbf{R}^{-T} = (\mathbf{R}^{-1})^T$,

$$\mathbf{R}^{-T}\mathbf{A}\mathbf{R}^{-1}\mathbf{y} = \lambda \mathbf{y}.$$

In this way the generalized eigenvalue problem is reduced to the special one for the matrix $\mathbf{R}^{-T}\mathbf{A}\mathbf{R}^{-1}$, which is also symmetric; its eigenvectors $\mathbf{y}$ still need to be multiplied by $\mathbf{R}^{-1}$ in order to obtain those of (2).

As an example we consider the vibrations of an elastic beam, clamped on the left, but free on the right. Let the bending stiffness be $JE(x)$ and the mass per unit length $M(x)$. Let the deflection of the beam at time t be described by $q(x,t)$; one thus has here a continuum of degrees of freedom, the functions $q_k(t)$ in (1) becoming the function $q(x,t)$, that is, the index k a continuous variable x.

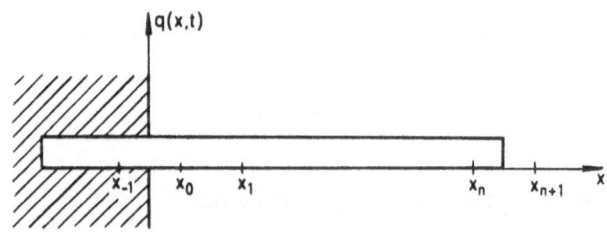

Figure 12.1. *Discretization for a beam clamped at one end*

The kinetic and potential energy are equal to

$$T = \frac{1}{2} \int_a^b M(x) \left[\frac{\partial q}{\partial t} \right]^2 dx, \quad U = \frac{1}{2} \int_a^b JE(x) \left[\frac{\partial^2 q}{\partial x^2} \right]^2 dx, \quad (4)$$

and to be discretized are these two integrals. For this purpose we select along the beam abscissas $x_k = x_0 + kh$ $(k = -1, 0, 1, \ldots, n + 1)$ as shown in Fig. 12.1. Then $\partial^2 q/\partial x^2$ is approximated in the usual way by $(q_{k-1} - 2q_k + q_{k+1})/h^2$ whereby the condition that the beam is clamped on the left can be expressed by $q_0 = q_{-1} = 0$. We furthermore use for the integrals the approximation

$$\int_a^b \phi(x)dx \approx h \sum_{k=0}^n \phi(x_k),$$

so that (4) becomes

$$T = \frac{h}{2} \sum_{k=0}^n M_k \dot{q}_k^2, \quad U = \frac{h}{2} \sum_{k=0}^n JE_k \left[\frac{q_{k-1} - 2q_k + q_{k+1}}{h^2} \right]^2.$$

In this way the original problem is now reduced to one with only finitely many degrees of freedom, though, of course, only approximately so. There now occurs, however, the fictitious deflection q_{n+1} in U, which is not present in T. But since the actual motion must take place such that $\int (T - U)dt$ becomes stationary, and q_{n+1} occurs only in the term $(q_{n-1} - 2q_n + q_{n+1})^2$, the integral can be stationary with respect to arbitrary variations $\delta q_{n+1}(t)$ only if this term vanishes, that is, if $q_{n-1}(t) - 2q_n(t) + q_{n+1}(t) \equiv 0$. If we still introduce the quantities $\gamma_k = JE_k/h^4$, then up to a common factor $h/2$,

$$T = \sum_{k=0}^n M_k \dot{q}_k^2,$$

$$(5)$$

$$U = \sum_{k=0}^{n-1} \gamma_k (q_{k-1}^2 - 4q_{k-1}q_k + 2q_{k-1}q_{k+1} + 4q_k^2 - 4q_k q_{k+1} + q_{k+1}^2).$$

The two matrices $\mathbf{A}$ and $\mathbf{B}$ occurring in the general equation (2) are now determined:

The matrix $\mathbf{B}$, which contains the coefficients of the quadratic form for T with respect to the $\dot{q}_k$, is a diagonal matrix:

$$\mathbf{B} = \begin{bmatrix} M_1 & & & & 0 \\ & M_2 & & & \\ & & \ddots & & \\ & & & \ddots & \\ 0 & & & & M_n \end{bmatrix}. \tag{6}$$

The matrix $\mathbf{A}$, which contains the coefficients of the quadratic form for U with respect to the q_k, is made up of the contributions of the individual terms of U, whereby the one corresponding to the index k furnishes the following contribution, which in turn we again represent in the form of the matrix:

$$
\begin{array}{ccccc}
1 & k-1 & k & k+1 & n \\
\downarrow & \downarrow & \downarrow & \downarrow & \downarrow
\end{array}
$$

$$
\mathbf{A}_k = \begin{bmatrix} 0 & & & & 0 \\ & \gamma_k & -2\gamma_k & \gamma_k & \\ & -2\gamma_k & 4\gamma_k & -2\gamma_k & \\ & \gamma_k & -2\gamma_k & \gamma_k & \\ 0 & & & & 0 \end{bmatrix}
\begin{array}{l}
\leftarrow 1 \\
\leftarrow k-1 \\
\leftarrow k \\
\leftarrow k+1 \\
\leftarrow n
\end{array}
$$

$(k = 2,3,\ldots, n-1)$. For $k=0$ and 1, respectively,

$$
\mathbf{A}_0 = \begin{matrix} \overset{1}{\downarrow} \\ \left[\begin{matrix} \gamma_0 & & \\ & & \\ & & 0 \end{matrix} \right] \end{matrix} \begin{matrix} \leftarrow 1 \\ \\ \end{matrix} \quad , \quad \mathbf{A}_1 = \begin{matrix} \overset{1}{\downarrow}\;\; \overset{2}{\downarrow} \\ \left[\begin{matrix} 4\gamma_1 & -2\gamma_1 & \\ -2\gamma_1 & \gamma_1 & \\ & & 0 \end{matrix} \right] \end{matrix} \begin{matrix} \leftarrow 1 \\ \leftarrow 2 \\ \end{matrix}
$$

(since q_0 and q_{-1} vanish identically). Then

$$\mathbf{A} = \mathbf{A}_0 + \mathbf{A}_1 + \cdots + \mathbf{A}_{n-1}.$$

There follows, in particular,

$$
\left.\begin{aligned}
a_{jj} &= \gamma_{j-1} + 4\gamma_j + \gamma_{j+1} \\
a_{j,j+1} &= a_{j+1,j} = -2\gamma_j - 2\gamma_{j+1} \\
a_{j,j+2} &= a_{j+2,j} = \gamma_{j+1}
\end{aligned}\right\} \quad (j = 1, 2, \ldots, n-2),
$$

$$a_{n-1,n-1} = \gamma_{n-2} + 4\gamma_{n-1}, \tag{7}$$

$$a_{n-1,n} = a_{n,n-1} = -2\gamma_{n-1},$$

$$a_{nn} = \gamma_{n-1}.$$

Knowledge of these matrices $\mathbf{A}$ and $\mathbf{B}$ now permits us to compute (approximately) the eigenfrequencies of the beam as the solution of the eigenvalue problem $(\mathbf{A} - \omega^2 \mathbf{B})\mathbf{z} = \mathbf{0}$. Both matrices automatically have become symmetric and positive definite.

Example. Let the beam have the form depicted in Fig. 12.2.

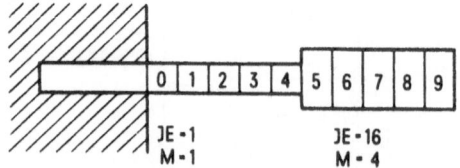

Figure 12.2. *Example of a beam clamped on the left*

Comparison with Fig. 12.1 shows that $n=9$, and by (7) and (6),

$$
A = \begin{bmatrix}
6 & -4 & 1 & & & & & & 0 \\
-4 & 6 & -4 & 1 & & & & & \\
1 & -4 & 6 & -4 & 1 & & & & \\
& 1 & -4 & 21 & -34 & 16 & & & \\
& & 1 & -34 & 81 & -64 & 16 & & \\
& & & 16 & -64 & 96 & -64 & 16 & \\
& & & & 16 & -64 & 96 & -64 & 16 \\
& & & & & 16 & -64 & 80 & -32 \\
0 & & & & & & 16 & -32 & 16
\end{bmatrix}, \tag{8}
$$

$$
B = \text{diag}\,(1,1,1,1,4,4,4,4). \tag{9}
$$

In order to reduce the generalized eigenvalue problem to a special one, one must now first compute the Cholesky decomposition $R^T R$ of B; here, it clearly gives

$$
R = \text{diag}\,(1,1,1,1,2,2,2,2).
$$

Then

$$\mathbf{R}^{-T}\mathbf{A}\mathbf{R}^{-1} = \begin{bmatrix} 6 & -4 & 1 & & & & & & 0 \\ -4 & 6 & -4 & 1 & & & & & \\ 1 & -4 & 6 & -4 & .5 & & & & \\ & 1 & -4 & 21 & -17 & 8 & & & \\ & & .5 & -17 & 20.25 & -16 & 4 & & \\ & & & 8 & -16 & 24 & -16 & 4 & \\ & & & & 4 & -16 & 24 & -16 & 4 \\ & & & & & 4 & -16 & 20 & -8 \\ 0 & & & & & & 4 & -8 & 4 \end{bmatrix} \tag{10}$$

is the matrix whose eigenvalues are (approximately) the squares of the frequencies of the beam. It must be observed, though, that only the smallest eigenvalues of this matrix actually correspond to frequencies of the beam.

§12.2. Extremal properties of eigenvalues

Let $\mathbf{A}$ be a symmetric matrix. We consider the quadratic form

$$Q(\mathbf{x}) = (\mathbf{x}, \mathbf{A}\mathbf{x}) = \sum_{i=1}^{n} \sum_{j=1}^{n} a_{ij} x_i x_j \tag{11}$$

as a function of the independent variables on the unit sphere, thus under the side condition $||\mathbf{x}||_2 = 1$.

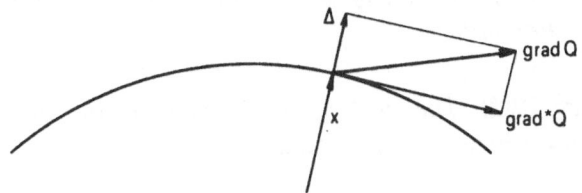

Figure 12.3. *Decomposition of* grad Q *into the radial and tangential component*

The gradient grad Q of $Q(\mathbf{x})$ equals $2\mathbf{Ax}$, since

$$\frac{\partial}{\partial x_k} \sum_{i=1}^{n} \sum_{j=1}^{n} a_{ij} x_i x_j = \sum_{i=1}^{n} a_{ik} x_i + \sum_{j=1}^{n} a_{kj} x_j.$$

The vector Δ in Fig. 12.3 is the projection of grad Q onto the position vector $\mathbf{x}$, so that $\Delta = 2(\mathbf{x}, \mathbf{Ax})\mathbf{x} = 2Q(\mathbf{x})\mathbf{x}$; thus,

$$\text{grad*}Q = 2\mathbf{Ax} - \Delta = 2(\mathbf{Ax} - Q(\mathbf{x})\mathbf{x}) \tag{12}$$

is the projection of the gradient onto the tangent plane. It is grad*Q, not grad Q, which is relevant for the variation of the function $Q(\mathbf{x})$ on the sphere. The extremal values of $Q(\mathbf{x})$ on the sphere indeed occur where grad*$Q = \mathbf{0}$. At these points, therefore,

$$\mathbf{Ax} = Q(\mathbf{x})\mathbf{x},$$

that is, $\mathbf{x}$ is an eigenvector of $\mathbf{A}$ and $Q(\mathbf{x})$ the associated eigenvalue.

Conversely, if $\mathbf{Ax} = \lambda\mathbf{x}$ and $||\mathbf{x}||_2 = 1$, then $Q(\mathbf{x}) = (\mathbf{Ax}, \mathbf{x}) = \lambda(\mathbf{x}, \mathbf{x}) = \lambda$, thus

$$\text{grad*}Q = 2(\mathbf{Ax} - Q(\mathbf{x})\mathbf{x}) = 2(\lambda\mathbf{x} - \lambda\mathbf{x}) = \mathbf{0},$$

so that Q is stationary at $\mathbf{x}$. We thus have:

Theorem 12.1. *The normalized eigenvectors and the eigenvalues of the symmetric matrix* $\mathbf{A}$ *are precisely the stationary points and associated function values, respectively, of the quadratic form* Q, *considered as a function on the unit sphere.*

Corollary 12.2. *The field of values of* $Q(\mathbf{x})$ *on the unit sphere is the closed interval between the smallest and the largest eigenvalue of* $\mathbf{A}$.

Examples. 1) The quadratic form associated with the matrix

$$\mathbf{A} = \begin{bmatrix} 6 & 3 & 1 \\ 3 & 2 & 1 \end{bmatrix}$$

has the following stationary points:

$$x = \pm\ [.860,\ .472,\ .194]^T, \qquad Q(x) = 7.873,$$
$$x = \pm\ [-.408,\ .408,\ .816]^T, \qquad Q(x) = 1,$$
$$x = \pm\ [-.306,\ .781,\ -.544]^T, \qquad Q(x) = .127.$$

The second, with value 1, is a saddle point (cf. Fig. 12.4).

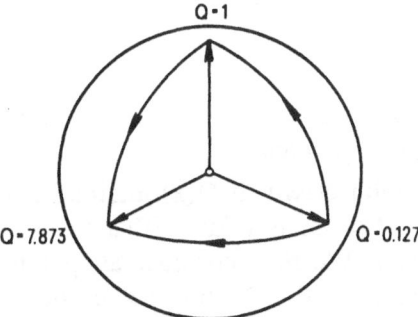

Figure 12.4. *To Example* 1

2) The matrix

$$A = \begin{bmatrix} 7 & 1 & 2 \\ 1 & 7 & -2 \\ 2 & -2 & 4 \end{bmatrix}$$

has the double eigenvalue $\lambda = 8$. Here, Q assumes its maximum 8 at all points of a certain great circle and the minimum 2 at the appertaining poles (cf. Fig. 12.5).

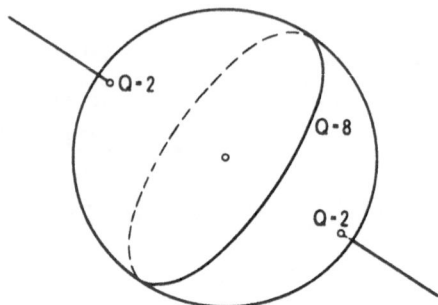

Figure 12.5. *To Example 2*

The eigenvectors and eigenvalues of **A** are also the solution of a second, similar extremal problem:

Let x_1 be the point at which $Q(x)$ attains its maximum on the unit sphere $||x||_2 = 1$. (We already know that x_1 is an eigenvector to the largest eigenvalue λ_1.) We now consider all points **x** on the sphere for which in addition $(x_1, x) = 0$. On this set of points we again seek the maximum of $Q(x)$; let it be assumed at x_2. We then consider the set of points defined by the conditions

$$(x, x) = 1, \quad (x_1, x) = (x_2, x) = 0$$

and thereon determine the maximum of $Q(x)$, giving the point x_3, etc. We so eventually obtain a complete orthogonal system $x_1, x_2, \ldots, x_n$.

One can show that these vectors x_k are again just eigenvectors of **A**, and that for the corresponding eigenvalues $\lambda_k = Q(x_k)$ one has $\lambda_1 \geq \lambda_2 \geq \cdots \geq \lambda_n$. For a symmetric matrix **A** there thus exists a complete orthogonal system of eigenvectors. If this is chosen as a new coordinate system, that is, if one puts

$$x = \sum_{k=1}^{n} \xi_k x_k,$$

then, because of $(x_k, Ax_\ell) = \lambda_\ell (x_k, x_\ell) = \lambda_\ell \, \delta_{k\ell}$,

$$Q(\mathbf{x}) = \sum_{k=1}^{n} \sum_{l=1}^{n} \xi_k \xi_l (\mathbf{x}_k, \mathbf{A}\mathbf{x}_l) = \sum_{k=1}^{n} \lambda_k \xi_k^2.$$

In the new system, the quadratic form Q is thus represented by the diagonal matrix diag $(\lambda_1, \ldots, \lambda_n)$ (*transformation to principal axes*).

Perturbation of the eigenvalues. The extremal property of eigenvalues, furthermore, yields a statement concerning the change of the eigenvalues of a symmetric matrix when its elements are perturbed in a certain way:

Let $\mathbf{A}$ and $\mathbf{B}$ be two symmetric matrices and $\mathbf{C} = \mathbf{B} - \mathbf{A}$ their difference, for which we assume that bounds α, β can be given for the field of values of the associated quadratic form (considered on the unit sphere):

$$\alpha \leq (\mathbf{x}, \mathbf{C}\mathbf{x}) \leq \beta \quad \text{for} \quad ||\mathbf{x}||_2 = 1. \tag{13}$$

Then for every normalized $\mathbf{x}$,

$$(\mathbf{x}, \mathbf{A}\mathbf{x}) + \alpha \leq (\mathbf{x}, \mathbf{B}\mathbf{x}) \leq (\mathbf{x}, \mathbf{A}\mathbf{x}) + \beta.$$

By Corollary 12.2, there now follows

$$\lambda_{\max}(\mathbf{A}) + \alpha \leq \lambda_{\max}(\mathbf{B}) \leq \lambda_{\max}(\mathbf{A}) + \beta.$$

An analogous statement holds for all eigenvalues:

Theorem 12.3. *If $\mathbf{A}$ and $\mathbf{B}$ are symmetric matrices with eigenvalues λ_k and μ_k, respectively, and if for $\mathbf{C} = \mathbf{B} - \mathbf{A}$ the estimate (13) holds, then each interval $[\lambda_k + \alpha, \lambda_k + \beta]$ contains (at least) one μ_k.*

Instead of a proof we give a plausibility argument: For the extreme eigenvalues, the assertion is obviously true (the maximum and minimum of the function $Q_A(\mathbf{x}) = (\mathbf{x}, \mathbf{A}\mathbf{x})$ (with $||\mathbf{x}||_2 = 1$) are changed at most by α and β, respectively). But the intermediate eigenvalues are saddle points of $Q_A(\mathbf{x})$, say of the kind depicted in Fig. 12.6, where the lowest top P of the pass, traversing it from W to E, has altitude λ_k (one has to climb at least that high up); but at the same time, P is also the lowest point on the path from N to S. Therefore, P is the lowest top of the pass on the way from valley to valley and at the same time the highest "bottom of the pass" on the way from mountain to mountain.

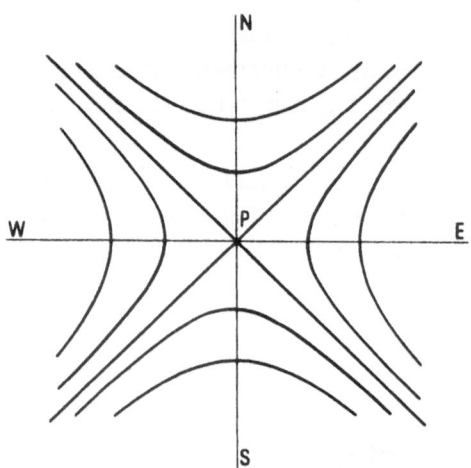

Figure 12.6. *Saddle point of the quadratic form* Q_A

If to Q_A one now adds a function with values between α and β, then we can think of construction debris being dumped over the whole terrain to a depth of at least α but no more than β. The passage from valley to valley via P then becomes higher by at most β, although by way of a detour one can perhaps find a lower maximum; at any rate, the new altitude of the pass is smaller than, or equal to, $\lambda_k + \beta$. Likewise, the path from mountain to mountain via P is elevated by at least α, thus the whole path is never lower than $\lambda_k + \alpha$. By way of a detour one perhaps finds a still higher "bottom of the pass", but the highest possible one definitely has an altitude of at least $\lambda_k + \alpha$. The eigenvalue μ_k of B, which corresponds to the lowest new top of the pass (and at the same time to the highest new "bottom of the pass"), therefore lies between $\lambda_k + \alpha$ and $\lambda_k + \beta$.

Theorem 12.3 states that the eigenvalues of a symmetric matrix are changed only a little by small symmetric perturbations, namely, at most by $\pm\varepsilon$, if all eigenvalues of the perturbation matrix $\mathbf{C}$ lie in the interval $[-\varepsilon,\varepsilon]$; this, in particular, holds true if the Schur norm (see §10.7, Eq. (56)) does not exceed ε:

$$\sum_{i=1}^{n} \sum_{j=1}^{n} c_{ij}^2 \le \varepsilon^2 .$$

Example. We consider the matrices

$$A = \begin{bmatrix} 19 & 7 & 7 & 0 \\ & 14 & -8 & 5 \\ & & 14 & -5 \\ \text{sym.} & & & 3 \end{bmatrix},$$

$$B = \begin{bmatrix} 18.9992 & 7.0008 & 6.9997 & .0005 \\ & 14.0010 & -8.0006 & 4.9996 \\ & & 14.0000 & -5.0007 \\ \text{sym.} & & & 3.0007 \end{bmatrix}.$$

A has the double eigenvalues .657281 and 24.342719. The Schur norm of the perturbation matrix $C = B - A$ equals .002474, consequently the eigenvalues of B must lie in the intervals

$$.654807 \le \mu \le .659755, \quad 24.340245 \le \mu \le 24.345193 .$$

B actually has the eigenvalues

$$.656240, \ .658381, \ 24.341959, \ 24.344320.$$

Perturbation of the eigenvectors. For the eigenvectors there is no analogous proposition. Indeed, a small change in the matrix elements can produce a large change in the eigenvectors. For example,

$$\begin{bmatrix} 1 & 10^{-5} & 0 \\ 10^{-5} & 1 & 10^{-5} \\ 0 & 10^{-5} & 1 \end{bmatrix}$$

has the eigenvectors (not normalized)

$$[1,0,-1]^T, \quad [1,\sqrt{2},1]^T, \quad [1,-\sqrt{2},1]^T,$$

while

$$
\begin{bmatrix}
1 & 10^{-5} & 10^{-5} \\
10^{-5} & 1 & 10^{-5} \\
10^{-5} & 10^{-5} & 1
\end{bmatrix}
$$

has the eigenvector $[1,1,1]^T$ corresponding to the eigenvalue $\lambda = 1 + 2_{10}{-5}$.

Nevertheless, there is a valid statement concerning the perturbation of an eigenvector, provided the associated eigenvalue is simple and sufficiently separated from the remaining eigenvalues:

Theorem 12.4. *Let* x *be a normalized eigenvector belonging to the eigenvalue* λ *of the symmetric matrix* A, *and let there be no further eigenvalue* λ' *of* A *with* $|\lambda - \lambda'| < t\varepsilon$, *where* $t > 2$. *In addition, let the Schur norm of the symmetric matrix* C *be no greater than* ε. *Then* $B = A + C$ *has a normalized eigenvector* y *with*

$$||x-y||_2 < \frac{1}{t-1}.$$

Proof. One can assume that A has been brought to diagonal form by an orthogonal transformation and that $\lambda = \lambda_1$, $x = e_1 = [1,0,\ldots,0]^T$. When C is transformed in the same way, the sum of the squares of the elements remains $\leq \varepsilon^2$. One thus has (in the new coordinate system)

$$
B = A + C =
\begin{bmatrix}
\lambda_1 + c_{11} & c_{12} & \cdots & c_{1n} \\
c_{21} & \lambda_2 + c_{22} & & c_{2n} \\
\vdots & & & \\
& & & \\
c_{n1} & c_{n2} & & \lambda_n + c_{nn}
\end{bmatrix}
$$

$$= \lambda_1 \mathbf{I} + \begin{bmatrix} c_{11} & \mathbf{q}^T \\ \mathbf{q} & \mathbf{P} \end{bmatrix},$$

with

$$\mathbf{P} = \begin{bmatrix} \lambda_2 - \lambda_1 + c_{22} & c_{23} & \cdots & c_{2n} \\ c_{32} & \lambda_3 - \lambda_1 + c_{33} & & c_{3n} \\ \vdots & & & \\ \vdots & & & \\ c_{n2} & c_{n3} & & \lambda_n - \lambda_1 + c_{nn} \end{bmatrix}, \quad \mathbf{q} = \begin{bmatrix} c_{12} \\ c_{13} \\ \vdots \\ \vdots \\ c_{1n} \end{bmatrix}.$$

By Theorem 12.3, $\mathbf{P}$ has no eigenvalue in the interval $|\mu| < (t-1)\varepsilon$, since by assumption $|\lambda_k - \lambda_1| \geq t\varepsilon \ (k = 2, \ldots, n)$ and

$$\sum_{i=2}^{n} \sum_{j=2}^{n} c_{ij}^2 \leq \varepsilon^2.$$

Now the eigenvectors of $\mathbf{B}$ are the same as those of

$$\begin{bmatrix} c_{11} & \mathbf{q}^T \\ \mathbf{q} & \mathbf{P} \end{bmatrix} = \mathbf{B} - \lambda_1 \mathbf{I}.$$

With $\mathbf{v}^T = [1 \mid \mathbf{z}^T]$ being such an eigenvector corresponding to the eigenvalue μ one obtains

$$c_{11} + \mathbf{q}^T \mathbf{z} = \mu - \lambda_1,$$

$$\mathbf{q} + \mathbf{P} \mathbf{z} = (\mu - \lambda_1) \mathbf{z}.$$

There follows:

$$
\begin{aligned}
||Pz||^2 &= ||(\mu-\lambda_1)\,z-q||^2 = (\mu-\lambda_1)^2||z||^2 - 2(\mu-\lambda_1)(z,q) + ||q||^2 \\
&= (\mu-\lambda_1)^2||z||^2 - 2(\mu-\lambda_1)(\mu-\lambda_1-c_{11}) + ||q||^2 \\
&= (\mu-\lambda_1)^2||z||^2 - (\mu-\lambda_1)^2 - (\mu-\lambda_1-c_{11})^2 + c_{11}^2 + ||q||^2.
\end{aligned}
$$

Since P has no eigenvalue smaller in modulus that $(t-1)\varepsilon$, one has $||Pz|| \ge (t-1)\,\varepsilon\,||z||$, and furthermore,

$$
||q||^2 + c_{11}^2 = \sum_{j=1}^{n} c_{1j}^2 \le \varepsilon^2.
$$

Therefore,

$$
(t-1)^2\varepsilon^2||z||^2 - (\mu-\lambda_1)^2||z||^2 \le \varepsilon^2 - (\mu-\lambda_1)^2 - (\mu-\lambda_1-c_{11})^2
$$
$$
\le \varepsilon^2 - (\mu-\lambda_1)^2.
$$

By Theorem 12.3, the matrix B has an eigenvalue μ with the property $|\mu-\lambda_1| \le \varepsilon$. For the corresponding eigenvector $[1\,|\,z^T]$, therefore,

$$
||z||^2 \le \frac{\varepsilon^2-(\mu-\lambda_1)^2}{(t-1)^2\varepsilon^2-(\mu-\lambda_1)^2} = \frac{1}{(t-1)^2}\,\frac{\varepsilon^2-(\mu-\lambda_1)^2}{\varepsilon^2-\dfrac{(\mu-\lambda_1)^2}{(t-1)^2}} < \frac{1}{(t-1)^2}\,,
$$

so long as only $(t-1)^2 > 1$, thus $t > 2$.

It is true that $v^T = [1,z^T]$ is not a normalized eigenvector, but for the corresponding normalized vector y the distance to the eigenvector x (here e_1) is even smaller (cf. Fig. 12.7), that is, we certainly have $||x-y|| < 1/(t-1)$, q.e.d.

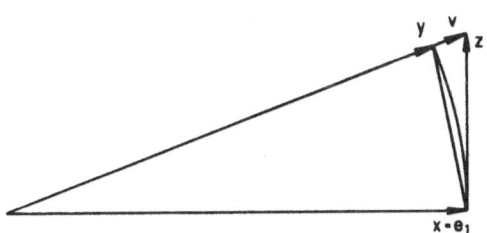

Figure 12.7. *To the proof of Theorem* 12.4

§12.3. The classical Jacobi method

Following Jacobi[1], the transformation to principal axes is carried out by means of an iterative process: Beginning with the matrix $A_0 = A$, one performs a sequence of "elementary" orthogonal transformations

$$A_1 = U_0^T A_0 U_0, \quad A_2 = U_1^T A_1 U_1 , \ldots$$

such that the matrices $A_0, A_1, A_2, \ldots$ become "more and more diagonal". As a measure for the deviation from a diagonal matrix one chooses the sum of the squares of all off-diagonal elements,

$$S_k = \sum_{i=1}^{n} \sum_{\substack{j=1 \\ j \neq i}}^{n} [a_{ij}^{(k)}]^2, \tag{14}$$

where $a_{ij}^{(k)}$ denote the elements of the matrix A_k. We thus require that $S_0 > S_1 > S_2 > \cdots$.

A_k is related to the original matrix A through

$$A_k = U_{k-1}^T U_{k-2}^T \ \cdots \ U_0^T A U_0 U_1 \ \cdots \ U_{k-1},$$

---

[1] Jacobi C.G.J.: Über ein leichtes Verfahren die in der Theorie der Säcularstörungen vorkommenden Gleichungen aufzulösen, *J. Reine Agnew. Math.* **30**, 51–94 (1846).

which, by means of the "accumulated" transformation matrix

$$\mathbf{V}_k = \mathbf{U}_0 \mathbf{U}_1 \cdots \mathbf{U}_{k-1} \tag{15}$$

(which is also orthogonal), can be written as

$$\mathbf{A}_k = \mathbf{V}_k^T \mathbf{A} \mathbf{V}_k. \tag{16}$$

If one succeeds in making S_k arbitrarily small, say $\leq \varepsilon$, then $\mathbf{A}_k$ deviates from a diagonal matrix $\mathbf{D}$ only by a matrix whose Schur norm is $\leq \varepsilon$: $||\mathbf{A}_k - \mathbf{D}|| \leq \varepsilon$. To the extent that this deviation is permissible on the basis of the perturbation theory (that is, Theorems 12.3 and 12.4), the transformation to principal axes is accomplished (within the desired accuracy) by $\mathbf{A} \to \mathbf{A}_k = \mathbf{V}_k^T \mathbf{A} \mathbf{V}_k$.

As to the choice of the matrices $\mathbf{U}_k$, one has complete freedom; they must only be orthogonal and reduce the sum of squares S_k. The simplest possibility is

$$\mathbf{U}_k = \mathbf{U}(p,q,\phi) = \begin{bmatrix} 1 & & & & & & & & & & 0 \\ & \ddots & & & & & & & & & \\ & & 1 & & & & & & & & \\ & & & \cos\phi & & & \sin\phi & & & & \\ & & & & 1 & & & & & & \\ & & & & & \ddots & & & & & \\ & & & & & & 1 & & & & \\ & & & -\sin\phi & & & \cos\phi & & & & \\ & & & & & & & & 1 & & \\ & & & & & & & & & \ddots & \\ 0 & & & & & & & & & & 1 \end{bmatrix} \begin{matrix} \\ \\ \\ \leftarrow p \\ \\ \\ \\ \leftarrow q \\ \\ \\ \\ \end{matrix} , \tag{17}$$

thus

$$u_{pp} = u_{qq} = \cos\phi, \quad u_{pq} = -u_{qp} = \sin\phi \quad (p < q).$$

($\mathbf{U}_k$ differs from the unit matrix only in these four elements.) A transformation

$$\mathbf{A}_k \rightarrow \mathbf{A}_{k+1} = \mathbf{U}_k^T \mathbf{A}_k \mathbf{U}_k \tag{18}$$

with an orthogonal matrix $\mathbf{U}_k = \mathbf{U}(p,q,\phi)$ of the form (17) is called a *Jacobi rotation with pivot element a_{pq} and angle of rotation ϕ.*

Denoting the elements of $\mathbf{A}_k$ briefly by a_{ij}, and those of $\mathbf{A}_{k+1}$ by $\hat{a}_{ij}$, the transformation (18) can be described elementwise as follows:

$$\hat{a}_{ij} = a_{ij} \quad (i \neq p,q \text{ and } j \neq p,q),$$

$$\left.\begin{aligned}
\hat{a}_{pj} &= a_{pj} \cos\phi - a_{qj} \sin\phi \\
\hat{a}_{qj} &= a_{pj} \sin\phi + a_{qj} \cos\phi
\end{aligned}\right\} \quad (j \neq p,q),$$

$$\left.\begin{aligned}
\hat{a}_{ip} &= a_{ip} \cos\phi - a_{iq} \sin\phi \\
\hat{a}_{iq} &= a_{ip} \sin\phi + a_{iq} \cos\phi
\end{aligned}\right\} \quad (i \neq p,q), \tag{19}$$

$$\hat{a}_{pq} = \hat{a}_{qp} = \tfrac{1}{2}(a_{pp} - a_{qq})\sin(2\phi) + a_{pq}\cos(2\phi),$$

$$\hat{a}_{pp} = a_{pp}\cos^2\phi - 2a_{pq}\sin\phi\cos\phi + a_{qq}\sin^2\phi,$$

$$\hat{a}_{qq} = a_{pp}\sin^2\phi + 2a_{pq}\sin\phi\cos\phi + a_{qq}\cos^2\phi.$$

There now follows:

α) $\hat{a}_{pj}^2 + \hat{a}_{qj}^2 = a_{pj}^2 + a_{qj}^2 \quad (j \neq p,q),$

β) $\hat{a}_{ip}^2 + \hat{a}_{iq}^2 = a_{ip}^2 + a_{iq}^2 \quad (i \neq p,q),$

γ) $\hat{a}_{ij}^2 = a_{ij}^2 \quad (i \neq p,q \text{ and } j \neq p,q).$

Denoting by $\sum_\alpha$, $\sum_\beta$, $\sum_\gamma$ the sums over the quantities occurring in the left- or (right-) hand members of these relations, whereby in $\sum_\gamma$ only terms with $i \neq j$ are to be considered, one evidently gets

$$S_k = \sum_{i=1}^{n} \sum_{j \neq i} a_{ij}^2 = \Sigma_\alpha + \Sigma_\beta + \Sigma_\gamma + a_{pq}^2 + a_{qp}^2,$$

$$S_{k+1} = \sum_{i=1}^{n} \sum_{j \neq i} \hat{a}_{ij}^2 = \Sigma_\alpha + \Sigma_\beta + \Sigma_\gamma + \hat{a}_{pq}^2 + \hat{a}_{qp}^2,$$

hence, because of symmetry,

$$S_{k+1} = S_k - 2a_{pq}^2 + 2\hat{a}_{pq}^2, \tag{20}$$

that is, the transformation (18) reduces the sum of the squares of all off-diagonal elements by $2a_{pq}^2 - 2\hat{a}_{pq}^2$. S_{k+1}, therefore, becomes minimal when a_{pq}^2 is as large as possible and $\hat{a}_{pq}^2$ becomes as small as possible.

Accordingly, one first chooses for p and q $(p < q)$ the row and column index, respectively, of the maximum modulus element of $\mathbf{A}_k$ above the diagonal, and then ϕ such that $\hat{a}_{pq} = 0$. This clearly entails

$$\tan(2\phi) = \frac{2a_{pq}}{a_{qq} - a_{pp}}, \tag{21}$$

which leaves four possibilities for ϕ in the interval $-\pi < \phi < \pi$. From among these four possible values one selects, as a matter of principle, the smallest: $|\phi| \leq \pi/4$.

One then carries out the transformation (18) and concurrently computes the matrix $\mathbf{V}_{k+1}$, which can be done, on the basis of (15), by

$$\mathbf{V}_{k+1} = \mathbf{V}_k \mathbf{U}_k;$$

in components, when v_{ij} are the elements of $\mathbf{V}_k$ and $\hat{v}_{ij}$ those of $\mathbf{V}_{k+1}$:

$$\left.\begin{aligned}
\hat{v}_{jp} &= v_{jp} \cos \phi - v_{jq} \sin \phi \\
\hat{v}_{jq} &= v_{jp} \sin \phi + v_{jq} \cos \phi
\end{aligned}\right\} \quad (j = 1, \ldots, n). \tag{22}$$

With this, a basic step is now completed.

In summary, the *classical Jacobi method* consists in executing the following operations for $k = 0,1,2, \ldots$:

1) Choice of the pivot element as the maximum modulus off-diagonal element of A_k.

2) Computation of the angle of rotation ϕ, or of the respective quantities $\cos \phi$ and $\sin \phi$ (only these are really needed).

3) Computation of the elements of the matrices A_{k+1}, V_{k+1} (i.e., the elements $\hat{a}_{ij}$, $\hat{v}_{ij}$ in the above notation) according to (19) and (22), respectively.

Convergence. The iteration can be terminated as soon as all off-diagonal elements of A_k are negligibly small; the only question is whether this ever happens. Now on the basis of (20), and our choice of ϕ,

$$S_{k+1} = S_k - 2a_{pq}^2.$$

Moreover, a_{pq}^2 as the square of the maximum modulus off-diagonal element is larger than or equal to the mean value $S_k/(n(n-1))$, hence

$$S_{k+1} \le S_k \left[1 - \frac{2}{n(n-1)} \right] ,$$

$$S_k \le S_0 \left[1 - \frac{2}{n(n-1)} \right]^k \to 0 \text{ as } k \to \infty . \tag{23}$$

This establishes convergence:

Theorem 12.5. *In the classical Jacobi method the sum S_k of the squares of all off-diagonal elements of A_k converges to 0 monotonically and (at least) linearly.*

§12.4. Programming considerations

In computational practice, a Jacobi rotation (18) is executed "in place", that is, a matrix element a_{ij} is stored as $a[i,j]$ regardless of whether it is an element of A_k or of A_{k+1}. Accordingly, such a step proceeds as follows:

1) Determination of p and q through a search of the maximum off-diagonal element. This search is considerably facilitated by keeping a list in which for each row one enters the location and magnitude of the maximum modulus element of that row[1].

2) Determination of $c = \cos \phi$ and $s = \sin \phi$. In the first place, one does not compute $\tan (2\phi)$, but

$$\cot (2\phi) = \frac{a_{qq} - a_{pp}}{2a_{pq}} .$$

(The denominator in this formula, in contrast to the numerator, cannot become 0, since one always works with the maximum modulus off-diagonal element a_{pq}.)

3) From the quantity $ct = \cot (2\phi)$ one finds $t = \tan (\phi)$ through solving the quadratic equation

$$t^2 + 2 \times ct \times t - 1 = 0.$$

Since we seek an angle ϕ with $|\phi| \leq \pi/4$, we must have $|t| \leq 1$; hence one takes the absolutely smaller root. In ALGOL notation,

$$t: = 1/(\text{abs}(ct) + \text{sqrt}(1 + ct \uparrow 2));$$

is already the absolute value of the smaller root; since the latter must have the same sign as ct, one has to add

if $ct < 0$ **then** $t: = - t;$

---

[1] Corbato F.J.: On the coding of Jacobi's method for computing eigenvalues and eigenvectors of real symmetric matrices, *J. Assoc. Comput. Mach.* **10**, 123–125 (1963).

Finally,

$$c := 1/\mathrm{sqrt}\,(1 + t \uparrow 2); \quad s := c \times t;$$

4) Since one works only with elements on or above the diagonal (and thereby reduces the computational work by almost 50%), an element $a[i,j]$ with $i > j$ occurring in the program must be replaced immediately by $a[j,i]$.

5) There are special formulas for the elements $\hat{a}_{pq}, \hat{a}_{pp}, \hat{a}_{qq}$. Namely, $\hat{a}_{pp} + \hat{a}_{qq} = a_{pp} + a_{qq}$, and

$$\begin{aligned}
\hat{a}_{pp} - a_{pp} &= (a_{qq} - a_{pp}) \sin^2\phi - 2a_{pq} \cos\phi \sin\phi \\
&= \tan\phi \, [(a_{qq} - a_{pp}) \cos\phi \sin\phi - 2a_{pq} \cos^2\phi] \\
&= -\tan\phi \, [\tfrac{1}{2} (a_{pp} - a_{qq}) \sin(2\phi) + a_{pq} \cos(2\phi) + a_{pq}] \\
&= -\tan\phi \, [\hat{a}_{pq} + a_{pq}].
\end{aligned}$$

Since $\hat{a}_{pq} = 0$ (by the choice of ϕ), one thus has

$$\begin{aligned}
\hat{a}_{pp} &= a_{pp} - a_{pq} \tan\phi, \\
\hat{a}_{qq} &= a_{qq} + a_{pq} \tan\phi, \\
\hat{a}_{pq} &= 0.
\end{aligned} \tag{24}$$

These formulae are much less susceptible to rounding errors than the original ones, and besides save computing time.

6) If the pivot element a_{pq} is very small in comparison to *both* diagonal elements a_{pp} and a_{qq}, that is, if in machine arithmetic

$$a_{pp} + a_{pq} = a_{pp} \quad \text{and} \quad a_{qq} + a_{pq} = a_{qq},$$

then the rotation is not carried out at all, and one simply puts $\hat{a}_{pq} = 0$. This, admittedly, is a falsification, but it is not larger than the one that would occur anyhow during the execution of the rotation on account of rounding errors.

Altogether one obtains for the rotation proper (after p, q, t, c, s have been determined), in observance of 4) and 5), the following piece of program:

$$h := a_{pq} \times t;$$
$$a_{pp} := a_{pp} - h;$$
$$a_{qq} := a_{qq} + h;$$
$$a_{pq} := 0;$$
for $j := 1$ **step** 1 **until** $p - 1$ **do**
begin
    $h := c \times a_{jp} - s \times a_{jq};$
    $a_{jq} := s \times a_{jp} + c \times a_{jq};$
    $a_{jp} := h$
end;
for $j := p + 1$ **step** 1 **until** $q - 1$ **do**
begin
    $h := c \times a_{pj} - s \times a_{jq};$
    $a_{jq} := s \times a_{pj} + c \times a_{jq};$
    $a_{pj} := h$
end;
for $j := q + 1$ **step** 1 **until** n **do**
begin
    $h := c \times a_{pj} - s \times a_{qj};$
    $a_{qj} := s \times a_{pj} + c \times a_{qj};$
    $a_{pj} := h$
end
for $j := 1$ **step** 1 **until** n **do**
begin
    $h := c \times v_{jp} - s \times v_{jq};$
    $v_{jq} := s \times v_{jp} + c \times v_{jq};$
    $v_{jp} := h$
end;

This completes *one* Jacobi step. The process is continued iteratively, and is terminated as soon as $a_{ij} = 0$ for all i, j with $i < j$, which through the measure in 6) is only accelerated. At the end, the a_{jj} are approximately the eigenvalues, and the columns of the matrix **V** the associated eigenvectors.

§12.5. The cyclic Jacobi method

Having to locate the off-diagonal element of maximum modulus in each step of the classical Jacobi method is somewhat awkward. To spare oneself this trouble, the following variant has been proposed when the method was rediscovered([1]) in 1952:

One chooses as pivot element a_{pq} in turn (rowwise from left to right) all elements above the diagonal and each time carries out a rotation which makes the pivot element equal to 0. After $n(n-1)/2$ rotations, all off-diagonal elements have each been "treated" once. One calls this a sweep. After one such sweep, not all off-diagonal elements, of course, are 0; rather, additional sweeps must be followed. Usually, 10 of them suffice.

One observes in practice that the off-diagonal elements at first decrease only slowly, but then become smaller faster and faster. The latter is due to the *quadratic convergence* of this *cyclic Jacobi method*, which we now propose to verify under the assumption of simple eigenvalues:

Let the moduli of all off-diagonal elements at the beginning of a sweep be less than or equal to ε, where ε is assumed small in comparison with the minimum difference δ of any two diagonal elements. First an auxiliary observation: we write for brevity $a = a_{pp}$, $b = a_{qq}$, $\varepsilon_0 = a_{pq}$, $\varepsilon_1 = a_{pj}$, $\varepsilon_2 = a_{qj}$, so that

$$
\mathbf{A} = \begin{bmatrix}
\ddots & & & & & & & \\
& a & \cdots & \varepsilon_0 & \cdots & \varepsilon_1 & \cdots & \\
& & \ddots & & & & & \\
& & & b & \cdots & \varepsilon_2 & \cdots & \\
& & & & \ddots & & & \\
& \text{sym.} & & & & \ddots & & \\
& & & & & & \ddots &
\end{bmatrix} .
$$

Then the angle of rotation ϕ is determined by

---

[1] Gregory R.T.: Computing eigenvalues and eigenvectors of a symmetric matrix on the ILLIAC, *Math. Tables Aids Comput.* **7**, 215–220 (1953).

$$\tan(2\phi) = \frac{2\varepsilon_0}{b-a} \ll 1,$$

from which

$$\tan \phi \approx \frac{\varepsilon_0}{b-a}, \quad \sin \phi \approx \phi, \quad \cos \phi \approx 1, \quad |\phi| \approx |\tan \phi| < \frac{\varepsilon_0}{\delta}.$$

Then,

$$\hat{\varepsilon}_1 \approx \varepsilon_1 - \varepsilon_2\phi, \quad |\hat{\varepsilon}_1 - \varepsilon_1| \leq \frac{\varepsilon_0\varepsilon_2}{\delta},$$

$$\hat{\varepsilon}_2 \approx \varepsilon_1\phi + \varepsilon_2, \quad |\hat{\varepsilon}_2 - \varepsilon_2| \leq \frac{\varepsilon_0\varepsilon_1}{\delta}.$$

With this, the change of the off-diagonal elements after one single rotation is estimated (asymptotically). During a complete sweep, however, each element is affected at most n times, hence can at most n times be enlarged by the product of two off-diagonal elements divided by δ. But for such a product one has, at each stage,

$$|\varepsilon_0\varepsilon_i| \leq \frac{1}{2}(\varepsilon_0^2 + \varepsilon_i^2) \leq \frac{1}{4} S < \frac{1}{4} n^2\varepsilon^2,$$

where S is the sum of squares of all off-diagonal elements defined in (14). (Note on account of (20) that S also here can never increase.) The total change of an element during the sweep, measured from the moment where as pivot element it has been rotated to 0, is thus smaller than $n^3\varepsilon^2/(4\delta)$. It can therefore not become larger than this bound; in particular, after the sweep, it is of the order $O(\varepsilon^2)$, q.e.d.

In spite of this quadratic convergence, one often hears complaints about the excessive computational work involved in the Jacobi method. In fact, 10 sweeps amount to about $5n^2$ Jacobi rotations, of which each requires $8n$ multiplications and $4n$ additions, which gives a total of about $60n^3$ arithmetic operations. For small n, this is not serious, but for large n one prefers other methods.

§12.6. The LR transformation

The large computational work for determining the eigenvalues of a matrix is felt to be particularly annoying when one is dealing with the following special case: *The matrix* A *is a symmetric positive definite band matrix* (i.e., $a_{ij} = 0$ if $|i - j| > m$), *and one needs only the* p *smallest eigenvalues* (p and the bandwidth ([1]) m are assumed small in comparison with the order n of the matrix).

Example. The eigenoscillations of a beam are described by the differential equation

$$\frac{\partial^2 u}{\partial t^2} = -\frac{\partial^4 u}{\partial x^4} \quad (0 \le x \le 1, \; -\infty < t < \infty). \tag{25}$$

If the beam is simply supported at each end, the boundary conditions are

$$u(0,t) = u(1,t) \equiv 0, \quad u_{xx}(0,t) = u_{xx}(1,t) \equiv 0. \tag{26}$$

Seeking the solution in the form $u(x,t) = e^{i\omega t} y(x)$, one obtains

$$y^{(4)} = \omega^2 y \; \text{ with } \; y(0) = y(1) = y''(0) = y''(1) = 0,$$

and from this, through discretization in the x-direction ($x_k = kh$ with $h = 1/n$), finally

$$\mathbf{Ay} = \omega^2 \mathbf{y},$$

with the $(n - 1 \times n - 1)$ - matrix

$$\mathbf{A} = n^4 \begin{bmatrix} 5 & -4 & 1 & & & & & & 0 \\ -4 & 6 & -4 & 1 & & & & & \\ 1 & -4 & 6 & -4 & 1 & & & & \\ & 1 & -4 & 6 & -4 & 1 & & & \\ & & \ddots & \ddots & \ddots & \ddots & \ddots & & \\ & & & \ddots & \ddots & 1 & -4 & 6 & -4 \\ 0 & & & & & & 1 & -4 & 5 \end{bmatrix}. \tag{27}$$

---

[1] Some authors refer to $2m + 1$ as bandwidth of **A**. (Translator's remark)

Here, the bandwidth is $m=2$, and one needs only the smallest eigenvalues, since only these, to some extent, are useful approximations to the eigenvalues of the continuous problem. (The latter are $\lambda_k = \pi^4 k^4$ $(k = 1,2, \ldots)$, while A has the eigenvalues $\lambda_k = 16n^4 \sin^4 (\pi k/(2n))$.)

In a case like that, the Jacobi method would have the following disadvantages:

a) In the course of the process, the whole matrix A is filled with nonzero numbers, which then have to be made 0 again.

b) One must compute all eigenvalues of A.

There exists, however, a method in which the many zeros in A, as well as the fact that only a few eigenvalues are needed, can be usefully exploited, namely the *LR transformation*:

Given an arbitrary symmetric positive definite matrix A, one applies to A an *LR step*: One decomposes $A_0 = A$ by Cholesky into $A_0 = R_0^T R_0$ and then computes $A_1 = R_0 R_0^T$. Thereupon, an *LR* step can be applied again to A_1, which gives A_2, etc. The kth step is:

$$A_k = R_k^T R_k \ \text{(by Cholesky)}, \ A_{k+1} = R_k R_k^T. \tag{28}$$

Theorem 12.6. *If $A = A_0$ is symmetric and positive definite, then for the matrices A_k generated iteratively by means of the LR transformation (28) one has*

$$\lim_{k \to \infty} A_k = \text{diag} (\lambda_1, \lambda_2, \ldots, \lambda_n),$$

where $\lambda_1, \ldots, \lambda_n$ are the eigenvalues of A.

Moreover, the eigenvalues in the limit matrix are usually ordered: $\lambda_1 \geq \lambda_2 \geq \cdots \geq \lambda_n$. There are exceptions, however: for example, if

$$A = \begin{bmatrix} 5 & 4 & 1 & 1 \\ 4 & 5 & 1 & 1 \\ 1 & 1 & 4 & 2 \\ 1 & 1 & 2 & 4 \end{bmatrix}, \ \text{then} \ \lim_{k \to \infty} A_k = \begin{bmatrix} 10 & 0 & 0 & 0 \\ 0 & 1 & 0 & 0 \\ 0 & 0 & 5 & 0 \\ 0 & 0 & 0 & 2 \end{bmatrix}.$$

Nevertheless, such convergence is unstable, and during numerical computation rounding errors cause the pattern to tip over, so that the limit matrix is still diag $(10, 5, 2, 1)$.

Proof of Theorem 12.6. a) Since

$$A_{k+1} = R_k R_k^T = R_k (R_k^T R_k) R_k^{-1} = R_k A_k R_k^{-1},$$

all matrices A_k are similar to one another. Therefore, if $\lim A_k$ is a diagonal matrix, then its diagonal elements are the eigenvalues of A.

b) If $s_\ell^{(k)}$ denotes the sum of the ℓ first diagonal elements of A_k:

$$s_\ell^{(k)} = \sum_{i=1}^{\ell} a_{ii}^{(k)},$$

then certainly (since A_k is positive definite),

$$0 < s_1^{(k)} < s_2^{(k)} < \cdots < s_{n-1}^{(k)} < s_n^{(k)},$$

where the trace $s = s_n^{(k)}$ is independent of k. Now $A_k = R_k^T R_k$ evidently means

$$a_{jj}^{(k)} = \sum_{i=1}^{j} (r_{ij}^{(k)})^2,$$

while from $A_{k+1} = R_k R_k^T$ there follows that

$$a_{ii}^{(k+1)} = \sum_{j=i}^{n} (r_{ij}^{(k)})^2.$$

Summing over j, respectively i, from 1 to ℓ, one gets

$$s_\ell^{(k)} = \sum_{j=1}^{\ell} \sum_{i=1}^{j} (r_{ij}^{(k)})^2 = \sum_{i=1}^{\ell} \sum_{j=i}^{\ell} (r_{ij}^{(k)})^2,$$

$$s_\ell^{(k+1)} = \sum_{i=1}^{\ell} \sum_{j=i}^{n} (r_{ij}^{(k)})^2 \,,$$

thus,

$$s_\ell^{(k+1)} - s_\ell^{(k)} = \sum_{i=1}^{\ell} \sum_{j=\ell+1}^{n} (r_{ij}^{(k)})^2 \geq 0. \tag{29}$$

Therefore, the sequence $s_\ell^{(k)}$ is monotone in k, and also bounded; it must thus converge. Consequently, by (29),

$$\lim_{k \to \infty} r_{ij}^{(k)} = 0 \text{ for } i \leq \ell, \ j > \ell.$$

Since this is true for each ℓ, there follows:

$$\lim_{k \to \infty} r_{ij}^{(k)} = 0 \text{ whenever } j > i.$$

Furthermore, the existence of $\lim_{k \to \infty} s_\ell^{(k)}$ implies immediately that of

$$\lim_{k \to \infty} (r_{\ell\ell}^{(k)})^2 = \lim_{k \to \infty} (s_\ell^{(k)} - s_{\ell-1}^{(k)}) \,.$$

Therefore, $\lim \mathbf{R}_k$ exists and is a diagonal matrix; finally

$$\lim_{k \to \infty} \mathbf{A}_k = \lim_{k \to \infty} (\mathbf{R}_k^T \mathbf{R}_k)$$

likewise becomes diagonal, q.e.d.

Speed of convergence. Convergence as such, in numerical analysis, does not mean a great deal yet. Indeed, a computational process is only then useful, in practice, when convergence is *good.* Here we propose to estimate convergence of the *LR* transformation under the assumption that $\mathbf{A}_k$ is already nearly diagonal:

$$\mathbf{A}_k = \begin{bmatrix} \lambda_1 & \varepsilon_{12} & \varepsilon_{13} & \cdots & \varepsilon_{1n} \\ & \lambda_2 & \varepsilon_{23} & & \varepsilon_{2n} \\ & & \cdot & \cdot & \\ & & & \cdot & \\ \text{sym.} & & & & \lambda_n \end{bmatrix}.$$

Then, with $\mathbf{R}_k = [r_{ij}]$,

$$r_{11} = \sqrt{\lambda_1}, \quad r_{12} = \varepsilon_{12} / \sqrt{\lambda_1}, \ldots, r_{1n} = \varepsilon_{1n} / \sqrt{\lambda_1},$$

$$r_{22} = \sqrt{\lambda_2 - r_{12}^2} = \sqrt{\lambda_2 - \varepsilon_{12}^2 / \lambda_1} \approx \sqrt{\lambda_2},$$

$$r_{23} = (\varepsilon_{23} - r_{12} r_{13}) / r_{22} \approx \varepsilon_{23} / \sqrt{\lambda_2}, \text{ etc.}$$

Altogether,

$$\mathbf{R}_k \approx \begin{bmatrix} \sqrt{\lambda_1} & \dfrac{\varepsilon_{12}}{\sqrt{\lambda_1}} & \dfrac{\varepsilon_{13}}{\sqrt{\lambda_1}} & \cdots & \dfrac{\varepsilon_{1n}}{\sqrt{\lambda_1}} \\[2ex] & \sqrt{\lambda_2} & \dfrac{\varepsilon_{23}}{\sqrt{\lambda_2}} & \cdots & \dfrac{\varepsilon_{2n}}{\sqrt{\lambda_2}} \\[2ex] & & \cdot & \cdot & \\ & & & \cdot & \\ 0 & & & & \sqrt{\lambda_n} \end{bmatrix},$$

and thus, neglecting quantities of the order ε^2:

$$
\mathbf{A}_{k+1} = \mathbf{R}_k \mathbf{R}_k^T \approx
\begin{bmatrix}
\lambda_1 & \varepsilon_{12}\left[\dfrac{\lambda_2}{\lambda_1}\right]^{1/2} & \varepsilon_{13}\left[\dfrac{\lambda_3}{\lambda_1}\right]^{1/2} & \cdots & \varepsilon_{1n}\left[\dfrac{\lambda_n}{\lambda_1}\right]^{1/2} \\
 & \lambda_2 & \varepsilon_{23}\left[\dfrac{\lambda_3}{\lambda_2}\right]^{1/2} & \cdots & \varepsilon_{2n}\left[\dfrac{\lambda_n}{\lambda_2}\right]^{1/2} \\
 & & & \cdot & \\
 & \text{sym.} & & \cdot & \\
 & & & & \lambda_n
\end{bmatrix}
.\tag{30}
$$

We therefore have:

Theorem 12.7. *If all eigenvalues of* **A** *are simple, then under the hypotheses of Theorem 12.6 one has, asymptotically: The* (i,j)-*element* $(j > i)$ *of the matrix* $\mathbf{A}_k$ *converges to 0 like* $(\lambda_j/\lambda_i)^{k/2}$ *as* $k \to \infty$.

From these considerations it already follows that convergence of the $\mathbf{A}_k$ to a diagonal matrix with unordered diagonal elements cannot be a stable convergence. One further recognizes that convergence is very bad when the relative difference of two eigenvalues is very small, as, say, in the case $\lambda_5 = 7.49$, $\lambda_6 = 7.47$. What, then, is the advantage of this method?

First advantage: If $\mathbf{A}_0$ is a band matrix, then so are $\mathbf{A}_1$, $\mathbf{A}_2$, ... Namely, if $a_{ij}^{(0)} = 0$ for $|i - j| > m$, then $r_{ij}^{(0)} = 0$ for $j < i$ and $j > i + m$, that is, $\mathbf{R}_0$ is a triangular matrix with bandwidth m. But then also, $\mathbf{A}_1 = \mathbf{R}_0 \mathbf{R}_0^T$ receives the same band form as $\mathbf{A}_0$. The same, of course, holds for $\mathbf{A}_2$, $\mathbf{A}_3$, etc.

This has the consequence that in a computational process one must only account for the elements inside the band; it suffices to store the matrices **A** and **R** as **array** $a, r\,[1{:}n,\ 0{:}m\,]$, where $a\,[i,j\,]$ means the matrix element $a_{i,i+j}$ $(= a_{i+j,i})$. The storage requirement, therefore, is reduced from n^2 to $(m + 1)n$, which means, for example, that a band matrix with $n = 2000$, $m = 5$ can still be stored in a medium-sized machine. Naturally, when this band storage is used, the *LR* method requires special programming.

The computational work for an *LR* step with dense matrix is about equal to $n^3/3$; for a band matrix it is reduced to about nm^2, thus in the above example from $2.667_{10}9$ to 50000, that is, by a factor of more than 50000.

Second advantage: *If one wants only some of the smallest eigenvalues, then comparatively few LR steps usually suffice.* These small eigenvalues, after a number of *LR* steps, assemble at the lower end of the diagonal. This goes rather fast, since the ratios λ_j/λ_i at the lower end of the spectrum are by nature small (provided there are no multiple eigenvalues). The matrices A_k therefore soon will show large off-diagonal elements only at the top, but at the bottom will already be diagonalized.

But even if, say, λ_n, λ_{n-1} and λ_{n-2} are nearly equal, this will only have the effect that at the lower right-hand corner a few off-diagonal elements develop which do not want to decrease, but which are distinctly separated in magnitude from the remaining off-diagonal elements. Then the lower end of the diagonal of A_k, for example, looks as follows:

$$
\begin{bmatrix}
\ddots & & \cdot & & \cdot & & \cdot & \\
 \cdot & \ddots & & \cdot & & \cdot & & \\
81.2351 & .0015 & -.0014 & .0001 & 0 & 0 & 0 & \cdot \\
 \cdot & 10.0259 & 1.2573 & -.0087 & -.0007 & 0 & 0 & \\
 \vdots & & 10.3592 & .0098 & .0003 & .0017 & 0 & \\
 \vdots & & & 4.3259 & .0238 & -.0142 & .0108 & \cdot \\
 \vdots & & & & 1.6928 & .1589 & .2725 & \\
 \vdots & \text{sym.} & & & & 1.6259 & -.0867 & \\
 \vdots & & & & & & 1.7235 &
\end{bmatrix}
$$

Here, the smallest eigenvalues can be easily obtained, for example by diagonalizing the 6×6 principal minor matrix at the lower right-hand corner by means of Jacobi's method. (There result the eigenvalues 11.4608, 8.9243, 4.3261, 1.9879, 1.7238, 1.3302.)

§12.7. The LR transformation with shifts

Since in the *LR* transformation the off-diagonal elements of the last column A_k converge to 0 like $(\lambda_n/\lambda_i)^{k/2}$ ($i = 1, \ldots, n-1$), one can accelerate this convergence by applying the transformation not to the matrix A, but to $A - tI$. Of course, one must have $t < \lambda_n$, since $A - tI$

must be positive definite.

The convergence of the element a_{in} then is given by

$$a_{in}^{(k)} = O\left[\left[\frac{\lambda_n - t}{\lambda_i - t} \right]^{k/2} \right], \tag{31}$$

and this becomes smaller the closer t moves to λ_n. The trick we have to accomplish, therefore, is to choose t as close to λ_n as possible, but in no case larger than this number.

The way one does this, in practice, is as follows: One first chooses d_0, carries out the decomposition $A_0 - d_0 I = R_0^T R_0$ and puts $A_1 = R_0^T R_0$; then one chooses d_1, decomposes $A_1 - d_1 I = R_1^T R_1$ (at this point, $t = d_0 + d_1$) and computes $A_2 = R_1 R_1^T$, etc. In the kth step one thus chooses a $d_k > 0$, but so, that it is smaller than the smallest eigenvalue of A_k. The total shift t in this way continually increases. A guide for this choice is provided by the smallest diagonal element, which, however, is an upper bound for $\lambda_{\min}(A_k)$. It thus can happen that d_k is chosen too large. This is signaled in the failure of the Cholesky decomposition of $A_k - d_k I$. This decomposition must then be repeated with a smaller d_k. One therefore uses a computational scheme somewhat as follows:

1. Choice of d_k; $s := 1$;
2. Cholesky decomposition $A_k - d_k I = R_k^T R_k$;
3. If the decomposition fails:
   $s := s + 1$; $d_k := d_k/2$;
   if $s > 3$ then $d_k := 0$;
   goto 2;
4. $A_{k+1} := R_k R_k^T$
5. $t := t + d_k$
6. **goto** 1;

Thus, t is here the sum of all shifts, so that A is similar to $A_k + tI$. If initially the Cholesky decomposition fails, one makes two more trials with reduced shifts, then sets $d_k = 0$. What is bad is only when the decomposition fails in the last case too, which can happen because of rounding errors.

There exists, however, a type of failure in the Cholesky decomposition which leads immediately to a safe, and usually also very sharp, lower bound for the smallest eigenvalue of A_k. Namely, if it is only the last diagonal element prior to taking the square root, thus

$$c = a_{nn}^{(k)} - \sum_{i=1}^{n-1} (r_{in}^{(k)})^2 - d_k,$$

which becomes negative, then $d_k + c$ is a lower bound for $\lambda_{\min}(A_k)$. Therefore, the decomposition *must* succeed (theoretically) if one repeats it with $d_k := d_k + c$.

This property of $d_k + c$ can be proved as follows: If the Cholesky decomposition has advanced through the $(n-1)$st row, then from the original quadratic form

$$\sum_{i=1}^{n} \sum_{j=1}^{n} a_{ij} x_i x_j - d \sum_{i=1}^{n} x_i^2$$

one has already subtracted the $n-1$ squares

$$\left[\sum_{j=\ell}^{n} r_{\ell j} x_j \right]^2, \quad \ell = 1, \ldots, n-1,$$

and there remains cx_n^2, where $\sqrt{c}$ then gives the last diagonal element r_{nn}. For the quadratic form associated with A, therefore,

$$Q(x) = \sum_{i=1}^{n} \sum_{j=1}^{n} a_{ij} x_i x_j = d \sum_{i=1}^{n} x_i^2 + \sum_{\ell=1}^{n-1} \left[\sum_{j=\ell}^{n} r_{\ell j} x_j \right]^2 + cx_n^2.$$

If now $c < 0$, then

$$Q(x) \geq d \sum_{i=1}^{n} x_i^2 + cx_n^2 \geq (d+c) \sum_{i=1}^{n} x_i^2; \quad \text{q.e.d.}$$

Numerical example. Consider the matrix

$$
A_0 = \begin{bmatrix} 101 & 10 & 1 \\ 10 & 11 & 1 \\ 1 & 1 & 1 \end{bmatrix} .
$$

The Cholesky decomposition with $d_0 = 1$ yields

$$
\begin{bmatrix} 100 & 10 & 1 \\ 10 & 10 & 1 \\ 1 & 1 & 0 \end{bmatrix} \rightarrow \begin{bmatrix} 10 & 1 & .1 \\ 0 & 3 & .3 \\ 0 & 0 & \sqrt{-.1} \end{bmatrix} ,
$$

that is, breaks down in the last step with the remainder element $c = 0 - (.1)^2 - (.3)^2 = -.1$. Therefore, $d_0 + c = .9$ is a lower bound for the smallest eigenvalue of A_0. Repeating the *LR* step with $d_0 = .9$ then gives

$$
A_1 = \begin{bmatrix} 101.10901 & 3.04509 & .00315 \\ & 9.19002 & .00939 \\ \text{sym.} & & .00099 \end{bmatrix} , \quad t = .9 ,
$$

A further step with $d_1 = .00099$ again produces a failure in the last diagonal element, this time with $c = -.00001$. The step, therefore, must be repeated with $d_1 = .00098$; in this way one finds (with a computing precision of 5 digits after the decimal point):

$$
A_2 = \begin{bmatrix} 101.19975 & .91342 & 0 \\ .91342 & 9.09735 & 0 \\ 0 & 0 & 0 \end{bmatrix} , \quad t = .90098.
$$

We observe:

1) 0 is an eigenvalue of A_2, hence $\lambda_3 = .90098$ an eigenvalue of A_0, and in fact guaranteed the smallest, since $A_2 = R_1 R_1^T$ cannot have negative eigenvalues.

2) The remaining eigenvalues of A_2 are those of the (2×2)-submatrix

$$\begin{bmatrix} 101.19975 & .91342 \\ .91342 & 9.09735 \end{bmatrix} .$$

3) One can carry out a *deflation*, that is, strike out the last row and column of A_2, and then continue with the *LR* transformation, whereby one can build on top of the t already obtained.

The next *LR* step with $d_2 = 9$ leads to

$$A_3 = \begin{bmatrix} 92.20880 & .02827 \\ .02827 & .08830 \end{bmatrix} , \quad t = 9.90098 .$$

Then, with $d_3 = .08830$, one first runs into a failure with $c = -.00001$, and with $d_3 = .08829$ finally obtains

$$A_3 = \begin{bmatrix} 92.12052 & 0 \\ 0 & 0 \end{bmatrix} , \quad t = 9.98927 .$$

A_0 therefore has the eigenvalues $\lambda_1 = 102.10979$, $\lambda_2 = 9.98927$, $\lambda_3 = .90098$.

§12.8. The Householder transformation

The computational work for determining the eigenvalues of a symmetric matrix A, also in the case where it is dense, can be significantly reduced by first transforming it to band form. For this purpose a special class of orthogonal matrices is useful:

Let w be a vector of length 1; we use it to form the matrix

$$H = I - 2ww^T, \tag{32}$$

that is, the symmetric matrix with elements

$$h_{ij} = \delta_{ij} - 2w_i w_j.$$

We first observe that

$$H^T H = H^2 = I - 4ww^T + 4(ww^T)(ww^T).$$

Here, $ww^T ww^T$ is the matrix with elements

$$\sum_{j=1}^{n} (w_i w_j)(w_j w_k) = w_i w_k \sum_{j=1}^{n} w_j^2 = w_i w_k ,$$

thus again the matrix ww^T. Therefore, $H^T H = I$, that is, H is orthogonal.

Now let A be subjected to the orthogonal transformation given by H:

$$H^T A H = H A H = (I - 2ww^T) A (I - 2ww^T)$$

$$= A - 2ww^T A - 2Aww^T + 4ww^T Aww^T.$$

Here, $ww^T Aww^T$, since $w^T A w$ is a scalar $Q(w)$, is equal to the matrix $Q(w)ww^T$, so that

$$H^T A H = A - 2w(Aw - Q(w)w)^T - 2(Aw - Q(w)w)w^T$$

$$= A - 2wg^T - 2gw^T , \tag{33}$$

where

$$g = Aw - Q(w)w. \tag{34}$$

Therefore, $B = H^T A H$ has the elements

$$b_{ij} = a_{ij} - 2w_i g_j - 2w_j g_i. \tag{35}$$

The g_j here are functions of the w_i; in view of $||w||_2 = 1$, one thus has exactly $n - 1$ degrees of freedom, which could be used, in principle, to satisfy $n - 1$ conditions, for example $b_{in} = 0$ for $i = 1, 2, \ldots, n - 1$. Then $\mathbf{B}$ would be of the form

$$\mathbf{B} = \begin{bmatrix} * & * & . & . & . & * & 0 \\ * & * & . & . & . & * & 0 \\ . & . & & & & . & . \\ . & . & & & & . & . \\ . & . & & & & . & . \\ * & * & . & . & . & * & 0 \\ 0 & 0 & . & . & . & 0 & \lambda_n \end{bmatrix}.$$

Unfortunately, this is not feasible. One can, however, solve a simpler problem by giving up one degree of freedom and trying, instead, to satisfy only $n - 2$ conditions:

$$w_n = 0 \quad \text{(1 fewer degree of freedom)},$$
$$b_{1n} = b_{2n} = \cdots = b_{n-2,n} = 0 \quad (n - 2 \text{ conditions}). \tag{36}$$

The relation (35) for $j = n$ then reduces to

$$b_{in} = a_{in} - 2g_n w_i$$

(which is to be equal to 0 for $i = 1, \ldots, n - 2$) with

$$g_n = \sum_{j=1}^{n-1} a_{jn} w_j. \tag{37}$$

From this, one obtains:

$$w_i = \frac{a_{in}}{2g_n}, \quad i = 1, \ldots, n-2,$$

$$w_{n-1} = \frac{a_{n-1,n} - b_{n-1,n}}{2g_n}.$$

$$(38)$$

For brevity, we henceforth write a and b in place of $a_{n-1,n}$ and $b_{n-1,n}$, respectively, so that $w_{n-1} = (a - b)/(2g_n)$. For $\mathbf{w}$ to be a vector of length 1, one must have

$$1 = \sum_{i=1}^{n-2} \frac{a_{in}^2}{4g_n^2} + \frac{a^2 - 2ab + b^2}{4g_n^2},$$

thus

$$4g_n^2 = \sum_{i=1}^{n-1} a_{in}^2 - 2ab + b^2. \qquad (39)$$

Furthermore, by (37) and (38),

$$g_n = \sum_{j=1}^{n-2} \frac{a_{jn}^2}{2g_n} + \frac{a(a - b)}{2g_n},$$

that is,

$$2g_n^2 = \sum_{j=1}^{n-1} a_{jn}^2 - ab. \qquad (40)$$

Comparison between (39) and (40) shows that

$$b^2 = \sum_{j=1}^{n-1} a_{jn}^2.$$

Consequently, one puts

$$b = \pm \sqrt{\sigma}, \quad g_n = \sqrt{\frac{\sigma - ab}{2}}, \quad \text{where} \quad \sigma = \sum_{j=1}^{n-1} a_{jn}^2 . \tag{41}$$

Of eminent importance, here, is the choice of the sign in the square root for b. If a and b have the same sign, the computation of g_n may be subject to cancellation, which could lead to an inaccurate vector $\mathbf{w}$. One thus chooses the sign of b opposite to the one of a:

$$b := \text{if } a > 0 \text{ then } - \text{sqrt}(\sigma) \text{ else sqrt}(\sigma);$$

Subsequently, g_n is determined by (41), and w_j by (38). Finally, one computes

$$z_k = \sum_{j=1}^{n-1} a_{kj} w_j, \quad k = 1, \ldots, n-1,$$

$$\tag{42}$$

$$Q = \sum_{k=1}^{n-1} w_k z_k$$

and, according to (34),

$$g_k = z_k - Q w_k, \quad k = 1, \ldots, n-1, \tag{43}$$

whereupon the actual transformation (35) can be carried out.

There are a few details, though, that still need to be noted:

1) After the transformation $\mathbf{A} \to \mathbf{B} = \mathbf{HAH}$ has been carried out, one has to be able, later on, to also transform back a vector $\mathbf{y}$: if $\mathbf{y}$ is eigenvector of $\mathbf{B}$, hence $\mathbf{By} = \lambda \mathbf{y}$, then $\mathbf{HAHy} = \lambda \mathbf{y}$, $\mathbf{AHy} = \lambda \mathbf{Hy}$, that is, $\mathbf{x} = \mathbf{Hy}$ is eigenvector of $\mathbf{A}$. One thus has to compute

$$\mathbf{x} = (\mathbf{I} - 2\mathbf{w}\mathbf{w}^T)\mathbf{y} = \mathbf{y} - 2\mathbf{w}\mathbf{w}^T\mathbf{y} ,$$

for which one first determines $c = \mathbf{w}^T\mathbf{y}$ and then $\mathbf{x} = \mathbf{y} - 2c\mathbf{w}$, requiring only a computing effort proportional to n.

2) Having transformed $\mathbf{A}$ into $\mathbf{A}^{(1)} = \mathbf{B} = \mathbf{HAH}$, which has the form

$$\mathbf{A}^{(1)} = \begin{bmatrix} & & & & 0 \\ & & & & \cdot \\ & \mathbf{A}_1 & & & \cdot \\ & & & & \cdot \\ & & & & 0 \\ \hline & & & & * \\ 0 & \cdot & \cdot & 0 & * & * \end{bmatrix} ,$$

the submatrix $\mathbf{A}_1$ is then treated in the same way. The transformation to be executed,

$$\mathbf{A}_1 \to \mathbf{B}_1 = \mathbf{H}_1 \mathbf{A}_1 \mathbf{H}_1$$

(with matrices of order $n - 1$), by virtue of the special form of the matrices $\mathbf{A}^{(1)}$ and

$$\mathbf{H}^{(1)} = \begin{bmatrix} & & & & 0 \\ & & & & \cdot \\ & \mathbf{H}_1 & & & \cdot \\ & & & & \cdot \\ & & & & 0 \\ & & & & 0 \\ \hline 0 & \cdot & \cdot & 0 & 0 & 1 \end{bmatrix} = \begin{bmatrix} * & \cdot & \cdot & \cdot & * & 0 & 0 \\ \cdot & & & & \cdot & \cdot & \cdot \\ \cdot & & & & \cdot & \cdot & \cdot \\ \cdot & & & & \cdot & \cdot & \cdot \\ * & \cdot & \cdot & \cdot & * & 0 & 0 \\ 0 & \cdot & \cdot & \cdot & 0 & 1 & 0 \\ 0 & \cdot & \cdot & \cdot & 0 & 0 & 1 \end{bmatrix} ,$$

however, has the same effect as the transformation of the corresponding full matrices,

$$\mathbf{A}^{(1)} \to \mathbf{B}^{(1)} = \mathbf{H}^{(1)} \mathbf{A}^{(1)} \mathbf{H}^{(1)} .$$

The result, therefore, is of the form

$$A^{(2)} = B^{(1)} = \begin{bmatrix} & & & & | & 0 & 0 \\ & & & & | & \cdot & \cdot \\ & & A_2 & & | & \cdot & \cdot \\ & & & & | & \cdot & \cdot \\ & & & & | & 0 & 0 \\ \rule{0pt}{0pt} & & & & \rfloor & * & 0 \\ 0 & \cdot\ \cdot\ \cdot & 0 & * & * & * \\ 0 & \cdot\ \cdot\ \cdot & 0 & 0 & * & * \end{bmatrix} .$$

Now one continues processing A_2, which thanks to the special form of $A^{(2)}$ and $H^{(2)}$ again causes no changes outside of A_2 in $A^{(2)}$, etc. After $n - 2$ steps, A is fully transformed to a symmetric tridiagonal matrix J:

$$J = U^T AU, \quad \text{where} \quad U = HH^{(1)} \ \cdots \ H^{(n-3)} . \qquad (44)$$

§12.9. Determination of the eigenvalues of a tridiagonal matrix

After the transformation of a symmetric matrix A to the tridiagonal form J, described in the preceding section, one still needs to determine the eigenvalues (and subsequently perhaps the eigenvectors) of J. For this, we use *Sylvester's law of inertia*: If A is symmetric, and X an arbitrary nonsingular matrix, then A and $X^T AX$ have the same number of positive eigenvalues as well as the same number of negative eigenvalues, and equally many that are 0.

If X is determined such that $X^T(J - tI)X$ is a diagonal matrix Q, the number of positive eigenvalues of $J - tI$ can thus be read off from the diagonal of Q and one knows, then, how many eigenvalues of J are greater than t. By carrying this out for different, suitably selected t, the eigenvalues of J can be estimated accurately.

We put:

$$
J = \begin{bmatrix} d_1 & e_1 & & & & 0 \\ e_1 & d_2 & e_2 & & & \\ & e_2 & d_3 & e_3 & & \\ & & \cdot & \cdot & \cdot & \\ & & & \cdot & \cdot & \cdot \\ 0 & & \cdot & & \cdot & \cdot \end{bmatrix}, \quad X^{-1} = \begin{bmatrix} 1 & x_1 & & & & 0 \\ & 1 & x_2 & & & \\ & & \cdot & \cdot & & \\ & & & \cdot & \cdot & \\ & & & & \cdot & x_{n-1} \\ 0 & & & & & 1 \end{bmatrix},
$$

$$
X^T(J - tI)X = Q = \operatorname{diag}(q_1, q_2, \ldots, q_n). \tag{45}
$$

Then we want the following to hold:

$$
J - tI = \begin{bmatrix} 1 & & & 0 \\ x_1 & 1 & & \\ & x_2 & 1 & \\ & & \cdot & \cdot \\ 0 & & x_{n-1} & 1 \end{bmatrix} \begin{bmatrix} q_1 & & & 0 \\ & q_2 & & \\ & & q_3 & \\ & & & \cdot \\ 0 & & & q_n \end{bmatrix} \begin{bmatrix} 1 & x_1 & & 0 \\ & 1 & x_2 & \\ & & \cdot & \cdot \\ & & & 1 & x_{n-1} \\ 0 & & & 1 \end{bmatrix}.
$$

From this, there follow the equations

$$
q_1 = d_1 - t, \quad q_1 x_1 = e_1,
$$

$$
q_2 + x_1^2 q_1 = d_2 - t, \quad q_2 x_2 = e_2, \text{ etc.,}
$$

in general:

$$
q_k + x_{k-1}^2 q_{k-1} = d_k - t, \quad q_k x_k = e_k, \quad k = 1, \ldots, n,
$$

with $x_0 = 0$, $q_0 = 1$. One can now eliminate the x_k and finds:

$$q_k = d_k - t - e_{k-1}^2/q_{k-1}, \quad k = 1, \ldots, n, \tag{46}$$

with $q_0 = 1$, $e_0 = 0$. (If a denominator q_{k-1} becomes 0 in this iteration, replace it, for example, by 10^{-100}.) *If now $m = m(t)$ of the values $q_1, q_2, \ldots, q_n$ in (46) become positive, then m eigenvalues of $J - tI$ are positive, thus m eigenvalues of J lie above t.*

Example. For the matrix

$$J = \begin{bmatrix} 1 & 1 & & 0 \\ 1 & 3 & 2 & \\ & 2 & 5 & 3 \\ 0 & & 3 & 7 \end{bmatrix} \tag{47}$$

and $t = 0, 1, \ldots, 10$ we obtain the results summarized in Table 12.1. One can see, for example, that the interval [1,2] contains the second-smallest eigenvalue of J. In order to localize it more accurately, one computes $m(t)$ for a sequence of further values of t, selected according to the bisection method; see Table 12.2.

Table 12.1. *Rough localization of the eigenvalues of the matrix (47)*

t	q_1	q_2	q_3	q_4	m
0	1.000000	2.000000	3.000000	4.000000	4
1	10^{-100}	$-_{10}100$	4.000000	3.750000	3
2	−1.000000	2.000000	1.000000	−4.000000	2
3	−2.000000	.500000	−6.000000	5.500000	2
4	−3.000000	−.666667	7.000000	1.714286	2
5	−4.000000	−1.750000	2.285714	−1.937500	1
6	−5.000000	−2.800000	.428571	−20.000000	1
7	−6.000000	−3.833333	−.956522	9.409090	1
8	−7.000000	−4.857143	−2.176471	3.135135	1
9	−8.000000	−5.875000	−3.319149	.711538	1
10	−9.000000	−6.888889	−4.419355	−.963504	0

In this way, through nesting of intervals, the eigenvalues of a symmetric tridiagonal matrix can be localized systematically and in a completely foolproof manner. It can even be shown that $m(t)$ also in numerical computation (that is, in the presence of rounding errors) cannot increase when t increases.

Initially, in practice, one begins with determining

$$a = \min_{1 \le i \le n} (d_i - |e_i| - |e_{i-1}|), \quad b = \max_{1 \le i \le n} (d_i + |e_i| + |e_{i-1}|).$$

According to the theorem of Gershgorin ([1]) all eigenvalues then

Table 12.2. *Determination of the second-smallest eigenvalue of the matrix (47) with the bisection method*

t	q_1	q_2	q_3	q_4	m
1.5	−.500000	3.500000	2.357143	1.681818	3
1.75	−.750000	2.583333	1.701613	−.039100	2
1.625	−.625000	2.975000	2.030462	.942511	3
1.6875	−.687500	2.767045	1.866915	.491712	3
1.71875	−.718750	2.672554	1.784555	.237975	3
1.734375	−.734375	2.627327	1.743165	.102603	3
1.7421875	−.742187	2.605181	1.722410	.032577	3
1.74609375	−.746094	2.594220	1.712017	−.003050	2
1.744140625	−.744141	2.599691	1.717215	.014816	3

lie in the interval $[a,b]$, thus $m(a) = n$, $m(b) = 0$. In order to compute a specific eigenvalue, say the pth one (from above), one immediately applies the bisection method to the function $m(t) - p + .5$([2]):

> **for** $t := (a + b)/2$ **while** $b \ne t \wedge a \ne t$ **do**
> **if** $m(t) \ge p$ **then** $a := t$ **else** $b := t$;

Here, $m(t)$ must be declared as an **integer procedure**; it carries out the recursion (46) and computes m.

---

[1] Gershgorin S.: Über die Abgrenzung der Eigenwerte einer Matrix, Bull. Acad. Sci. USSR Cl. Sci. Math. Nat. **6**, 749–754 (1931). (Editors' remark)

[2] Note that the machine-independent stopping criterion used here leads automatically to the maximum attainable accuracy. (Editors' remark)

Notes to Chapter 12

Careful implementation of some of the techniques described in this chapter appear in Wilkinson & Reinsch [1971, Part II]. This handbook eventually led to the widely used, and easily accessible, collection of Fortran routines called EISPACK. The original guide to this package (see Smith et al. [1976]) addresses the solution of "small" eigenvalue problems, that is, problems whose matrix can easily be stored in the fast memory of the user's computer system. It further emphasizes problems with *dense* matrices. These are matrices that do not have such a small number of nonzero entries that it would be worthwhile to exploit their sparsity. (See the notes to §§12.6–12.8 for a discussion of *sparse* matrices.) The eigenvalue problem for symmetric *band* matrices, the *generalized eigenvalue problem* (cf. Eq. (2) of §12.1) and the *singular value decomposition,* are dealt with in the EISPACK extension, see Garbow et al. [1977].

§12.2 Additional information about the location of eigenvalues of a symmetric matrix A can be had from knowledge of eigenvalues of neighboring (symmetric) matrices B, as well as from the eigenvalues of leading principal submatrices of A. For results along these lines, see Parlett [1980, Ch. 10]. A typical use of such results is in devising termination criteria for iterative methods for computing eigenvalues. A recent monograph on perturbation theory is Bhatia [1987], which addresses not only symmetric, but also normal – and even arbitrary – matrices.

§12.5 Quadratic convergence of the cyclic Jacobi method was first proved by Henrici [1958]. Refined (quadratic) convergence estimates were subsequently obtained by Schönhage [1961], who also showed that the classical Jacobi method (cf. §12.3) still converges "quadratically" (suitably defined!) if multiple eigenvalues are present, as long as they do not have multiplicities exceeding 2. The *general cyclic* Jacobi method consists of sweeps in which all off-diagonal elements are annihilated exactly once in some fixed, but otherwise arbitrary, order. Wilkinson [1962] has established quadratic convergence also for this version of Jacobi's method and showed that multiple eigenvalues improve convergence rates rather than reducing them.

§12.6–12.8 The advantage of the *LR* transformation of preserving the shape of banded matrices is shared by *QR transformations* in which at each stage the matrix A_k is decomposed into a product $Q_k R_k$ of an orthogonal matrix Q_k and an upper triangular matrix R_k, whereupon $A_{k+1} = R_k Q_k = Q_k^T A_k Q_k$. The superdiagonal (i,j)-element of A_k then converges to zero like $(\lambda_j/\lambda_i)^k$ $(j > i)$ as $k \to \infty$. This can be speeded up by suitable shifts, in the same way as described in §12.7 for the *LR* method. In the case of tridiagonal (symmetric) matrices, shift strategies have been designed that lead not only to guaranteed convergence, but indeed to extremely fast convergence – faster than cubic, in general! See Parlett [1980, Ch. 8] for a nice treatment of these techniques. Some applications favor the use of lower triangular matrices, L_k, in place of the R_k, giving rise to *QL transformations*.

Band matrices are an example of *sparse* matrices, i.e., matrices in which the number of nonzero elements is small compared to the total number of elements. While band matrices exhibit a regular pattern of sparsity, there sometimes occur sparse matrices whose pattern of nonzero elements is quite irregular. In such cases, none of the methods discussed here would be particularly suitable, since the similarity transformations employed would ruin sparsity. On the other hand, one is less likely to be interested in *all* eigenvalues, in such cases, but only in *some*, usually a few of the absolutely largest or smallest. This is particularly so for matrices of very large order. It is natural, then, to look for methods that make use only of matrix-times-vector operations Av, where A is the given (sparse) matrix and v an arbitrary vector. In this way, the sparsity pattern of A – however irregular – can be taken into account by an efficient programming of these matrix operations. A prototype of such a method is the *power method* (called v. Mises-Geiringer iteration in §13.2 of Chapter 13) in which one keeps multiplying an arbitrary initial vector by the matrix A. This method usually converges (often rather slowly!) to the absolutely largest eigenvalue and associated eigenvector. There are variants of this method that use not one, but several, vectors – for example, Rutishauser's own *simultaneous iteration method* – that can be successfully used to compute several of the largest eigenvalues and corresponding eigenvectors. An elegant implementation of this method is the algorithm *ritzit* in Wilkinson & Reinsch [1971, Contribution II/9], not included, incidentally, in EISPACK.

Another method is *Lanczos's algorithm*. In principle, Lanczos's techniques are more efficient than simultaneous iterations, but it is not easy to implement them so that their promise is realized. In the early 1950's the Lanczos algorithm was viewed as a technique for reducing a matrix to tridiagonal form. In 1971/72, C.C. Paige showed that it is better to terminate the algorithm early and obtain approximations to a few of the larger eigenvalues. The algorithm is described in Wikinson [1965], but is analyzed more thoroughly in Parlett [1980, Ch. 13], Golub & Van Loan [1989], and Cullum & Willoughby [1985]. The latter work includes a program.

§12.9 The method of *inverse iteration* (proposed by H. Wielandt) is often used in conjunction with the bisection method in order to compute the eigenvector associated with a just computed eigenvalue. For descriptions, the reader may consult Wilkinson [1965], Stewart [1973], and Parlett [1980]. The method of bisection is attractive for any matrix with a small bandwidth, not just for tridiagonal matrices.

References

Bhatia, R. [1987]: *Perturbation Bounds for Matrix Eigenvalues*, Pitman Research Notes in Mathematics Series **162**, Longman Scientific & Technical, New York.

Cullum, J.K. and Willoughby, R.A. [1985]: *Lanczos Algorithms for Large Symmetric Eigenvalue Computations*, Vols. I and II, Birkhäuser, Basel.

Garbow, B.S., Boyle, J.M., Dongarra, J.J. and Moler, C.B. [1977]: *Matrix Eigensystem Routines – EISPACK Guide Extension*, Lecture Notes Comp. Sci. **51**, Springer, New York.

Golub, G.H. and Van Loan, C.F. [1989]: *Matrix Computations*, 2nd ed., The Johns Hopkins University Press, Baltimore.

Henrici, P. [1958]: On the speed of convergence of cyclic and quasicyclic Jacobi methods for computing eigenvalues of Hermitian matrices, *J. Soc. Indust. Appl. Math.* **6**, 144–162.

Parlett, B.N. [1980]: *The Symmetric Eigenvalue Problem,* Prentice-Hall Series in Computational Mathematics, Prentice-Hall, Englewood Cliffs, N.J.

Schönhage, A. [1961]: Zur Konvergenz des Jacobi-Verfahrens, *Numer. Math.* **3**, 374–380.

Smith, B.T., Boyle, J.M., Dongarra, J.J., Garbow, B.S., Ikebe, Y., Klema, V.C. and Moler, C.B. [1976]: *Matrix Eigensystem Routines – EISPACK Guide,* Lecture Notes Comp. Sci. **6**, 2nd ed., Springer, New York.

Stewart, G.W. [1973]: *Introduction to Matrix Computations,* Academic Press, New York.

Wilkinson, J.H. [1962]: Note on the quadratic convergence of the cyclic Jacobi process, *Numer. Math.* **4**, 296–300.

Wilkinson, J.H. [1965]: *The Algebraic Eigenvalue Problem,* Clarendon Press, Oxford. [Paperback edition, 1988].

Wilkinson, J.H. and Reinsch, C. [1971]: *Linear Algebra,* Handbook for Automatic Computation, Vol. II, Springer, New York.

CHAPTER 13

The Eigenvalue Problem For Arbitrary Matrices

§13.1. Susceptibility to errors

The determination of the eigenvalues of nonsymmetric matrices is much more difficult, if for no other reason than the fact that for such matrices a concept analogous to the quadratic form is missing, and consequently, there are no extremal properties either. In accordance with these facts, the statement that eigenvalues are changed only a little by small perturbations in the matrix elements is also no longer valid.

Example. The matrix

$$
A = \begin{bmatrix}
0 & 1 & 0 & 0 & 0 \\
0 & 0 & 1 & 0 & 0 \\
0 & 0 & 0 & 1 & 0 \\
0 & 0 & 0 & 0 & 1 \\
0 & 0 & 0 & 0 & 0
\end{bmatrix}
$$

has the eigenvalues $\lambda_1 = \lambda_2 = \lambda_3 = \lambda_4 = \lambda_5 = 0$. A small perturbation gives rise to the matrix

$$
B = \begin{bmatrix}
0 & 1 & 0 & 0 & 0 \\
0 & 0 & 1 & 0 & 0 \\
0 & 0 & 0 & 1 & 0 \\
0 & 0 & 0 & 0 & 1 \\
10^{-5} & 0 & 0 & 0 & 0
\end{bmatrix},
$$

for which,

$$\mathbf{B} \begin{bmatrix} 10000 \\ 1000 \\ 100 \\ 10 \\ 1 \end{bmatrix} = \begin{bmatrix} 1000 \\ 100 \\ 10 \\ 1 \\ .1 \end{bmatrix},$$

hence $\lambda = .1$ is an eigenvalue.

This susceptibility to perturbations, however, occurs not only for multiple eigenvalues, which by their very nature, as we know, lead to a dangerous situation, but also for distinctly separated eigenvalues. It turns out that such a situation occurs also when for two eigenvalues $\lambda_1 \neq \lambda_2$ the angle between the corresponding eigenvectors x_1, x_2 is small.

To prove this, we consider a matrix $\mathbf{A}$ with n pairwise distinct eigenvalues $\lambda_1, \ldots, \lambda_n$, with eigenvectors $x_1, \ldots, x_n$, and further with eigenvectors $y_1, \ldots, y_n$ of $\mathbf{A}^T$, assuming $||x_i||_2 = 1$, but $x_i^T y_j = \delta_{ij}$, and thus $||y_i||_2 \geq 1$. (Such a normalization is possible.) Perturbation of $\mathbf{A}$ by (a matrix) Δ has the effect that λ_1 and x_1 are perturbed by quantities μ and ξ, respectively:

$$(\mathbf{A} + \Delta)(x_1 + \xi) = (\lambda_1 + \mu)(x_1 + \xi). \tag{1}$$

Using $\mathbf{A}x_1 = \lambda_1 x_1$, and neglecting all quantities which are small of second order, we obtain

$$\mathbf{A}\xi + \Delta x_1 = \lambda_1 \xi + \mu x_1,$$

$$(\mathbf{A} - \lambda_1 \mathbf{I})\xi = (\mu \mathbf{I} - \Delta)x_1.$$

Left-multiplication by y_1^T, on account of

$$y_1^T \mathbf{A} = (\mathbf{A}^T y_1)^T = \lambda_1 y_1^T,$$

yields

$$0 = \mathbf{y}_1^T(\mu\mathbf{I} - \Delta)\mathbf{x}_1,$$

that is,

$$\mu = \frac{\mathbf{y}_1^T \Delta \mathbf{x}_1}{\mathbf{y}_1^T \mathbf{x}_1} = \mathbf{y}_1^T \Delta \mathbf{x}_1 . \tag{2}$$

Furthermore, from $\mathbf{y}_1^T\mathbf{x}_1 = 1$, $\mathbf{y}_1^T\mathbf{x}_2 = 0$, there first follows $\mathbf{y}_1^T (\mathbf{x}_1 - \mathbf{x}_2) = 1$, and from this,

$$||\mathbf{y}_1|| \geq \frac{1}{||\mathbf{x}_1 - \mathbf{x}_2||} = \frac{1}{\sqrt{2 - 2\mathbf{x}_1^T\mathbf{x}_2}} , \tag{3}$$

showing that $||\mathbf{y}_1||$ is very large when $\mathbf{x}_1$ and $\mathbf{x}_2$ are nearly parallel. It then follows from (2) that μ may exceed the norm of the perturbation matrix Δ by a large amount (namely by the factor $||\mathbf{y}_1||$).

Example. If the eigenvalues of the matrix

$$\begin{bmatrix}
12 & 11 & 10 & 9 & 8 & 7 & 6 & 5 & 4 & 3 & 2 & 1 \\
11 & 11 & 10 & 9 & 8 & 7 & 6 & 5 & 4 & 3 & 2 & 1 \\
   & 10 & 10 & 9 & 8 & 7 & 6 & 5 & 4 & 3 & 2 & 1 \\
   &   & 9 & 9 & 8 & 7 & 6 & 5 & 4 & 3 & 2 & 1 \\
   &   &   & 8 & 8 & 7 & 6 & 5 & 4 & 3 & 2 & 1 \\
   &   &   &   & 7 & 7 & 6 & 5 & 4 & 3 & 2 & 1 \\
   &   &   &   &   & 6 & 6 & 5 & 4 & 3 & 2 & 1 \\
   &   &   &   &   &   & 5 & 5 & 4 & 3 & 2 & 1 \\
   &   & 0 &   &   &   &   & 4 & 4 & 3 & 2 & 1 \\
   &   &   &   &   &   &   &   & 3 & 3 & 2 & 1 \\
   &   &   &   &   &   &   &   &   & 2 & 2 & 1 \\
   &   &   &   &   &   &   &   &   &   & 1 & 1
\end{bmatrix}$$

are computed on a machine with a 60 bit mantissa, one obtains for the two smallest of them (rounded to 6 digits after the decimal point):

$$\lambda_{11} = .049689, \quad \lambda_{12} = .030945 .$$

For the corresponding eigenvectors

$$
\mathbf{x}_{11} =
\begin{bmatrix}
0 \\
0 \\
0 \\
0 \\
.000014 \\
-.000218 \\
.002207 \\
-.015922 \\
.082412 \\
-.294454 \\
.655782 \\
-.690071
\end{bmatrix}
, \quad
\mathbf{x}_{12} =
\begin{bmatrix}
0 \\
0 \\
0 \\
-.000004 \\
.000045 \\
-.000433 \\
.003324 \\
-.020117 \\
.092891 \\
-.308605 \\
.658624 \\
-.679656
\end{bmatrix}
$$

one has

$$\mathbf{x}_{11}^T \mathbf{x}_{12} = .999777.$$

Therefore, in computations with a shorter mantissa, one immediately obtains relatively large errors for these two eigenvalues, the bound on the right in (3) being 47.35.

§13.2. Simple vector iteration

A "well established" method for dealing with the eigenvalue problem $\mathbf{Ax} = \lambda\mathbf{x}$ is the "simple" or *v. Mises-Geiringer vector iteration*: Starting with a vector $\mathbf{x}_0$, one constructs iteratively a sequence $\mathbf{x}_1, \mathbf{x}_2, \ldots$ according to

$$\mathbf{x}_k = \mathbf{Ax}_{k-1}.$$

Then, evidently,

$$\mathbf{x}_k = \mathbf{A}^k \mathbf{x}_0, \tag{4}$$

although the iteration vectors are usually not formed by means of the powers $\mathbf{A}^k$.

Consider now the generating function

$$\mathbf{x}(z) = \sum_{k=0}^{\infty} z^{-(k+1)} \mathbf{x}_k \tag{5}$$

of the vector sequence $\mathbf{x}_0, \mathbf{x}_1, \mathbf{x}_2, \ldots$. This is a vector-valued function of the complex variable z. It must first be examined, for which z this series converges. If $||\mathbf{A}|| = r$, then $||\mathbf{x}_k|| \leq r ||\mathbf{x}_{k-1}||$, hence $||\mathbf{x}_k|| \leq r^k ||\mathbf{x}_0||$, so that the series converges for $|z| > r$ and represents there a vector-valued analytic function. One is justified, therefore, in even writing

$$\mathbf{x}(z) = \sum_{k=0}^{\infty} z^{-(k+1)} \mathbf{A}^k \mathbf{x}_0 = \left[\sum_{k=0}^{\infty} z^{-(k+1)} \mathbf{A}^k \right] \mathbf{x}_0.$$

$\sum z^{-(k+1)} \mathbf{A}^k$ is the so-called Neumann series, which also converges for $|z| > r$ and represents there the matrix $(z\mathbf{I} - \mathbf{A})^{-1}$. Thus,

$$\mathbf{x}(z) = (z\mathbf{I} - \mathbf{A})^{-1} \mathbf{x}_0. \tag{6}$$

Now $(z\mathbf{I} - \mathbf{A})^{-1} \mathbf{x}_0$ is a vector-valued rational function of z, which vanishes for $z \to \infty$, and whose poles are obviously just the eigenvalues of $\mathbf{A}$. Moreover, the denominator of this rational function is a polynomial (^{1}) of degree $m \leq n$ (n = order of the matrix $\mathbf{A}$), since there certainly exist

---

[1] If $\mathbf{x}_0$ is in general position relative to a normal basis for $\mathbf{A}$ (i.e., has only nonzero coefficients in this basis), this polynomial is called *minimal polynomial*. (A *normal basis* is a basis with respect to which the linear operator defined by the matrix $\mathbf{A}$ assumes the Jordan normal form.) If $\mathbf{A}$ has an eigenvalue to which there belong several linearly independent eigenvectors (several "boxes" in the Jordan normal form), then $m < n$. (Editors' remark)

$m + 1 \leq n + 1$ constants $a_0, a_1, \ldots, a_m$ such that $a_m = 1$ and $a_0 x_0 + a_1 x_1 + \cdots + a_m x_m = 0$. Multiplication of this relation by $\mathbf{A}^k$ yields

$$a_0 x_k + a_1 x_{k+1} + \cdots + a_m x_{k+m} = 0, \quad k = 0, 1, \ldots .$$

There follows

$$(a_0 + a_1 z + \cdots + a_m z^m) x(z) = \sum_{k=0}^{\infty} z^{-(k+1)} (a_0 x_k + \cdots + a_m x_{k+m})$$
$$+ y(z) = y(z),$$

where $y(z)$ is a polynomial in z of degree $< m$ in which all terms with nonnegative powers of z are collected. Consequently,

$$x(z) = \frac{1}{a_0 + a_1 z + \cdots + a_m z^m} \, y(z). \tag{7}$$

For this rational function, however, we have an expansion in partial fractions, which in the case of simple eigenvalues has the form

$$x(z) = \sum_{j=1}^{m} \frac{1}{z - \lambda_j} \, c_j. \tag{8}$$

On the other hand, if for example $\lambda_1 = \lambda_2 = \lambda_3$ is a triple pole ([2]), then the three terms for $j = 1, 2, 3$ are replaced by the combination

$$\frac{1}{z - \lambda_1} \, c_1 + \frac{1}{(z - \lambda_1)^2} \, c_1' + \frac{1}{(z - \lambda_1)^3} \, c_1'' .$$

---

[2] For the Jordan normal form of $\mathbf{A}$ this means: The largest of the "boxes" with eigenvalue λ_1 has dimension 3. (Editors' remark)

Expanding the right-hand side of (8) in descending powers of z, one obtains

$$\sum_{j=1}^{m} \frac{1}{z - \lambda_j} \, \mathbf{c}_j = \sum_{j=1}^{m} \mathbf{c}_j \sum_{k=0}^{\infty} \frac{\lambda_j^k}{z^{k+1}} = \sum_{k=0}^{\infty} \frac{1}{z^{k+1}} \sum_{j=1}^{m} \mathbf{c}_j \lambda_j^k,$$

so that, by a comparison of coefficients, there follows

$$\mathbf{x}_k = \sum_{j=1}^{m} \mathbf{c}_j \lambda_j^k \tag{9}$$

(with fixed vectors $\mathbf{c}_j$ not depending on k). In the case $\lambda_1 = \lambda_2 = \lambda_3$, on the other hand, since

$$\frac{1}{(z - \lambda_1)^2} = \sum_{k=1}^{\infty} \frac{k \lambda_1^{k-1}}{z^{k+1}}, \quad \frac{1}{(z - \lambda_1)^3} = \sum_{k=2}^{\infty} \frac{\binom{k}{2} \lambda_1^{k-2}}{z^{k+1}},$$

one gets

$$\mathbf{x}_k = \mathbf{c}_1 \lambda_1^k + k \mathbf{c}_1' \lambda_1^{k-1} + \binom{k}{2} \mathbf{c}_1'' \lambda_1^{k-2} + \sum_{j=4}^{m} \mathbf{c}_j \lambda_j^k. \tag{10}$$

If now $|\lambda_1| > |\lambda_2| \geq |\lambda_3| \geq \cdots$, thus, λ_1 is a simple dominant eigenvalue, then in the relation

$$\frac{1}{\lambda_1^k} \mathbf{x}_k = \mathbf{c}_1 + \sum_{j=2}^{m} \mathbf{c}_j \left[\frac{\lambda_j}{\lambda_1} \right]^k \tag{11}$$

derived from (9), the sum converges to 0 as $k \to \infty$, that is, $\mathbf{x}_k$ converges in direction toward $\mathbf{c}_1$, the convergence in fact being linear with convergence factor $|\lambda_2/\lambda_1|$. But what is the meaning of $\mathbf{c}_1$? By (6),

$$A\mathbf{x}(z) = z\mathbf{x}(z) - \mathbf{x}_0.$$

Here we substitute for $\mathbf{x}(z)$ the partial fraction expansion (8):

$$\sum_{j=1}^{m} \frac{1}{z - \lambda_j} \mathbf{A}\mathbf{c}_j = -\mathbf{x}_0 + \sum_{j=1}^{m} \frac{z}{z - \lambda_j}\mathbf{c}_j = -\mathbf{x}_0 + \sum_{j=1}^{m} \mathbf{c}_j + \sum_{j=1}^{m} \frac{\lambda_j}{z - \lambda_j}\mathbf{c}_j,$$

$$\sum_{j=1}^{m} \frac{1}{z - \lambda_j} (\mathbf{A}\mathbf{c}_j - \lambda_j\mathbf{c}_j) = \sum_{j=1}^{m} \mathbf{c}_j - \mathbf{x}_0.$$

It follows that the right-hand side must be $\mathbf{0}$, since it is constant, and the left-hand side tends to $\mathbf{0}$ as $z \to \infty$. But for a rational function which is identically zero, also the residues must vanish:

$$\mathbf{A}\mathbf{c}_j - \lambda_j\mathbf{c}_j = \mathbf{0}.$$

The $\mathbf{c}_j$ are thus eigenvectors. Hence:

Theorem 13.1. *The sequence of iteration vectors $\mathbf{x}_k$ generated by (4) converges in direction to an eigenvector $\mathbf{c}_1$ belonging to the eigenvalue λ_1 of $\mathbf{A}$ of maximum modulus, as $k \to \infty$, provided λ_1 is a simple dominant eigenvalue* ([3]).

If, on the other hand, there are three eigenvalues of maximum modulus, which either coincide ([4]) or merely have the same absolute value, then according to (10) one has, up to terms of order $O((\lambda_4/\lambda_1)^k)$, as $k \to \infty$,

$$\frac{1}{\lambda_1^k}\mathbf{x}_k = \mathbf{c}_1 + k \frac{1}{\lambda_1} \mathbf{c}_1' + \binom{k}{2} \frac{1}{\lambda_1^2} \mathbf{c}_1'' \tag{12}$$

or else, according to (11),

---

[3] It must be assumed, more precisely, that $\mathbf{x}_0$ is not orthogonal to the eigenvector of $\mathbf{A}^T$ belonging to λ_1. In practice, however, rounding errors may produce convergence (in direction) to $\mathbf{c}_1$ even in such a case of orthogonality.

[4] More precisely: if $\lambda_1 = \lambda_2 = \lambda_3$ and one has the situation described in footnote ([2]). (Editors' remark)

$$\frac{1}{\lambda_1^k} x_k = c_1 + \left[\frac{\lambda_2}{\lambda_1} \right]^k c_2 + \left[\frac{\lambda_3}{\lambda_1} \right]^k c_3, \tag{13}$$

that is, the x_k lie in the three-dimensional subspace ([5]) spanned by c_1, c_1', c_1'' and c_1, c_2, c_3, respectively. Four successive vectors x_k, x_{k+1}, x_{k+2}, x_{k+3}, therefore, are practically linearly dependent when k is large. This can be recognized by orthonormalizing these vectors from left to right. The linear dependence then manifests itself in a collapse of the length of a vector during orthonormalization. In the case of three eigenvalues with equal moduli, this will happen with x_{k+3}, in case of a simple dominant eigenvalue, however, already with x_{k+1}.

As a first by-product of such a collapse during the orthogonalization of x_{k+p}, one obtains the solution of the minimum problem

$$\min_{a_0, \dots, a_{p-1}} \; ||x_{k+p} + a_{p-1} x_{k+p-1} + \cdots + a_1 x_{k+1} + a_0 x_k||_2, \tag{14}$$

where the minimum is noticeably small. Now, from

$$\sum_{j=0}^{p} a_j x_{k+j} = 0$$

(with $a_p = 1$) there would follow

$$\sum_{j=0}^{p} a_j \lambda_\ell^{\,j} = 0, \quad \ell = 1, \dots, p, \tag{15}$$

as is seen immediately by substituting (12) or (13) and recalling the linear independence of c_1, c_2, c_3, ... and c_1, c_1', c_1'', ..., respectively. One thus has an algebraic equation whose roots are the p eigenvalues of A of maximum modulus ([6]).

---

[5] One can show that the m vectors $c_j^{(k)}$ ($k = 0, \dots, m_j - 1$, $\sum m_j = m$) occurring in the partial fraction expansion of $x(z)$ are linearly independent, if x_0 is in general position relative to a normal basis. (Editors' remark).

[6] These roots occur here with the same multiplicity as in the minimal polynomial of A. The multiplicity in the characteristic polynomial (i.e., as eigenvalue) can be larger. (Editors' remark)

Secondly, the orthonormalization process produces an orthonormal system of p vectors $y_1, \ldots, y_p$ which approximately (with a deviation of $O((\lambda_{p+1}/\lambda_p)^k)$) span the same subspace as the eigenvectors $c_1, \ldots, c_p$. If we now succeed in transforming these vectors $y_1, \ldots, y_p$ by means of an orthogonal transformation into the first p coordinate vectors, and if B denotes the matrix which in the new system defines the same linear transformation as A does in the old system, then the subspace spanned by $e_1, \ldots, e_p$ is nearly invariant under B. The matrix B thus has the form

$$B = \begin{bmatrix} B_1 & \vdots & B_2 \\ \cdots & \cdots & \cdots \\ B_4 & \vdots & B_3 \end{bmatrix}, \tag{16}$$

where B_1 is a $p \times p$-matrix whose eigenvalues are the p eigenvalues of A of maximum modulus and B_4 is a $(n-p) \times p$-matrix with *small* elements ([7]), so that B practically decomposes into B_1 and B_3 (B_3 has "normal" elements). Since (15) yields the eigenvalues of B_1, our problem is thus reduced to the one of determining the eigenvalues of B_3.

But how does one transform $y_1, \ldots, y_p$ into $e_1, \ldots, e_p$? A sequence of Jacobi rotations (cf. §12.3, Eq. (17))

$$U(n-1, n, \phi_n), \ U(n-2, n-1, \phi_{n-1}), \ldots, \ U(1, 2, \phi_2)$$

allows us to annihilate in succession the nth, $(n-1)$st $, \ldots,$ 2nd component of y_1. For example, with $s = \sin \phi_n$, $c = \cos \phi_n$, one has

$$U^T(n-1, n, \phi_n) y_1 = \begin{bmatrix} 1 & & & & & 0 \\ & 1 & & & & \\ & & \ddots & & & \\ & & & \ddots & & \\ & & & & 1 & \\ & & & & & c & -s \\ 0 & & & & & s & c \end{bmatrix} \begin{bmatrix} y_{11} \\ y_{12} \\ \vdots \\ \vdots \\ \vdots \\ y_{1n} \end{bmatrix} = \begin{bmatrix} y_{11} \\ y_{12} \\ \vdots \\ \vdots \\ * \\ 0 \end{bmatrix},$$

---

[7] In another manuscript of the author it is shown that (in exact arithmetic) the submatrix B_4 has nonzero elements only in its last column. (Editors' remark)

provided $cy_{1n} + sy_{1,n-1} = 0$, that is, $\cot \phi_n = -y_{1,n-1}/y_{1n}$. In this way, finally,

$$U^T(1,2,\phi_2)U^T(2,3,\phi_3) \cdots U^T(n-1,n,\phi_n)y_1 = e_1,$$

and at the same time,

$$U^T(1,2,\phi_2)U^T(2,3,\phi_3) \cdots U^T(n-1,n,\phi_n)y_j = \begin{bmatrix} 0 \\ * \\ \cdot \\ \cdot \\ \cdot \\ * \end{bmatrix} = y'_j,$$

since the orthogonality of the y_j is not destroyed by the rotations. Furthermore, with suitable ϕ'_j $(j = 3, \ldots, n)$, one achieves

$$U^T(2,3,\phi'_3)U^T(3,4,\phi'_4) \cdots U^T(n-1,n,\phi'_n)y'_2 = e_2.$$

These rotations no longer affect e_1. The process can evidently be continued until finally

$$U^T(p,p+1,\phi_{p+1}^{(p-1)})U^T(p+1,p+2,\phi_{p+2}^{(p-1)}) \cdots U^T(n-1,n,\phi_n^{(p-1)})y_p^{(p-1)} = e_p.$$

During each rotation, the matrix A participates in the transformation, that is, each time one forms $U^T(j-1,j,\phi_j^{(k)})AU(j-1,j,\phi_j^{(k)})$. In this manner one obtains, at the end, the form (16).

Examples. 1) For the matrix

$$A = \begin{bmatrix} 1 & 1 & 1 \\ 3 & 2 & 1 \\ 6 & 3 & 1 \end{bmatrix},$$

with the eigenvalues $\lambda_1 = 5.28799. \ldots$, $\lambda_2 = -1.42107. \ldots$, $\lambda_3 = .13307. \ldots$, one expects relatively good convergence. With $x_0 = [1,1,1]^T$

one gets

$$
\mathbf{x}_1 = \begin{bmatrix} 3 \\ 6 \\ 10 \end{bmatrix}, \quad
\mathbf{x}_2 = \begin{bmatrix} 19 \\ 31 \\ 46 \end{bmatrix}, \quad
\mathbf{x}_3 = \begin{bmatrix} 96 \\ 165 \\ 253 \end{bmatrix}, \quad
\mathbf{x}_4 = \begin{bmatrix} 514 \\ 871 \\ 1324 \end{bmatrix},
$$

$$
\mathbf{x}_5 = \begin{bmatrix} 2709 \\ 4608 \\ 7021 \end{bmatrix}, \quad
\mathbf{x}_6 = \begin{bmatrix} 14338 \\ 24364 \\ 37099 \end{bmatrix}.
$$

Orthogonalizing $\mathbf{x}_6$ relative to $\mathbf{x}_5$ yields $\lambda_1 = 5.285733$ as solution of

$$
\min_{\lambda} \ ||\mathbf{x}_6 - \lambda\mathbf{x}_5||
$$

and (in 7-digit computation)

$$
\mathbf{x}_6 - \lambda_1\mathbf{x}_5 = \begin{bmatrix} 18.95 \\ 7.34 \\ -12.13 \end{bmatrix}, \quad r_{22} = ||\mathbf{x}_6 - \lambda_1\mathbf{x}_5|| = 23.66675,
$$

which, in comparison with $||\mathbf{x}_6|| = 46642.46$, is very small. The predicted collapse of the length of $\mathbf{x}_6$ during orthogonalization thus materialized.

2) The matrix

$$
\mathbf{A} = \begin{bmatrix} 0 & -1 & 1 \\ 1 & 9 & -1 \\ -1 & 1 & 10 \end{bmatrix},
$$

on the other hand, has two dominant eigenvalues, which are conjugate to one another:

$$
\lambda_{1,2} = 9.3932451 \ldots \pm i\, .8693946 \ldots, \quad \lambda_3 = .2135098 \ldots .
$$

One expects, therefore, that linear dependence occurs only between three consecutive iteration vectors.

Starting again with $x_0 = [1,1,1]^T$, one obtains here

$$x_1 = \begin{bmatrix} 0 \\ 9 \\ 10 \end{bmatrix}, \quad x_2 = \begin{bmatrix} 1 \\ 71 \\ 109 \end{bmatrix}, \quad x_3 = \begin{bmatrix} 38 \\ 531 \\ 1160 \end{bmatrix}, \quad x_4 = \begin{bmatrix} 629 \\ 3657 \\ 12093 \end{bmatrix},$$

$$x_5 = \begin{bmatrix} 8436 \\ 21449 \\ 123958 \end{bmatrix}, \quad x_6 = \begin{bmatrix} 102509 \\ 77519 \\ 1252593 \end{bmatrix}.$$

To be orthogonalized are now x_4, x_5, x_6 (cf. §§5.3, 5.4):

$$r_{11} = ||x_4|| = 12649.50,$$

$$y_1 = \frac{1}{r_{11}} x_4 = [.04972529, .2891023, .9560062]^T,$$

$$r_{12} = (y_1, x_5) = 125125.0,$$

$$x_5 - r_{12}y_1 = [2214.123, -14724.93, 4337.700]^T,$$

$$r_{22} = ||x_5 - r_{12}y_1|| = 15509.40,$$

$$y_2 = \frac{1}{r_{22}} (x_5 - r_{12}y_1) = [.1427601, -.9494197, .2796820]^T,$$

$$r_{13} = (y_1, x_6) = 1224995,$$

$$x_6 - r_{13}y_1 = [41595.77, -276629.9, 81490.00]^T,$$

$$r_{23} = (y_2, x_6 - r_{13}y_1) = 291363.8,$$

$$x_6 - r_{13}y_1 - r_{23}y_2 = [.64, -3.40, .79]^T,$$

$$r_{33} = ||\mathbf{x}_6 - r_{13}\mathbf{y}_1 - r_{23}\mathbf{y}_2|| = 3.548760.$$

As expected, $\mathbf{x}_6$ has been shortened drastically during orthogonalization. Thus, $p = 2$, and to obtain the coefficients a_0 and a_1 in (14), it follows from

$$\mathbf{x}_4 = r_{11}\mathbf{y}_1, \quad \mathbf{x}_5 = r_{12}\mathbf{y}_1 + r_{22}\mathbf{y}_2,$$
$$\mathbf{x}_6 - r_{13}\mathbf{y}_1 - r_{23}\mathbf{y}_2 = \mathbf{x}_6 + a_1\mathbf{x}_5 + a_0\mathbf{x}_4$$
$$= \mathbf{x}_6 + (a_0 r_{11} + a_1 r_{12})\mathbf{y}_1 + a_1 r_{22}\mathbf{y}_2$$

that one needs only to solve the system of equations

	a_0	a_1	1
$0 =$	r_{11}	r_{12}	r_{13}
$0 =$	0	r_{22}	r_{23}

(17)

(The value of the minimum in (14) is equal to r_{33}.) One finds

$$a_0 = 88.98668, \quad a_1 = -18.78627.$$

The approximations for the dominant eigenvalues are finally obtained as roots of the equation (15), which in this case is quadratic, and one finds

$$\lambda_{1,2} = 9.393135 \pm .8693043i.$$

If now $\mathbf{y}_1$ is transformed into $\mathbf{e}_1$, by first multiplying by

$$\mathbf{U}_1^T = \begin{bmatrix} 1 & 0 & 0 \\ 0 & .2894603 & .9571900 \\ 0 & -.9571900 & .2894603 \end{bmatrix},$$

and then by

$$\mathbf{U}_2^T = \begin{bmatrix} .0497253 & .9987629 & 0 \\ -.9987629 & .0497253 & 0 \\ 0 & 0 & 1 \end{bmatrix},$$

y_2 changes into $y_2' = [0,-.1429367, .9897318]^T$ and $\mathbf{A}$ into

$$\mathbf{U}_2^T \mathbf{U}_1^T \mathbf{A} \mathbf{U}_1 \mathbf{U}_2 = \begin{bmatrix} 9.8916928 & 1.1602063 & -.6600471 \\ -.1752530 & .0245189 & -1.2810561 \\ 1.2134986 & 1.3086107 & 9.0837869 \end{bmatrix}.$$

It remains to transform y_2' into $\mathbf{e}_2$ through multiplication by

$$\mathbf{U}_3^T = \begin{bmatrix} 1 & 0 & 0 \\ 0 & .1429367 & -.9897318 \\ 0 & .9897318 & .1429367 \end{bmatrix}.$$

This finally transforms $\mathbf{A}$ into

$$\mathbf{B} = \begin{bmatrix} 9.9816928 & .8191056 & 1.0539480 \\ -1.2260882 & 8.8947992 & -2.5896532 \\ 0 & .0000134 & .2135060 \end{bmatrix}. \qquad (18)$$

Here one can read off directly an approximation to the third eigenvalue:

$$\lambda_3 = .2135060.$$

Refinement. Once $\mathbf{A}$, by Jacobi rotations as described above, has been brought to the form (16) (where $\mathbf{B}_4$ has only small elements), one can follow up with an additional transformation with a nonorthogonal matrix of the form

$$\mathbf{T} = \begin{bmatrix} \mathbf{I} & \vdots & \mathbf{O} \\ \cdots & \vdots & \cdots \\ \mathbf{X} & \vdots & \mathbf{I} \end{bmatrix}, \quad \text{with} \quad \mathbf{T}^{-1} = \begin{bmatrix} \mathbf{I} & \vdots & \mathbf{O} \\ \cdots & \vdots & \cdots \\ -\mathbf{X} & \vdots & \mathbf{I} \end{bmatrix}. \qquad (19)$$

Then,

$$\mathbf{T}^{-1} \mathbf{B} \mathbf{T} = \begin{bmatrix} \mathbf{B}_1 + \mathbf{B}_2 \mathbf{X} & \vdots & \mathbf{B}_2 \\ \cdots \cdots \cdots \cdots \cdots \cdots & \vdots & \cdots \cdots \cdots \cdots \\ \mathbf{B}_4 - \mathbf{X} \mathbf{B}_1 + & \vdots & \\ & \vdots & \mathbf{B}_3 - \mathbf{X} \mathbf{B}_2 \\ \mathbf{B}_3 \mathbf{X} - \mathbf{X} \mathbf{B}_2 \mathbf{X} & \vdots & \end{bmatrix}.$$

In order that the submatrix at the lower left becomes the zero matrix, one must have, in first approximation,

$$\mathbf{B}_4 - \mathbf{X}\mathbf{B}_1 + \mathbf{B}_3\mathbf{X} = \mathbf{O}, \tag{20}$$

since $\mathbf{X}$ can be expected to also have only small elements. In this way, one obtains for the $p(n-p)$ unknown elements of the matrix $\mathbf{X}$ the same number of linear equations:

$$b_{ij}^{(4)} = \sum_{\ell=1}^{p} x_{i\ell} b_{\ell j}^{(1)} - \sum_{k=p+1}^{n} b_{ik}^{(3)} x_{kj} \quad (i = p+1, \ldots, n; \; j = 1, \ldots, p).$$

By means of the Kronecker symbol, one can write these also in the form

$$b_{ij}^{(4)} = \sum_{k=p+1}^{n} \sum_{\ell=1}^{p} (\delta_{ik} b_{\ell j}^{(1)} - b_{ik}^{(3)} \delta_{\ell j}) x_{k\ell}$$
$$(i = p+1, \ldots, n; \; j = 1, \ldots, p). \tag{21}$$

This system, with coefficient matrix

$$\mathbf{M} = [m_{ijk\ell}], \quad \text{where} \quad m_{ijk\ell} = \delta_{ik} b_{\ell j}^{(1)} - b_{ik}^{(3)} \delta_{\ell j}$$
$$(i,k = p+1, \ldots, n; \; j,\ell = 1, \ldots, p), \tag{22}$$

is often very large; for example, when $n = 50$, $p = 4$, it already contains 184 equations.

Nevertheless, under certain simple conditions, which are usually satisfied in practice, $\mathbf{M}$ is nonsingular; in fact, the following theorem holds, which we state without proof:

Theorem 13.2. *If in the matrix $\mathbf{B}$ of the form (16) the eigenvalue of $\mathbf{B}_1$ of smallest modulus is still greater than the eigenvalue of $\mathbf{B}_3$ of largest modulus, then the matrix $\mathbf{M}$ defined by (22) is nonsingular. In addition, the solution $\mathbf{X}$ of (20) (and (21)) can also be found by means of the iteration*

$$\mathbf{X}_0 = \mathbf{O}, \quad \mathbf{X}_{k+1} = \mathbf{B}_4\mathbf{B}_1^{-1} + \mathbf{B}_3\mathbf{X}_k\mathbf{B}_1^{-1} \quad (k = 0,1,2, \ \ldots \ ). \qquad (23)$$

Example. We apply this refinement to (18). Then $n = 3$, $p = 2$, so that the system (21) consists of $p(n - p) = 2$ equations:

	x_1	x_2	1
0 =	9.67818	−1.22609	0
0 =	0.81911	8.68129	0.0000134

The solution is:

$$x_1 = 1.932_{10}{-}7, \quad x_2 = 15.253_{10}{-}7,$$

so that **B** has yet to be transformed by

$$\mathbf{T} = \begin{bmatrix} 1 & 0 & 0 \\ 0 & 1 & 0 \\ 1.932_{10}{-}7 & 15.253_{10}{-}7 & 1 \end{bmatrix},$$

which leads to

$$\mathbf{T}^{-1}\mathbf{BT} = \begin{bmatrix} 9.8916930 & .8191072 & 1.0539480 \\ -1.2260887 & 8.8947953 & -2.5896532 \\ -1.2_{10}{-}9 & -1.6_{10}{-}9 & .2135097 \end{bmatrix}.$$

From this matrix one reads off the improved approximation $\lambda_3 = $ .2135097.

Notes to Chapter 13

Great progress has been made in the treatment of small-order matrices since this chapter was written. The approach that turned out to be successful avoids trying to obtain the Jordan canonical form. It also avoids computation of the eigenvalues, one by one, using modifications of the simple iteration described in §13.2. Instead, it basically relies on Schur's lemma: Every square complex matrix is unitarily similar to an upper triangular matrix, $A = PSP^H$, $PP^H = I$. The method has two phases, i) reduction to upper Hessenberg form H ($h_{ij} = 0$ if $i > j + 1$) by a finite sequence of Householder transformations (as described in §12.8 of Chapter 12), ii) reduction of H to upper triangular form S by a sequence of elementary unitary matrices. For more information, see Stewart [1973] or Wilkinson [1965]. Phase ii) uses QR transformations (see notes to §§12.6–12.8 of Chapter 12). In principle, phase ii) requires infinitely many transformations, but in practice, for a matrix of order n, it requires fewer than $2n$ QR transformations to reduce H to S.

The search for reliable methods to extract a few eigenvalues from a large, sparse, nonsymmetric matrix goes on. See Parlett [1984] and Saad [1989] for recent surveys.

References

Parlett, B.N. [1984]: The software scene in the extraction of eigenvalues from sparse matrices, *SIAM J. Sci. Statist. Comput.* **5**, 590–603.

Saad, Y. [1989]: Numerical solution of large nonsymmetric eigenvalue problems, *Comput. Phys. Comm.* **53**, 71–90.

Stewart, G.W. [1973]: *Introduction to Matrix Computations*, Academic Press, New York.

Wilkinson, J.H. [1965]: *The Algebraic Eigenvalue Problem*, Clarendon Press, Oxford. [Paperback edition, 1988].

APPENDIX

An Axiomatic Theory of Numerical Computation with an Application to the Quotient-Difference Algorithm

Editor's Foreword

H. Rutishauser occupied himself with the topic of this appendix already in 1968 (report [21] in the bibliography to the appendix) and discussed part of this material in a course held during the spring semester of 1969. Later, however, the content and text were revised by Rutishauser and significantly extended. He also intended to present the very interesting third chapter on "Finite Arithmetic" at a meeting in Oberwolfach which took place in November, 1970, shortly after his death. It is not known, however, in which form he wanted to publish the whole work, which does not require for the reader to have any extensive previous knowledge, but which exceeds the usual length of a journal article. What is certain is only that the work remained unfinished. At least seven chapters were contemplated, but the manuscript breaks off in the fifth, entitled "Forcing Coincidence". Fortunately, the text nevertheless is fairly well rounded. It contains in the first two chapters an introduction to the qd-algorithm in a partly novel exposition. The third chapter, as already mentioned, gives an axiomatic approach to numerical computation. The principal goal is not an examination of completeness and independence of the axiomatic system, but rather the possibility of proving for an algorithm that it never fails in spite of the presence of rounding errors. The discussions of the qd-algorithm and its stationary form in the fourth and fifth chapters then also point into the same direction.

The text of the appendix agrees over long stretches word for word with the handwritten manuscript of H. Rutishauser. In a few places, however, theorems and proofs were formulated a bit more accurately and with more details. A larger reorganization was necessary only in the third chapter, where the statements now contained in Theorems A13, A14 and

A16 have been regrouped. The editors, in addition, prepared the bibliography and inserted the references thereto (which were left open in the manuscript).

Vancouver, B.C., February, 1976   M. Gutknecht

Introduction

§A1.1. The eigenvalues of a qd-row

An important area of application of the *qd-algorithm*[1] is the computation of the eigenvalues of a tridiagonal matrix. By a trivial (diagonal) similarity transformation, such a matrix almost always can be brought into the form

$$
A = \begin{bmatrix}
q_1 & -q_1 & & & & & 0 \\
-e_1 & q_2 + e_1 & -q_2 & & & & \\
& -e_2 & q_3 + e_2 & -q_3 & & & \\
& & & \ddots & \ddots & \ddots & \\
& & & & \ddots & \ddots & -q_{n-1} \\
0 & & & & & -e_{n-1} & q_n + e_{n-1}
\end{bmatrix}, \quad (1)
$$

in which only $2n-1$ independent quantities occur. These are collected in a *qd-row*

---

[1] The *quotient-difference algorithm* (briefly *qd-algorithm*) in principle is a computational method for the determination of the poles of a meromorphic function, but has many other applications. It is due to H. Rutishauser [8-12]. (The report [14], among other things, contains [8-11] in partly revised form. [21] represents a precursor of the unfinished work printed here.) (Editors' remark)

$$Z = \{q_1, e_1, q_2, e_2, \ldots, e_{n-1}, q_n\}. \tag{2}$$

By the eigenvalues of Z one means the eigenvalues of the matrix $\mathbf{A}$ associated with (2) according to (1). This terminology suggests itself very naturally, since the computation of the eigenvalues is accomplished exclusively with the help of data structures of the form (2).

The present work deals with the computation of the eigenvalues of a qd-row (2), particular attention being paid to the sequential reliability of the numerical process (that is, the process contaminated by rounding errors). It is possible to prove in an important special case that the computational process, even when perturbed by rounding errors, must run its course without any mishaps, and must furnish approximately the correct eigenvalues.

§A1.2. The progressive form of the qd-algorithm

The determination of the eigenvalues $\lambda_1, \ldots, \lambda_n$ of a qd-row (2) is effected in principle by means of an iterative process which consists of infinitely many steps of the following kind: A *progressive qd-step* is defined by the following computational algorithm([1]):

> **comment** it is assumed that $e_n = 0$;
> $q'_1 := q_1 + e_1$;
> **for** $k := 2$ **step** 1 **until** n **do**
> **begin**   (3)
> $\quad e'_{k-1} := (e_{k-1}/q'_{k-1}) \times q_k$;
> $\quad q'_k := (q_k - e'_{k-1}) + e_k$;
> **end** for k;

Provided that these operations are executable (which presupposes q'_1, $q'_2, \ldots, q'_{n-1} \neq 0$), (3) produces a new qd-row $Z' = \{q'_1, e'_1, q'_2, \ldots, q'_n\}$, which is expressed symbolically by

---
[1] Computational algorithms are given as pieces of ALGOL programs in which declarations and input and output operations are omitted, and indices, etc. are written in a form not permissible in ALGOL. Lower indices are true indices, while upper indices, primes, asterisks, etc. distinguish quantities which during the computation are stored in the same register. (Editors' remark)

$$Z \xrightarrow{\;0\;} Z', \tag{4}$$

the number 0 indicating that one is dealing with a *qd*-step *without shift*. It is to be noted that the type of parenthesizing prescribed in (3) is of crucial importance for the numerical execution (see Ch. A4).

The matrix associated with the row Z' is

$$\mathbf{A}' = \begin{bmatrix}
q'_1 & -q'_1 & & & & & 0 \\
-e'_1 & q'_2 + e'_1 & -q'_2 & & & & \\
& -e'_2 & q'_3 + e'_2 & -q'_3 & & & \\
& & \cdot & \cdot & \cdot & & \\
& & & \cdot & \cdot & \cdot & \\
& & & & \cdot & \cdot & -q'_{n-1} \\
0 & & & & & -e'_{n-1} & q'_n + e'_{n-1}
\end{bmatrix}$$

On the basis of the rhombus rules $q'_k + e'_{k-1} = q_k + e_k$, $q'_k e'_k = q_{k+1} e_k$, which follow from (3), it transpires that $\mathbf{A}'$ is diagonally similar[2] to the matrix

$$\mathbf{B} = \begin{bmatrix}
q_1 + e_1 & -q_2 & & & & & 0 \\
-e_1 & q_2 + e_2 & -q_3 & & & & \\
& -e_2 & q_3 + e_3 & -q_4 & & & \\
& & & \cdot & \cdot & \cdot & \\
& & & & \cdot & \cdot & \cdot \\
& & & & & \cdot & -q_n \\
0 & & & & & -e_{n-1} & q_n
\end{bmatrix}, \tag{5}$$

---

[2] The diagonal matrix $\mathbf{D}$ with $\mathbf{B} = \mathbf{D}^{-1}\mathbf{A}'\mathbf{D}$ has the diagonal elements $d_1 = 1$, $d_k = \prod_{i=1}^{k-1} e'_i/e_i = \prod_{i=1}^{k-1} q_{i+1}/q'_i$ $(k = 2, \ldots, n)$. (Editors' remark)

which in turn is similar to **A** in (1), because **A = XY**, **B = YX** with

$$
X = \begin{bmatrix}
q_1 & & & & & & 0 \\
-e_1 & q_2 & & & & & \\
& -e_2 & q_3 & & & & \\
& & & \cdot & \cdot & & \\
& & & & \cdot & \cdot & \\
& & & & & \cdot & \\
0 & & & & & -e_{n-1} & q_n
\end{bmatrix}, \quad
Y = \begin{bmatrix}
1 & -1 & & & & & 0 \\
& 1 & -1 & & & & \\
& & 1 & -1 & & & \\
& & & \cdot & \cdot & & \\
& & & & \cdot & \cdot & \\
& & & & & \cdot & -1 \\
0 & & & & & & 1
\end{bmatrix}.
$$

We thus have:

Theorem A1. *If the operations* (3) *are executable, then Z and Z′ have the same eigenvalues.*

The computational process (4) is now continued iteratively:

$$
Z \xrightarrow{\ 0\ } Z'; \; Z' \xrightarrow{\ 0\ } Z''; \; Z'' \xrightarrow{\ 0\ } Z''' ; \dots ,
$$

which produces an infinite sequence of qd-rows, all having the same eigenvalues, for which under certain conditions

$$
\lim_{j \to \infty} Z^{(j)} = \{\lambda_1, 0, \lambda_2, 0, \lambda_3, 0, \dots, 0, \lambda_n\}, \tag{6}
$$

that is, the q_k-values tend to the eigenvalues λ_k as the iteration progresses (see [14], Ch. I)([3]).

---

[3] A detailed convergence proof is given in [4], §7.6. [21] contains a simple convergence proof for the special case of positive qd-rows (cf. §A1.4). (Editors' remark)

§A1.3. The generating function of a qd-row

One associates with the qd-row (2) as generating function the finite continued fraction

$$f(z) = \frac{1}{z-} \; \frac{q_1}{1-} \; \frac{e_1}{z-} \; \frac{q_2}{1-} \; \frac{e_2}{z-} \; \cdots \; \frac{e_{n-1}}{z-} \; \frac{q_n}{1} \tag{7}$$

(see [14, 20]). (7) represents a rational function with a denominator of degree n; its poles are also the eigenvalues of (2). The qd-algorithm therefore also permits the calculation of the poles of a rational function; if they are simple, one can in this way even compute the residues([1]).

Between the generating function $f(z)$ of Z and $f'(z)$ of Z' there holds the relation([2])

$$f'(z) = \frac{zf(z) - 1}{q_1}, \tag{8}$$

which was used in [22] to prove the convergence of the qd-algorithm.

§A1.4. Positive qd-rows

The computational algorithm (3), and hence also its iterative continuation, is numerically endangered because of the possibility of one of the denominators q'_{k-1} vanishing or almost vanishing. There exists, however, a special case in which this danger (even in numerical computation) does not arise, namely the case when all elements of the row Z are positive:

Definition. *A qd-row* $Z = \{q_1, e_1, q_2, \ldots, q_n\}$ *is called* **positive** *(in symbols:* $Z > 0$*) if*

---

[1] The author had the intention to describe this in a 7th chapter of this work. He dealt with this problem already briefly in [14], Ch. II, §10, and in [22]. (Editors' remark)
[2] f' here does *not* denote the derivative of f. (Editors' remark)

$$q_k > 0 \quad (k = 1, \ldots, n),$$
$$e_k > 0 \quad (k = 1, \ldots, n-1). \tag{9}$$

One then has by [6], Ch. 9, Theorems 1 and 5:

Theorem A2. *The eigenvalues of a positive qd-row are all real, positive, and simple.*

Proof. In the case of a positive qd-row the associated matrix (1) is diagonally similar to

$$
\mathbf{H} =
\begin{bmatrix}
q_1 & \sqrt{q_1 e_1} & & & & 0 \\
\sqrt{q_1 e_1} & q_2 + e_1 & \sqrt{q_2 e_2} & & & \\
& \cdot & \cdot & \cdot & & \\
& & \cdot & \cdot & \cdot & \\
& & & \cdot & \cdot & \sqrt{q_{n-1} e_{n-1}} \\
0 & & & & \sqrt{q_{n-1} e_{n-1}} & q_n + e_{n-1}
\end{bmatrix}
\tag{10}
$$

(with all square roots positive), and this matrix admits a Cholesky decomposition $\mathbf{H} = \mathbf{R}^T \mathbf{R}$ with

$$
\mathbf{R} =
\begin{bmatrix}
\sqrt{q_1} & \sqrt{e_1} & & & 0 \\
& \sqrt{q_2} & \sqrt{e_2} & & \\
& & \cdot & \cdot & \\
& & & \cdot & \sqrt{e_{n-1}} \\
0 & & & & \sqrt{q_n}
\end{bmatrix},
\tag{11}
$$

where all q_k, e_k are positive; q.e.d.([1])

For positive rows one now obtains the following important fact:

Theorem A3. *If the qd-row Z is positive, then the qd-rows $Z^{(j)}$ generated from it by the progressive qd-algorithm, that is, by*

$$Z \xrightarrow{\;0\;} Z', \quad Z' \xrightarrow{\;0\;} Z'', \quad Z'' \xrightarrow{\;0\;} Z''', \ldots,$$

are likewise positive, and one has unconditionally:

$$\lim_{j \to \infty} Z^{(j)} = \{\lambda_1, 0, \lambda_2, 0, \ldots, \lambda_n\}, \tag{12}$$

where $\lambda_1 > \lambda_2 > \cdots > \lambda_n > 0$ are the eigenvalues of Z.

Proof. (a) By (3) and (9),

$$q_1' = q_1 + e_1 > e_1 > 0,$$
$$e_1' = q_2(e_1/q_1') < q_2, \quad \text{as well as} \quad e_1' > 0,$$
$$q_2' = (q_2 - e_1') + e_2 > e_2 > 0,$$
$$e_2' = q_3(e_2/q_2') < q_3, \quad \text{as well as} \quad e_2' > 0,$$
$$\vdots$$
$$q_n' = (q_n - e_{n-1}') > 0.$$

Therefore, $Z' > 0$, and likewise $Z'' > 0$, etc.

b) For the convergence of

$$\lim_{j \to \infty} q_k^{(j)} = \ell_k \quad \text{and} \quad \lim_{j \to \infty} e_k^{(j)} = 0, \tag{13}$$

see, for example, [21]. One has, moreover, $\ell_1 > \ell_2 > \cdots > \ell_n$.

---

[1] The eigenvalues are simple, since **H** is a symmetric tridiagonal matrix with nonzero side diagonal elements. Cf. Theorem 4.9 in [24]. (Editors' remark)

c) For sufficiently large j, the matrix $\mathbf{H}^{(j)}$ formed with the elements of $Z^{(j)}$ in analogy to (10), differs from $\mathrm{diag}(\ell_1, \ell_2, \ldots, \ell_n)$ by an arbitrarily small amount. Consequently, also the eigenvalues of $\mathbf{H}^{(j)}$, which by Theorem A1 continue to be equal to $\lambda_1, \lambda_2, \ldots, \lambda_n$, differ from $\ell_1, \ell_2, \ldots, \ell_n$ by as little as one likes, q.e.d.

§A1.5. Speed of convergence of the qd-algorithm

For practical computation, a convergence statement like (13), phrased in general terms, is not yet sufficient; rather, one ought to have some information concerning the speed of convergence. Now from (13) and the computational rule (3), however, it follows that the $e_k^{(j)}$ converge to zero linearly as $j \to \infty$, more precisely:

$$e_k^{(j)} = O\left\{ \left[\frac{\lambda_{k+1}}{\lambda_k} \right]^j \right\} \qquad (k = 1, 2, \ldots, n-1). \qquad (14)$$

From this, and (3), one then also obtains the convergence behavior of the $q_k^{(j)}$,

$$q_k^{(j)} - \lambda_k = O(s^j), \quad \text{with} \quad s = \max\left\{ \frac{\lambda_{k+1}}{\lambda_k}, \frac{\lambda_k}{\lambda_{k-1}} \right\}. \qquad (15)$$

(Here one has to put $\lambda_0 = \infty$, $\lambda_{n+1} = 0$.)

The convergence of the qd-algorithm, therefore, can be very slow, namely if two eigenvalues λ_p and λ_{p+1} lie very close together. According to (15), though, convergence is then slow only for these two eigenvalues; for the remaining ones, it may still be fast.

Example 1. Let $Z = \{9, 1, 1000, 1, 9\}$. We display the qd-rows Z, Z', Z'', etc. in the form of a qd-scheme [14]:

$Z \rightarrow$	9				
Z'		1			
$\searrow$	10		1000		
Z''		100		1	
$\searrow$	110		901		9
$Z^{(3)}$		819.09091		.009989	
$\searrow$	929.09091		81.919079		8.990011
$Z^{(4)}$		72.220245		.001096	
$\searrow$	1001.3112		9.699931		8.988915
$Z^{(5)}$		.699614		.001016	
$\searrow$	1002.0108		9.001332		8.987899
	·	.006285		.001014	
	·		8.996062		8.986885
$Z^{(20)}$	·			.001013	
$\searrow$	1002.0171				8.985872
	·	$<10^{-30}$			
	·		9.010972		
$Z^{(50)}$	·			.000972	
$\searrow$	1002.0171				8.970946
		$<10^{-90}$			
			9.037557		
				.000776	
					8.944556

(Here, $\lambda_1 = 1002.0171$, $\lambda_2 = 9.087030$, $\lambda_3 = 8.895860$.)

In addition, one must observe that (14) and (15) are merely asymptotic statements. It may well take a long time, in certain cases, until they finally become effective, so that a very large number of *qd*-steps may be necessary, even if convergence eventually becomes quite fast.

Example 2. $Z = \{1, .001, 2, .001, 4, .001, 8, .001, 16\}$. Since the eigenvalues here are approximately 16, 8, 4, 2, 1, the *qd*-algorithm ought to converge like a geometric series with ratio .5. In reality, however, one has:

$$Z^{(10)} = \{1.580, .266, 2.224, .821, 3.290, 1.408, 6.499, 1.286, 13.631\},$$

$$Z^{(20)} = \{10.196, 1.507, 9.643, .458, 4.383, .028, 2.026, .001, 1.000\},$$

$$Z^{(30)} = \{15.989, 7_{10}-3, 8.006, 8_{10}-4, 4.002, 5_{10}-5, 2.001, 2_{10}-6, .999\}.$$

Only starting with $Z^{(30)}$, where the relative errors of the q_k are less than $10-3$, does the law (15) apply; the errors in $Z^{(40)}$, in fact, are less than $10-6$, those in $Z^{(50)}$ less than $10-9$.

§A1.6. The qd-algorithm with shifts

To the extent that the slow convergence demonstrated in §A1.5 is caused by excessively large quotients λ_{k+1}/λ_k, there exists the possibility to influence these quotients (and with them the convergence) by a shift of the origin of the λ-plane.

Such a shift can be realized if one succeeds in transforming the relation (8) between the generating functions f and f' into

$$f'(z - v) = \frac{zf(z) - 1}{q_1}. \tag{16}$$

Then the poles of f', which, as we know, are also eigenvalues of Z', are indeed diminished by v, so that the further convergence behavior is determined by the quotients $\lambda'_{k+1}/\lambda'_k = (\lambda_{k+1} - v)/(\lambda_k - v)$.

The algorithm for the modified (in the sense of (16)) qd-step (*a progressive qd-step with shift* v) is given by:

```
comment it is assumed that eₙ = 0;
q₁' := (q₁ - v) + e₁;
for k := 2 step 1 until n do
begin                                                        (17)
    e'ₖ₋₁ := (eₖ₋₁/q'ₖ₋₁) × qₖ;
    q'ₖ := ((qₖ - e'ₖ₋₁) - v) + eₖ;
end for k;
```

We denote this process formally by

$$Z \xrightarrow{\text{v}} Z', \tag{18}$$

where it is assumed, of course, that Z' exists, that is, no divisions by 0 occur in (17). One then has:

Theorem A4. *The eigenvalues of the row Z' generated by $Z \xrightarrow{\text{v}} Z'$ are smaller than those of Z by the amount of the shift v.*

Proof. With the row Z' is associated the matrix

$$A' = \begin{bmatrix} q_1' & -q_1' & & & \\ -e_1' & q_2' + e_1' & -q_2' & & \\ & \cdot & \cdot & \cdot & \\ & & \cdot & \cdot & \\ & & & \cdot & -q_{n-1}' \\ & & & -e_{n-1}' & q_n' + e_{n-1}' \end{bmatrix}$$

It can be transformed by means of the computing rule (17), analogously as in §A1.2, into the similar matrix

$$C = \begin{bmatrix} q_1 + e_1 - v & -q_2 & & & \\ -e_1 & q_2 + e_2 - v & -q_3 & & \\ & -e_2 & q_3 + e_3 - v & -q_4 & \\ & & \cdot & \cdot & \cdot \\ & & & \cdot & \cdot & -q_n \\ & & & & -e_{n-1} & q_n - v \end{bmatrix} = B - vI,$$

where B is the matrix (5), which is similar to A; q.e.d.

The effects of such shifts in the Example 1 of §A1.5 become quite noticeable. The qd-scheme containing the chain $Z \xrightarrow{8} Z' \xrightarrow{.8} Z'' \xrightarrow{.09} Z''' \xrightarrow{0} Z^{(4)} \xrightarrow{0} \ldots \xrightarrow{0} Z^{(7)}$ indeed is

					$\sum v_i$
9					
	1				
2		1000			
	500		1		
501.2		493		9	
	491.8196329		.0182556		0
992.9296		.3986227		.9817444	
	.1974465		.0449606		8
993.1271		.1561368		.1367838	
	.0000310		.0393878		8.8
993.1271		.1954935		.0073961	
	0		.0014901		8.89
993.1271		.1969837		.0059059	
	0		.0000447		8.89
993.1271		.1970284		.0058612	
	0		.0000013		8.89
		.1970297		.0058599	
			0		8.89
		.0058599			
					8.89

(On the right are indicated the accumulated shifts.)

By taking into account the shifts 8, .8, .09, one thus obtains the three eigenvalues

$$\lambda_1 = 1002.0171, \quad \lambda_2 = 9.0870297, \quad \lambda_3 = 8.8958599.$$

§A1.7. Deflation after the determination of an eigenvalue

For the matrix $\mathbf{A}$ associated with a *qd*-row in accordance with (1), one clearly has

$$
\mathbf{A}
\begin{bmatrix}
1 \\
1 \\
1 \\
\cdot \\
\cdot \\
\cdot \\
1
\end{bmatrix}
=
\begin{bmatrix}
0 \\
0 \\
\cdot \\
\cdot \\
0 \\
q_n
\end{bmatrix}.
\tag{19}
$$

Thus, when $q_n = 0$, $\mathbf{A}$ is singular, that is:

Theorem A5. *A qd-row* $Z = \{q_1, e_1, q_2, \ldots, e_{n-1}, 0\}$ *has 0 as an eigenvalue.*

If, therefore, by a sequence of *qd*-steps (with suitable choice of the shifts $v_0, v_1, \ldots$)

$$
Z \xrightarrow{v_0} Z' \xrightarrow{v_1} Z'' \xrightarrow{v_2} Z''' \longrightarrow \ldots ,
$$

one succeeds in obtaining a row $Z^{(j)}$ whose last element $q_n^{(j)} = 0$, then 0 is an eigenvalue of $Z^{(j)}$, and therefore, by Theorem A4,

$$
\lambda_n = v_0 + v_1 + \cdots + v_{j-1}
\tag{20}
$$

an eigenvalue of the given row Z. In practice, (20) of course holds only approximately.

Example 3. For $Z = \{4, 3, 3, 2, 2, 1, 1\}$ one obtains with the shifts $v_0 = .3$, $v_1 = .02$, $v_2 = .002$, $v_3 = .000548$ (in 6-digit computation) the following *qd*-scheme:

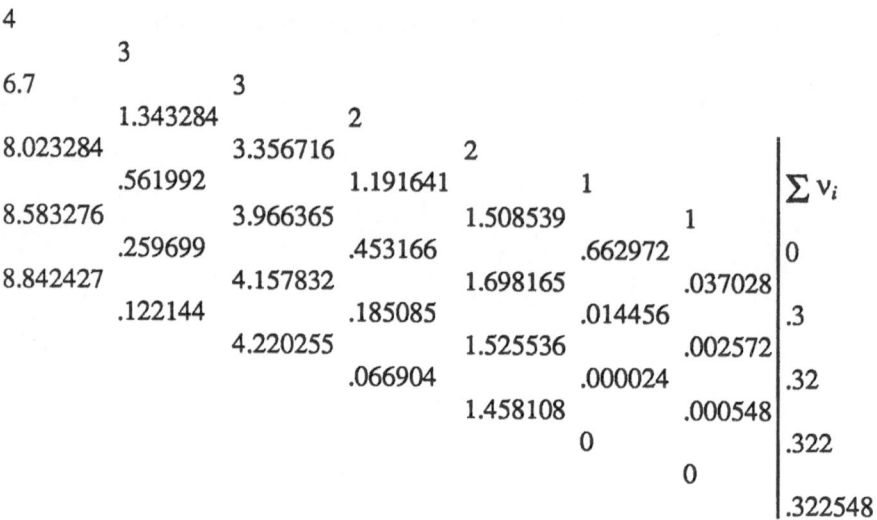

					$\sum v_i$
4					
	3				
6.7		3			
	1.343284		2		
8.023284		3.356716		2	
	.561992		1.191641		1
8.583276		3.966365		1.508539	1
	.259699		.453166		.662972	0
8.842427		4.157832		1.698165		.037028
	.122144		.185085		.014456	.3
	4.220255		1.525536		.002572	.32
		.066904		.000024		.322
		1.458108		.000548	.322
		0			0	.322548

(The last two zeros, of course, are correct only within the computing precision.)

Since the sum of the shifts is .322548, the eigenvalues $\lambda_k^{(4)}$ of $Z^{(4)}$ are smaller by this amount than those of Z; but according to Theorem A5, $\lambda_4^{(4)} = 0$, hence $\lambda_4 = .322548$ (the exact value is $.32254769\ldots$).

To determine further eigenvalues of $Z^{(4)}$, one exploits the fact that for a qd-row $Z_0 = \{q_1,\ e_1,\ q_2, \ldots, e_{n-2},\ q_{n-1},\ 0,\ 0\}$ the associated matrix is

$$
\mathbf{A}_0 = \begin{bmatrix}
q_1 & -q_1 & & & & \\
-e_1 & q_2 + e_1 & -q_2 & & & \\
& \cdot & \cdot & \cdot & & \\
& & \cdot & \cdot & \cdot & \\
& & & \cdot & \cdot & \cdot \\
& & & -e_{n-2} & q_{n-1} + e_{n-2} & -q_{n-1} \\
& & & & 0 & 0
\end{bmatrix}.
$$

Here, one eigenvalue is 0, while the remaining $n - 1$ eigenvalues obviously coincide with the eigenvalues of the reduced matrix

$$A_1 = \begin{bmatrix} q_1 & -q_1 & & & & \\ -e_1 & q_2 + e_1 & -q_2 & & & \\ & \cdot & \cdot & \cdot & & \\ & & \cdot & \cdot & \cdot & \\ & & & \cdot & \cdot & \cdot & \\ & & & & & -q_{n-2} \\ & & & & -e_{n-2} & q_{n-1} + e_{n-2} \end{bmatrix}$$

This, however, is precisely the matrix associated with the qd-row $Z_1 = \{q_1, e_1, q_2, e_2, \ldots, e_{n-2}, q_{n-1}\}$; one thus has the following rule:

Theorem A6. *A qd-row of the form*

$$Z_0 = \{q_1, e_1, q_2, \ldots, q_{n-1}, 0, 0\}$$

has the eigenvalue 0; the remaining $n - 1$ eigenvalues of Z_0 are the eigenvalues of

$$Z_1 = \{q_1, e_1, q_2, \ldots, q_{n-1}\}.$$

Deleting the two zeros at the end of the qd-row Z_0, that is, the transition from Z_0 to Z_1, is called *deflation*.

Thus, in Example 3 one would obtain from

$$Z_0^{(4)} = \{8.842427, .122144, \ldots, 1.458108, 0, 0\}$$

by deflation the row

$$Z_1^{(4)} = \{8.842427, .122144, 4.220255, .066904, 1.458108\},$$

and from it determine further eigenvalues, for example, according to

$$Z_1^{(4)} \xrightarrow{\ 1\ } Z_1^{(5)} \xrightarrow{\ .4\ } Z_1^{(6)} \xrightarrow{\ .02\ } Z_1^{(7)} \xrightarrow{\ .003\ } Z_1^{(8)} \xrightarrow{\ .000213\ } Z_1^{(9)}:$$

				$\sum v_i$
8.842427				
	.122144			
7.964571		4.220255		
	.064721		.066904	
7.629292		3.222438		1.458108
	.027337		.030273	.322548
7.636629		2.825374		.427835
	.010114		.004584	1.322548
7.643743		2.799844		.023251
	.003705		.000038	1.722548
7.647235		2.793177		.003213
	.001353		.000000	1.742548
		2.791611		.000213
		0		1.745548
			0	
				1.745761

Note that in the sum of the shifts executed up until now, those executed before the deflation have to be included. One thus has now $\sum v_i = 1.745761$, $Z_1^{(9)} = \{7.647235, .001353, 2.791611, 0, 0\}$, that is, $\lambda_3 = 1.745761$.

Choice of Shifts

§A2.1. Effect of the shift v on Z'

For reasons mentioned in §§A1.5, A1.6, only the qd-algorithm *with shifts* has any practical significance for the determination of the eigenvalues of a positive qd-row. The correct choice of the shifts, however, is also crucial for a successful completion of the task.

According to [20], §3, the positive qd-rows enjoy particularly favorable numerical properties. One therefore aims at choosing the shifts v_0, $v_1, v_2, \ldots$ in the iterative sequence

$$ Z \xrightarrow{v_0} Z' \xrightarrow{v_1} Z'' \xrightarrow{v_2} Z''' \ldots $$

in such a way that also Z', Z'', Z''', $\ldots$ remain positive, so that this property is not lost. In this connection, the following is relevant.

Theorem A7. *If the qd-row Z is positive, then the row Z' generated from it by $Z \xrightarrow{v} Z'$ is also positive precisely if $v < \lambda_n$ (λ_n = smallest eigenvalue of Z).*

Proof. a) If $v \geq \lambda_n$, then Z' cannot be positive, since otherwise the computational algorithm (A1, 17) would certainly be executable and by Theorems A4 and A2, therefore, $\lambda'_n = \lambda_n - v > 0$, in contradiction to $v \geq \lambda_n$.

b) As v increases from 0 monotonically and continuously, the following holds for $Z \xrightarrow{v} Z'$ and for the elements $q'_k(v)$, $e'_k(v)$, which depend on v: By Theorem A3, $q'_1(0)$, $e'_1(0), \ldots, q'_n(0)$ are positive; furthermore, on the basis of the computational algorithm (A1, 17):

$$q_1'(v) = q_1 - v + e_1$$ decreases monotonically,

$$e_1'(v) = (e_1/q_1'(v))q_2$$ increases monotonically, so long as

q_1' remains positive, hence

$$q_2'(v) = q_2 - e_1'(v) - v + e_2$$ decreases monotonically, etc.,

until finally $\qquad\qquad (1)$

$$q_n'(v) = q_n - e_{n-1}'(v) - v$$ decreases monotonically, so long

as none of the values $q_1'(v)$, $q_2'(v)$,

$\ldots$, $q_{n-1}'(v)$ becomes negative .

Therefore, under the last condition mentioned, all $q_k'(v)$ are decreasing monotonically, and the $e_k'(v)$ increase monotonically. From $q_\ell'(v) \downarrow 0$ (ℓ fixed, $\ell < n$), however, there follows $e_\ell'(v) \uparrow \infty$ and thus $q_{\ell+1}'(v) \downarrow -\infty$. Of all q_k', therefore, only q_n' can vanish, without another q_k' becoming negative; that is, with increasing v it is $q_n'(v)$ which first attains the value 0. There thus exists a $v = v_0 > 0$ such that $q_n'(v_0) = 0$, but $Z' > 0$ for $v < v_0$. Consequently, $\lambda_k' = \lambda_k - v > 0$ for all k and $v < v_0$, hence $\lambda_k \geq v_0$ ($k = 1, 2, \ldots, n$); but since $q_n'(v_0) = 0$, one gets $\lambda_n' = 0$ from Theorem A5, hence $\lambda_n = v_0$. By Theorem A2, the eigenvalues of Z are real and simple, thus $\lambda_1 > \lambda_2 > \cdots > \lambda_{n-1} > \lambda_n = v_0$, q.e.d.

With this, the question as to the appropriate choice of v is answered: One should choose the shift always below the smallest eigenvalue of the qd-row, to which the shift is applied, and in fact should choose it as closely below as possible, as was illustrated, for example, in Example 3 of §A1.7.

Now, unfortunately, this rule cannot be applied without knowledge of λ_n; methods must therefore be developed which permit an independent determination of v (see §A2.3).

An important fact, which can facilitate the choice of v, is given in the following

Theorem A8. *If the qd-row Z is positive, and $v \leq \lambda_n$, then for the row Z' obtained from Z by $Z \overset{v}{\longrightarrow} Z'$ one has:*

$$
\left.\begin{array}{l}
q_k' > e_k \\[2mm]
e_k' < q_{k+1}
\end{array}\right\} \quad k = 1, 2, \ldots, n-1, \tag{2}
$$

$$
q_n' \geq 0.
$$

Proof. If $\nu \geq 0$, it follows from (A1, 17) and Theorem A7 that

$$
q_n' = q_n - e_{n-1}' - \nu \left\{\begin{array}{ll}
> 0 & \text{for } \nu < \lambda_n, \\[2mm]
= 0 & \text{for } \nu = \lambda_n,
\end{array}\right.
$$

hence, in every case, $e_{n-1}' < q_n$. Furthermore, $q_{n-1} - e_{n-2}' - \nu + e_{n-1} = q_{n-1}' > e_{n-1}$, thus $e_{n-2}' < q_{n-1}$, etc., until $e_1' < q_2$ and $q_1' > e_1$. For $\nu < 0$, the assertion follows from the monotonicity property (1) of the q_k' and e_k', q.e.d.

Thus, for example, if one applies to

$$
Z = \{5, 10, 7, 5, 8, 3, 9, 1, 10\}
$$

a *qd*-step with $\nu = 3$, which yields $q_1' = 12$, $e_1' = 5.8333333$, $q_2' = 3.1666666$ ($< e_2 = 5$), then Theorem A8 tells us that one cannot have $\nu \leq \lambda_n$, that is, that the shift is chosen too large and would produce a non-positive Z'.

§A2.2. Semipositive qd-rows

Definition. *A qd-row* $Z = \{q_1, e_1, q_2, e_2, \ldots, q_n\}$ *with*

$$
\left.\begin{array}{l}
q_k > 0 \\[2mm]
e_k > 0
\end{array}\right\} \quad k = 1, 2, \ldots, n-1, \tag{3}
$$

$$
q_n = 0
$$

is called semipositive (in symbols: $Z \geq 0$).

By Theorem A5, every semipositive row has the eigenvalue 0, but since in this case the associated (according to (A1, 10)) matrix H is positive semidefinite, being decomposable with (A1, 11), one has:

Theorem A9. *The eigenvalues of a semipositive qd-row are*

$$\lambda_1 > \lambda_2 > \lambda_3 > \cdots > \lambda_{n-1} > \lambda_n = 0.$$

Furthermore, from the proof of Theorem A7, there follows:

Theorem A10. *With $v = \lambda_n$ (and only then) the step $Z \xrightarrow{v} Z'$ produces from a positive row Z a semipositive row Z'.*

It would be ideal if, by a single qd-step, one could obtain from Z a semipositive row Z', since then one immediately would have $\lambda'_n = 0$, hence $\lambda_n = v$, and one could obtain the remaining eigenvalues as follows:

Apply to $Z' \geq 0$ a qd-step $Z' \xrightarrow{0} Z''$, whereby, as in the proof of Theorem A3,

$$q''_1 > e'_1 > 0,$$
$$e''_1 < q'_2 \text{ (and } e''_1 > 0),$$
$$q''_2 > e'_2 > 0,$$
$$e''_2 < q'_3 \text{ (and } e''_2 > 0),$$
$$\cdot$$
$$\cdot \tag{4}$$
$$\cdot$$
$$q''_{n-1} > e'_{n-1} > 0, \text{ but now}$$
$$e''_{n-1} = q'_n(e'_{n-1}/q''_{n-1}) = 0,$$
$$q''_n = q'_n - e''_{n-1} = 0.$$

Thus, $Z'' = \{q''_1, e''_1, \ldots, q''_{n-1}, 0, 0\}$; by deflation according to §A1.7, and on account of (4), one obtains from this again a positive row,

$$Z''_1 = \{q''_1, e''_1, \ldots, q''_{n-1}\},$$

which is then further processed in the same way.

Now in practical computation, owing to rounding errors, it is usually impossible to obtain a semipositive row in *one* step, even when λ_n is known accurately. For the row

$$Z = \{100, 1, 200, 1, 300, 1, 400, 1, 500\},$$

for example, the smallest eigenvalue is $\lambda_5 = 99.00985285.\ldots$. Working consistently with 3 decimals after the decimal point, λ_5 can be approximated at best by 99.009 or 99.010. With $v = 99.010$ one obtains a negative value $q_4' = -1257$, and with 99.009 the value of q_5' does not become small; one succeeds only with additional steps: the sequence

$$Z \xrightarrow{\ 99.009\ } Z' \xrightarrow{\ 0\ } Z'' \xrightarrow{\ 0\ } Z''' \xrightarrow{\ .001\ } Z^{(4)}$$

yields the *qd*-scheme:

```
100
        1
1.991           200
       100.452         1
102.443          1.539        300
        1.509        194.932        1                          │ Σ vᵢ
103.952         194.962        7.059       400                 │
        2.830         7.056        56.665        1             │ 0
106.781         199.190       56.666       245.326        500  │
        5.279         2.008       245.322       2.038          │ 99.009
                195.918       299.980       2.042      398.953 │
                 3.075         1.670       398.172            │ 99.009
                        298.574       398.544        .781     │
                         2.229          .780                  │ 99.009
                                397.094        .001           │
                                         0                    │ 99.009
                                                  0           │
                                                              │ 99.010
```

Thanks to the rounding errors, it was still possible, here, to carry out a step with $v = .001$ and to thus increase the sum of shifts to 99.010. In higher precision, this would not be possible.

§A2.3. Bounds for λ_n

In order to adhere to the rule "v is to be chosen closely below λ_n", one needs, of course, information concerning the approximate location of λ_n. In cases where such information is not sufficiently accurate, one must resort to the nesting procedure described in the next section.

According to [20], every q_k-value is already an upper bound for the smallest eigenvalue of the row $Z = \{q_1, e_1, \ldots, q_n\}$. More precisely, the following holds:

Theorem A11. *For a positive qd-row, define the quantities $F_1 = 1$ and*

$$F_k = 1 + \frac{e_{k-1}}{q_{k-1}}\left[1 + \frac{e_{k-2}}{q_{k-2}}\left[1 + \cdots \left[1 + \frac{e_1}{q_1}\right] \cdots \right]\right], \quad k = 2, \ldots, n. \quad (5)$$

Then

$$\lambda_n < sup = \min_{1 \le k \le n} \frac{q_k}{F_k}. \qquad (6)$$

Proof. a) With the quantities F_k defined in (5) one clearly has

$$\overline{d}_k = \frac{q_k}{F_k} = \frac{q_k}{1 + \dfrac{e_{k-1}}{q_{k-1}} F_{k-1}} = \frac{q_k}{1 + \dfrac{e_{k-1}}{\overline{d}_{k-1}}}$$

(with $\overline{d}_1 = q_1$). Constructing, however, $Z \xrightarrow{\ 0\ } Z'$, and with the elements of Z' the quantities $d_k = q'_k - e_k$ ($k = 1, \ldots, n$), one has by (A1, 3):

$$d_1 = q_1$$

and for $k > 1$,

$$d_k = q_k - e'_{k-1} = q_k - \frac{q_k e_{k-1}}{q'_{k-1}}$$

$$= q_k \frac{q'_{k-1} - e_{k-1}}{q'_{k-1}} = q_k \frac{d_{k-1}}{d_{k-1} + e_{k-1}} = \frac{q_k}{1 + \dfrac{e_{k-1}}{d_{k-1}}}.$$

The d_k and $\overline{d}_k$ thus satisfy the same recursion formula with the same initial values; therefore,

$$d_k = q'_k - e_k = \frac{q_k}{F_k}. \tag{7}$$

b) For the qd-row $Z'' = \{q''_1, e''_1, \ldots, q''_n\}$ obtained by a qd-step $Z \xrightarrow{v>0} Z''$ we have on the basis of the monotonicity property (1): $q''_k < q'_k$, $e''_k > e'_k$, so long as $q''_1, q''_2, \ldots, q''_{k-1}$ are positive. With the shift $v = d_j$ (j fixed), therefore, the following is true: either at least one of the $q''_1, q''_2, \ldots, q''_{j-1}$ is negative, or $q''_j = q_j - d_j - e''_{j-1} + e_j = q_j - q_j + e'_{j-1} - e''_{j-1} + e_j < e_j$, contrary to the statement (2) of Theorem A8. Consequently, we must have $d_j > \lambda_n$, hence also min $d_j > \lambda_n$, q.e.d.

It is to be noted that Theorem A11 follows also from the fact that F_k/q_k is the kth diagonal element of the matrix $\mathbf{H}^{-1}$ ($\mathbf{H}$ is defined in (A1, 10)); the line of proof above will allow us later to also make statements concerning the effect of rounding errors upon (6).

Since $\mathbf{H}^{-1}$ is a positive definite matrix, the trace is an upper bound for λ_n^{-1}, and therefore

$$\lambda_n > inf = \frac{1}{\displaystyle\sum_{k=1}^{n} \frac{F_k}{q_k}} \geq \frac{sup}{n}. \tag{8}$$

Remark. For the quantities d_k in (7), Bauer and Reinsch [23] give the recursion formula

$$d_k = \frac{q_k}{q'_{k-1}} d_{k-1}, \tag{9a}$$

which of course is equivalent to the formula

$$d_k = \frac{q_k}{1 + \dfrac{e_{k-1}}{d_{k-1}}} \tag{9b}$$

used in the proof. Likewise, an expression for the quantity *inf* in (8) can already be found in the cited paper.

§A2.4. A formal algorithm for the determination of eigenvalues

By a formal algorithm (cf. §1.1) one means a computational procedure which attains the desired goal if one disregards the limitations of finite arithmetic (rounding errors, limited number range).

The desired goal, here, is to use *qd*-steps (with shifts) to obtain a semipositive *qd*-row $Z^{(j)}$; how one proceeds from there is then outlined in §A1.7. Strictly speaking, only a *qd*-row $Z^{(j)}$ with a negligibly small $q_n^{(j)}$ is achievable; according to [20], the error due to neglecting $q_n^{(j)}$ can be estimated.

Since the smallest eigenvalue λ_n must still lie below the smallest *q*-element $q_{\min}$, one starts the algorithm with $v_0 = q_{\min}/2$. The complete procedure then consists in the following (under the assumption that the given row Z is positive):

1. The first step $Z \xrightarrow{v_0} Z'$ is executed with $v_0 = q_{\min}/2$.

2. If the step $Z^{(j)} \xrightarrow{v_j} Z^{(j+1)}$ leads to a row $Z^{(j+1)} > 0$, the step "succeeds" and one puts $v_{j+1} := v_j/2; \ j := j + 1$.

3. If the step $Z^{(j)} \xrightarrow{v_j} Z^{(j+1)}$ yields an element $q_k^{(j+1)} \leq 0$, the step "fails" (since $v_j \geq \lambda_n^{(j)}$); it must be terminated immediately and repeated with $v_j := v_j/2$ ([1]).

---

[1] An exception is the case $q_k^{(j+1)} > 0$, $k = 1, \dots, n - 1$, $q_n^{(j+1)} = 0$ of a semipositive row $Z^{(j+1)}$. (Editors' remark)

4. The shifts are accumulated as follows: if $Z^{(j)} \xrightarrow{\;v_j\;} Z^{(j+1)}$ succeeds (and only then), one puts $w_{j+1} := w_j + v_j$ (one begins with $w_0 = 0$). w_j then always indicates by how much the eigenvalues of Z and $Z^{(j)}$ differ.

It will be shown now that the computational process thus defined *theoretically* achieves the desired goal:

To begin with, we prove that

$$w_j < \lambda_n \le w_j + 2v_j, \tag{10}$$

where v_j, w_j denote the individual and accumulated shift, respectively, at the beginning of the step $Z^{(j)} \xrightarrow{\;v_j\;} Z^{(j+1)}$ (also in the case of a repetition): Initially, $w_0 = 0$, $\lambda_n > 0$, $v_0 = q_{min}/2$, thus $2v_0 = q_{min} \ge \lambda_n$. The relation (10) can thus be used as an induction hypothesis.

a) If the step $Z^{(j)} \xrightarrow{\;v_j\;} Z^{(j+1)}$ succeeds, then $w_{j+1} = w_j + v_j$, $v_{j+1} = v_j/2$, and because of $\lambda_n^{(j)} > v_j$ thus $w_{j+1} < \lambda_n^{(j)} + w_j = \lambda_n$, while $w_{j+1} + 2v_{j+1} = w_{j+1} + v_j = w_j + 2v_j \ge \lambda_n$.

b) If, however, $Z^{(j)} \xrightarrow{\;v_j\;} Z^{(j+1)}$ fails, one has $v_j \ge \lambda_n^{(j)}$, thus $w_j < \lambda_n \le w_j + v_j$; after the operation $v_j := v_j/2$ the relation (10) therefore continues to be valid.

Now in each step, whether it succeeds or not, v_j is halved, so that

$$v_j \to 0, \quad \lambda_n^{(j)} \to 0, \quad w_j \to \lambda_n \quad \text{as} \quad j \to \infty, \tag{11}$$

while always $\lambda_k^{(j)} = \lambda_k - w_j > \lambda_k - \lambda_n$. For the generating function, this means, according to the considerations in [20], §2, that

$$f^{(j)}(z) = \sum_{k=1}^{n} \frac{c_k^{(j)}}{z + w_j - \lambda_k},$$

where

$$c_k^{(j)} = c_k^{(0)} \; \frac{\prod\limits_{\ell=0}^{j-1} (\lambda_k - w_\ell)}{\prod\limits_{\ell=0}^{j-1} q_1^{(\ell)}} \; , \tag{12}$$

so that by (11),

$$\frac{c_n^{(j)}}{c_k^{(j)}} \to 0 \quad (k = 0, 1, \ldots, n-1).$$

On the basis of [14], §I.10, this finally implies that also $q_n^{(j)} \to 0$ and $e_{n-1}^{(j)} \to 0$.

Naturally, this theoretical convergence $q_n^{(j)} \to 0$, $e_{n-1}^{(j)} \to 0$ does not mean much for practical computation; therefore, we must examine more closely the arithmetic employed, in order to arrive at a useful algorithm.

Finite Arithmetic

§A3.1. The basic sets

Numerical processes, always, can only be executed in a finite (inexact) arithmetic; as a matter of fact, the attribute "numerical" precisely means that one is dealing with an inexact procedure.

In the domain $\mathbf{R}$ of the real numbers, the arithmetic operations $+$, $-$, $\times$, $/$ are defined exactly, as are also the six order relations $>$, $\geq$, $<$, $\leq$, $=$, $\neq$. On the other hand, a finite arithmetic is characterized by a nonempty finite subset $\mathfrak{G} \subset \mathbf{R}$ in which are defined the *numerical* operations $\tilde{+}$, $\tilde{-}$, $\tilde{\times}$, $\tilde{/}$ (as approximations to the exact operations). These operations, which are applicable only to elements of $\mathfrak{G}$, produce as results again elements of $\mathfrak{G}$, or else the singular value Ω (that is, "undefined"). The union of $\mathfrak{G}$ and $\{\Omega\}$ is denoted by $\bar{\mathfrak{G}}$ Contrary to the arithmetic operations, the order relations $>$, $\geq$, $<$, $\leq$, $=$, $\neq$ are defined exactly in $\mathfrak{G}$.

The structure of the set $\mathfrak{G}$ is determined by the following axioms:

I_1: *To each $x \in \mathbf{R}$ there is associated uniquely an element $\tilde{x} \in \bar{\mathfrak{G}}$*

In this map $\mathbf{R} \to \bar{\mathfrak{G}}$ many x, of course, can be mapped into the same element $z \in \bar{\mathfrak{G}}$ but:

I_2: *For each $z \in \mathfrak{G}$ the set $\mathfrak{P}(z) = \{x \mid x \in \mathbf{R}, \tilde{x} = z\}$ is connected.*

The set $\mathfrak{O} = \{x \mid x \in \mathbf{R}, \tilde{x} = \Omega\}$ of real numbers to which the element Ω is assigned, that is, which are not representable in the arithmetic, is called the *overflow domain* of the arithmetic. We require:

I_3: *The complementary set* $\mathfrak{D} = \mathbf{R} - \mathfrak{D} = \{x \mid x \in \mathbf{R}, \ \tilde{x} \in \mathfrak{S}\}$ *is connected.*

I_4: $x \in \mathfrak{S} \Rightarrow \tilde{x} = x$, *that is, the elements of* $\mathfrak{S}$ *are their own representers.*

One can formulate I_4 also as $z \in \mathfrak{P}(z)$ $(z \in \mathfrak{S})$.

As a consequence of the Axioms I one obtains a certain *monotonicity property* of the map $\mathbf{R} \to \bar{\mathfrak{S}}$

Theorem A12. *If* $x,y \in \mathfrak{D}$, *then*

$$
\begin{aligned}
x < y & \Rightarrow \tilde{x} \leq \tilde{y}, \\
x = y & \Rightarrow \tilde{x} = \tilde{y}, \\
x > y & \Rightarrow \tilde{x} \geq \tilde{y}.
\end{aligned}
\tag{1}
$$

Proof. The second statement is a direct consequence of I_1; it also means that for $\tilde{x} \neq \tilde{y}$ the sets $\mathfrak{P}(\tilde{x})$ and $\mathfrak{P}(\tilde{y})$ must be disjoint. Since by definition, $x \in \mathfrak{P}(\tilde{x})$ and $y \in \mathfrak{P}(\tilde{y})$, and by I_4, in addition, $\tilde{x} \in \mathfrak{P}(\tilde{x})$, $\tilde{y} \in \mathfrak{P}(\tilde{y})$, it follows from I_2 that for $x < y$ there cannot hold $\tilde{x} > \tilde{y}$, and vice versa; q.e.d.

We further require the elements 0 and 1 to be in $\mathfrak{S}$, and also want symmetry of $\mathfrak{S}$ with respect to the origin:

II_1: $\tilde{0} = 0$,
II_2: $\tilde{1} = 1$,
II_3: $(-x)^{\tilde{}} = -\tilde{x}$.

(From II_3 there also follows the existence of an exact unitary operation "$-$" in $\mathfrak{S}$.)

The set $\mathfrak{U} = \{x \mid x \in \mathbf{R}, \ \tilde{x} = 0\}$ is called the *underflow domain* of the arithmetic.

§A3.2. Properties of the arithmetic

III: For each pair $a, b \in \mathfrak{C}$ and each operator $\circ = +, -, \times, /$ there is defined the operation

$$c = a \; \widetilde{\circ} \; b \quad \text{with} \quad c \in \overline{\mathfrak{C}}$$

Thus, either the result c is again in $\mathfrak{C}$, or $c = \Omega$; the latter simply means that the operation $a \; \widetilde{\circ} \; b$ is undefined. In particular, of course, one always has $a \; \widetilde{/} \; 0 = \Omega$.

A first group of axioms imposes the commutativity of addition and multiplication (it is always assumed that $a, b, \in \mathfrak{C}$):

IV_1: $a \; \widetilde{+} \; b = b \; \widetilde{+} \; a,$

IV_2: $a \; \widetilde{\times} \; b = b \; \widetilde{\times} \; a.$

(For example, if $a \; \widetilde{+} \; b = \Omega$, then also $b \; \widetilde{+} \; a = \Omega$.)

However, associativity and distributivity are out of the question; after all, it need not even be true that $(a \; \widetilde{+} \; b) \; \widetilde{-} \; b = a$. On the other hand, we can require:

IV_3: *If $a \geq b \geq 0$, then $(a \; \widetilde{-} \; b) \; \widetilde{+} \; b = a$.*

(In fact, IV_3 is a property which in floating-point arithmetic is usually valid.)

A second group of axioms imposes the sign-symmetry of certain operations:

$$V_1: \quad a \; \widetilde{-} \; b = a \; \widetilde{+} \; (-b) = - (b \; \widetilde{-} \; a),$$
$$V_2: \quad (-a) \; \widetilde{\times} \; b = a \; \widetilde{\times} \; (-b) = -(a \; \widetilde{\times} \; b),$$
$$V_3: \quad (-a) \; \widetilde{/} \; b = a \; \widetilde{/} \; (-b) = -(a \; \widetilde{/} \; b).$$

From V_1 and IV_1 there follows, in addition[1]:

---

[1] Using V_1 twice, and then IV_1, one indeed obtains $(-a) \; \widetilde{-} \; (-b) = -(b \; \widetilde{-} \; (-a)) = -(b \; \widetilde{+} \; (-(-a))) = -(b \; \widetilde{+} \; a) = -(a \; \widetilde{+} \; b).$

$$V_4: \quad (-a) \mathbin{\widetilde{\mp}} (-b) = -(a \mathbin{\widetilde{\mp}} b).$$

Furthermore, the following properties can be derived(^{2}):

Theorem A13. *If $a \in \mathfrak{G}$, then*

$$a \mathbin{\widetilde{=}} a = 0,$$
$$a \mathbin{\widetilde{\mp}} 0 = a \mathbin{\widetilde{=}} 0 = a, \tag{2}$$
$$a \mathbin{\widetilde{\times}} 0 = 0.$$

§A3.3. Monotonicity of the arithmetic

Sequential reliability of a program (see §1.1) can in effect be achieved and proved only if the arithmetic operations exhibit certain monotonicity properties:

Suppose that $a, b, c, d \in \mathfrak{G}$ and $0 \le a \le b$, $0 \le c \le d$. Then the following is to hold:

$$VI_1: \quad a \mathbin{\widetilde{\mp}} c \le b \mathbin{\widetilde{\mp}} d,$$
$$VI_2: \quad a \mathbin{\widetilde{\times}} c \le b \mathbin{\widetilde{\times}} d,$$
$$VI_3: \quad a \mathbin{\widetilde{=}} d \le b \mathbin{\widetilde{=}} c,$$
$$VI_4: \quad a \mathbin{\widetilde{/}} d \le b \mathbin{\widetilde{/}} c.$$

The $\le$-signs in the hypotheses and in the assertions are not meant to be coherent, that is, it may be possible, for example, that $a < b$, $c < d$, and yet $a \mathbin{\widetilde{\times}} c = b \mathbin{\widetilde{\times}} d$. Furthermore, the $\le$-sign is to have the additional

---

^{2} For the proof one needs the following axioms:

1) For $a \mathbin{\widetilde{=}} a = 0$: V_1 (with $a = b$), II_1.
2) For $a \mathbin{\widetilde{\mp}} 0 = a$: IV_3 (with $a = b \ge 0$), assertion 1), IV_1, V_4 (with $a < 0$, $b = 0$), II_1, II_3.
3) For $a \mathbin{\widetilde{=}} 0 = a$: V_1 (with $b = 0$), II_1, assertion 2).
4) For $a \mathbin{\widetilde{\times}} 0 = 0$: V_2 (with $b = 0$).

(Editors' remark)

meaning that when the value on the left-hand side is Ω, then also the right-hand side must have this value.

With the aid of these axioms, one now obtains

Theorem A14([1]). *If $a, b \in \mathfrak{G}$, then*

$$b \geq 0 \;=>\; a \tilde{\mp} b \geq a, \tag{3a}$$

$$a \geq b \;=>\; a \overset{\sim}{-} b \geq 0, \tag{3b}$$

$$a \overset{\sim}{-} b \geq 0 \;=>\; a \geq b, \tag{3c}$$

$$a \overset{\sim}{-} b > 0 \;=>\; a > b, \tag{3d}$$

$$a \tilde{\mp} b > a \;=>\; b > 0. \tag{3e}$$

For $a, b, c \in \mathfrak{G}$ with $a, b, c \geq 0$, one has furthermore: If $a \overset{\sim}{-} b \geq c$, then either $a \overset{\sim}{-} c > b$ or $b \tilde{\mp} c = a$.

Proof. If $a, b \geq 0$, then (3a) and (3b) follow at once from the axioms VI_1, VI_3 and Theorem A13. If $a < 0$, $b \geq 0$, one applies, besides II_3, in turn IV_1, V_1, VI_3, V_1, IV_1 and Theorem A13:

$$a \tilde{\mp} b = b \tilde{\mp} a = b \overset{\sim}{-} (-a) \geq 0 \overset{\sim}{-} (-a) = 0 \tilde{\mp} a = a \tilde{\mp} 0 = a.$$

To prove (3b), one still needs to treat two cases: If $b < 0 \leq a$, then V_1, VI_1 and Theorem A13 yield

$$a \overset{\sim}{-} b = a \tilde{\mp} (-b) \geq a \tilde{\mp} 0 = a \geq 0.$$

If $b \leq a \leq 0$, one uses V_1, IV_1, V_1, VI_3 and Theorem A13:

---

[1] The original contains only the inequalities (3a) and (3b), with a proof valid only in the case $a, b \geq 0$, and the second part of the theorem is stated with the stronger hypothesis $a, b, c > 0$ and the weaker assertion $a \overset{\sim}{-} b \geq c \;=>\; \{a \overset{\sim}{-} c \geq b$ or $b \tilde{\mp} c = a\}$ (in the proof, $a, b \geq 0$, $c > 0$ would suffice). Later, in Chapter A4, however, also the inequalities (3c), (3d) and (3e) will be used. The first, which is nontrivial, immediately allows one to weaken the hypotheses of the second part. Accordingly, Theorem A14 and its proof are given here in an extended form. (Editors' remark)

$$a \simeq b = a \mathbin{\widetilde{\mp}} (-b) = (-b) \mathbin{\widetilde{\mp}} a = (-b) \simeq (-a) \geq (-a) \simeq (-a) = 0.$$

Similar case distinctions are needed to prove (3c) by contradiction, assuming $a < b$. If $0 \leq a < b$, there first follows from VI_3 and Theorem A13 that $a \simeq b \leq 0$. If we had $a \simeq b = 0$, then IV_3, V_1, IV_1 and Theorem A13 would give the contradiction

$$b = (b \simeq a) \mathbin{\widetilde{\mp}} a = [-(a \simeq b)] \mathbin{\widetilde{\mp}} a = 0 \mathbin{\widetilde{\mp}} a = a \mathbin{\widetilde{\mp}} 0 = a.$$

If $a < 0 < b$, then two applications of V_1 lead to $a \simeq b = -(b \simeq a) = -[b \mathbin{\widetilde{\mp}} (-a)]$, and the expression in brackets, by VI_1 and Theorem A13, is not smaller than b, thus $a \simeq b \leq -b < 0$, contrary to the assumption. Finally, if $a < b \leq 0$, one first obtains from IV_1 and V_1 that $a \simeq b = (-b) \simeq (-a)$, and an appeal to the first case shows that this again is negative.

Implication (3d) is now an immediate consequence of (3c) and the first equation of Theorem A13. It could also be proved at once by contradiction from (3b), with the help of V_1. Analogously, to prove (3e), one assumes $b \leq 0$ and deduces from V_1 and IV_1 that $a \mathbin{\widetilde{\mp}} b = -[(-a) \mathbin{\widetilde{\mp}} (-b)]$; the expression in brackets, according to (3a), is not smaller than $-a$, hence $a \mathbin{\widetilde{\mp}} b \leq a$, contradicting the assumption.

For the second part of the theorem, we first note that $a \simeq b \geq c \geq 0$ implies $a \geq b$, by (3c). The axioms IV_3, VI_1 and IV_1 then give $a = (a \simeq b) \mathbin{\widetilde{\mp}} b \geq c \mathbin{\widetilde{\mp}} b = b \mathbin{\widetilde{\mp}} c$. In particular, $a \geq c$ and $a \simeq c \geq 0$, cf. (3b). If we now had $a \simeq c \leq b$, then likewise $a = (a \simeq c) \mathbin{\widetilde{\mp}} c \leq b \mathbin{\widetilde{\mp}} c$. That is, $a \simeq b \geq c$ is compatible with $a \simeq c \leq b$ only in the case $a = b \mathbin{\widetilde{\mp}} c$; q.e.d.

As an example for the second part of the theorem, consider $a = 1.01$, $b = 9.74_{10}-1$, $c = 3.70_{10}-2$. Then, in 3-digit floating-point arithmetic,

$$a \simeq b = 1.01 - .97 = .04 = 4.00_{10}-2 > c,$$

$$a \simeq c = 1.01 - .04 = .97 = 9.70_{10}-1 < b,$$

$$b \mathbin{\widetilde{\mp}} c = 9.74_{10}-1 + .37_{10}-1 = 1.01 = a.$$

§A3.4. Precision of the arithmetic

A further characteristic of the arithmetic is its precision, which we introduce through the following axioms:

VII_1: *There exists a smallest number* $\theta > 0$ *having the property that for all* $x \in \mathbf{R}, a \in \mathfrak{G}, a > 0$:

$$\tilde{x} = a \implies |x - a| \le \theta |a|.$$

VII_2: *There exists a largest number* $\vartheta > 0$ *having the property that for all* $x \in \mathbf{R}, a \in \mathfrak{G}, a > 0$:

$$\tilde{x} \ne a \implies |x - a| \ge \vartheta |a|.$$

These constants ϑ and θ, characteristic for the arithmetic, can also be defined by([1])

$$\vartheta = \min \{ |b/a| \mid a \in \mathfrak{G}, \ a \ne 0, \ (a + b)^\sim \ne a \},$$
$$\theta = \max \{ |b/a| \mid a \in \mathfrak{G}, \ a \ne 0, \ (a + b)^\sim = a \}. \tag{4}$$

While $\vartheta > \theta$ is feasible, in practice one always has $\vartheta < \theta$; in fact, often $\vartheta << \theta$, but in this case (say, when $\vartheta = {}_{10}-5\theta$) the arithmetic in question must be rated as *unbalanced*, although no precise criteria are imposed in this regard. At any rate, $\theta/\vartheta \approx$ basis of the number system is technically realizable and is also satisfactory for practical purposes.

So far, θ and ϑ describe only the "density" of the set. We define now for each $a > 0$, $a \in \mathfrak{G}$, a "predecessor" a^- and a "successor" a^+ with the property that among all elements of $\mathfrak{G}$ only a has the property $a^- < a < a^+$ (for the smallest positive number, $a^- = 0$, for the largest number, $a^+ = \Omega$). With the set $\mathfrak{G}^+ \subset \mathfrak{G}$ defined as $\mathfrak{G}^+ = \{a \mid a > 0, a^- \ne 0, a^+ \ne \Omega\}$, the following holds:

---

[1] The fact that the axioms *VII* are valid also for negative a follows from II_3.

Theorem A15. *One has*

$$\vartheta \le \min_{a \in \mathfrak{S}^+} \min \left\{ \frac{a^+ - a}{a} , \frac{a - a^-}{a} \right\} \tag{5}$$

and either

$$\theta < \max_{a \in \mathfrak{S}^+} \max \left\{ \frac{a^+ - a}{a}, \frac{a - a^-}{a} \right\} \tag{6}$$

or there is an $a \in \mathfrak{S}$ *with* $a^+ = \Omega$ *or* $a^- = 0$ *such that either* $(a + \theta a)^\sim = a$ *or* $(a - \theta a)^\sim = a$.

Proof. a) For $x = a^+$ one has $\tilde{x} \ne a$ (Axiom I_4), thus $|a^+ - a| \ge \vartheta |a|$, and likewise $|a^- - a| \ge \vartheta |a|$ for all $a \in \mathfrak{S}^+$. This implies (5).

b) If $\max \{ |b/a| \mid a \in \mathfrak{S}, \ a \ne 0, \ (a + b)^\sim = a \}$ is attained for $a = a_1, \ b = b_1$ (where, because of II_3, it may be assumed that $a_1 > 0$), then $a_1 + b_1 < a_1^+, \ a_1 + b_1 > a_1^-$, thus, if $a_1 \in \mathfrak{S}^+$,

$$\theta = \left| \frac{b_1}{a_1} \right| < \max \left\{ \frac{a_1^+ - a_1}{a_1}, \frac{a_1 - a_1^-}{a_1} \right\} \le \max_{a \in \mathfrak{S}^+} \max \left\{ \frac{a^+ - a}{a}, \frac{a - a^-}{a} \right\}, \text{ q.e.d.}$$

As to the rounding errors in arithmetic operations, we first require the multiplicative operations $\times, /$ to satisfy:

VIII_1: $\quad a \tilde{\times} b = (a \times b)^\sim$,

VIII_2: $\quad a \tilde{/} b = (a/b)^\sim \quad (b \ne 0)$.

(These equations are meant to include also such statements as $a \tilde{\times} b = \Omega$ if $a \times b \in \mathfrak{D}$.)

On the basis of these axioms one now obtains properties for multiplication and division which are analogous to those in Theorem A13 and

Theorem A14 for addition and subtraction(^{2}):

Theorem A16. *If a* $\in$ $\mathfrak{G}$, *then*

$$a \tilde{\times} 1 = a \,\tilde{/}\, 1 = a,$$
$$a \,\tilde{/}\, a = 1 \;\; if \;\; a \neq 0. \tag{7}$$

If a, b $\in$ $\mathfrak{G}$, *then*

$$a \geq 0, \; 0 \leq b \leq 1 \;\Rightarrow\; a \,\tilde{\times}\, b \leq a,$$
$$a \geq b > 0 \;\Rightarrow\; a \,\tilde{/}\, b \geq 1. \tag{8}$$

However, from $a, b \in \mathfrak{G}$ and $a \tilde{\times} b = 0$ it does not follow, of course, that $a = 0$ or $b = 0$.

For the additive operations $+$, $-$ properties analogous to those in VIII_1, VIII_2 cannot be demanded; indeed, an arithmetic in which also $a \,\tilde{\pm}\, b = (a \pm b)^\sim$, we would call *optimal*. However, this property will not be required, but only

VIII_3: *a* $\tilde{\pm}$ *b* = $(a_1 \pm b_1)^\sim$, *where* a_1, b_1 *are quantities* (*not necessarily in* $\mathfrak{G}$) *not further specified except that* $\tilde{a}_1 = a$, $\tilde{b}_1 = b$.

With this, the rounding errors in arithmetic operations can now be estimated; namely:

Theorem A17. *One has*

$$|(a \,\tilde{\pm}\, b) - (a \pm b)| \leq \theta(|a| + |b| + |a \,\tilde{\pm}\, b|),$$
$$|(a \,\tilde{\times}\, b) - (a \times b)| \leq \theta\,|a \,\tilde{\times}\, b|, \tag{9}$$
$$|(a \,\tilde{/}\, b) - (a/b)| \leq \theta\,|a \,\tilde{/}\, b|.$$

Proof. The statements concerning $\times$ and $/$ are direct consequences of Axioms VII_1 and VIII_1, VIII_2. For addition and subtraction, we first

---

^{2} For the proof, one requires the following axioms:

  1) For $a \,\tilde{\times}\, 1 = a$: $VIII_1, I_4$.
  2) For $a \,\tilde{/}\, 1 = a$: $VIII_2, I_4$.
  3) For $a \,\tilde{/}\, a = 1$: $VIII_2, II_2$.

To prove the remaining two assertions, one needs in addition VI_2 and VI_4, respectively. (Editors' remark)

note that

$$|a_1 - a| \le \theta |a|, \quad |b_1 - b| \le \theta |b|,$$

hence

$$|(a_1 \pm b_1) - (a \pm b)| \le \theta(|a| + |b|).$$

On the other hand,

$$|(a_1 \pm b_1)^\sim - (a_1 \pm b_1)| \le \theta |(a_1 \pm b_1)^\sim|; \quad \text{q.e.d.}$$

§A3.5. Underflow and overflow control

It should be evident that in any computational process overflow must be prevented under all circumstances; and underflow also, in most cases. What is controversial is only how to proceed. We shall rely on three assumptions:

IX_1: *There exists a constant $\Gamma > 0$ such that*
$$\Gamma \,\tilde{/}\, x \,\tilde{/}\, y \ne 0 \implies x \,\tilde{\times}\, y \ne \Omega.$$

IX_2: *For all $c, x \in \mathfrak{S}$ one has $(\Gamma \,\tilde{/}\, c) \,\tilde{\times}\, x \ne 0 \implies c \,\tilde{/}\, x \ne \Omega$.*

IX_3: *(Maehly's rule) $(\theta^4)^\sim \ne 0$.*

The last requirement guarantees an exponent range in the floating-point representation which has a reasonable relationship to the computing precision (number of digits in the mantissa). The first two axioms allow a safe test on overflow.

Example. Normalization of a vector. Overflow is threatened if the squares of the components, multiplied by n, yield overflow. Let $\Gamma = 2$. One may program as follows:

```
max := 0;
for k := 1 step 1 until n do
    if abs(a[k]) > max then max := abs(a[k]);
    if (if max > 1 then 2/n/max/max = 0 else false) then
        go to measures;
```

(Then, after the **label** *measures*, one would have to make provisions against overflow.)

Influence of Rounding Errors

§A4.1. Persistent properties of the qd-algorithm

Normally, the properties of a numerical method are considerably altered by rounding errors. A property of a computational process, on the other hand, is called *persistent* if it remains preserved also when the process is carried out in finite arithmetic in the sense of Ch. A3. As it turns out, the *qd*-algorithm is precisely one of the algorithms that exhibits a number of persistent properties; this only, to be sure, if the sequencing of the operations in the algorithms (A1, 3) and (A1, 17) is properly observed.

Theorem A18. *If the qd-row $Z = \{q_1, e_1, \ldots, q_n\}$ is positive or semipositive, then for the row Z' computed numerically by $Z \xrightarrow{\ 0\ } Z'$,*

$$q_k' > 0,\ e_k' \geq 0 \quad (k = 1, 2, \ldots, n-1),$$
$$q_n' \geq 0. \tag{1}$$

Proof. By (A1, 3), (A3, 3), (A3, 8) and Axioms *IV* and *VI* one has:

$$q_1' = q_1 \mathbin{\widetilde{\mp}} e_1 \geq e_1 > 0,$$

$$e_1' = (e_1 \mathbin{\widetilde{/}} q_1') \mathbin{\widetilde{\times}} q_2 \leq q_2, \text{ but also } e_1' \geq 0,$$

$$q_2' = (q_2 \mathbin{\widetilde{\simeq}} e_1') \mathbin{\widetilde{\mp}} e_2 \geq e_2 > 0,$$

$$e_2' = (e_2 \; \tilde{/} \; q_2') \; \tilde{\times} \; q_3 \leq q_3, \text{ but also } e_2' \geq 0,$$

.

.

.

$$e_{n-1}' = (e_{n-1} \; \tilde{/} \; q_{n-1}') \; \tilde{\times} \; q_n \leq q_n, \text{ but also } e_{n-1}' \geq 0,$$

$$q_n' = (q_n \; \tilde{=} \; e_{n-1}') \geq 0, \quad \text{q.e.d.}$$

Note that the e'-values may become 0 because of underflow, and furthermore, e_{n-1}' and q_n' certainly become 0, if Z was semipositive. (Then, according to §A1.7, deflation is possible.) Thus, Z' no longer needs to be positive or semipositive; but, since $q_1', q_2', \ldots, q_{n-1}'$ are certainly positive, the step $Z \xrightarrow{\;0\;} Z'$ is definitely executable.

There remains, to be sure, the possibility of an overflow in one of the operations $(q_k \; \tilde{=} \; e_{k-1}') \; \tilde{\mp} \; e_k$. We show later under very weak conditions that this can be excluded.

Furthermore, Theorem A8 also is essentially persistent:

Theorem A19. *If the qd-row Z is positive and the row Z' obtained numerically from Z by $Z \xrightarrow{\;v\;} Z'$ (with $v > 0$, $v \in \mathfrak{G}$) is positive or semipositive, then*

$$\left. \begin{array}{c} q_k' > e_k \\[2mm] e_k' < q_{k+1} \end{array} \right\} \quad k = 1, 2, \ldots, n-1. \tag{2}$$

Proof. By (A1, 17) and Theorem A14 there first follows from $q_n' \geq 0$ that $(q_n \; \tilde{=} \; e_{n-1}') \; \tilde{=} \; v = q_n' \geq 0$, thus $q_n \; \tilde{=} \; e_{n-1}' \geq v > 0$, which according to Theorem A13 is only possible if $q_n > e_{n-1}'$. This, however, implies $(e_{n-1} \; \tilde{/} \; q_{n-1}') \; \tilde{\times} \; q_n = e_{n-1}' < q_n$, thus by Theorem A16, $e_{n-1} \; \tilde{/} \; q_{n-1}' < 1$, and further $e_{n-1} < q_{n-1}'$. Likewise, from $((q_{n-1} \; \tilde{=} \; e_{n-2}') \; \tilde{=} \; v) \; \tilde{\mp} \; e_{n-1} = q_{n-1}' > e_{n-1}$ one concludes $q_{n-1} > e_{n-2}'$, etc., until $e_1' < q_2$ and $q_1' > e_1$, q.e.d.

Another property which is persistent is the monotonicity property (A2, 1) of the q'_k and e'_k with respect to changes in the shift v:

Theorem A20. *Let the qd-row Z be positive and with $0 \le v_2 \le v_1$ ($v_1, v_2 \in \mathfrak{C}$) suppose that one computes from it numerically the rows Z' and Z'' by*

$$Z \xrightarrow{\ v_1\ } Z', \ Z \xrightarrow{\ v_2\ } Z''.$$

Then the following holds: if Z' is still positive, then

$$\begin{aligned} q''_k &\ge q'_k & (k = 1, \dots, n), \\ 0 \le e''_k &\le e'_k & (k = 1, \dots, n-1). \end{aligned} \tag{3}$$

Proof. By virtue of the Axioms *VI* one has

$$q'_1 = (q_1 \mathbin{\widetilde{-}} v_1) \mathbin{\widetilde{+}} e_1 \le (q_1 \mathbin{\widetilde{-}} v_2) \mathbin{\widetilde{+}} e_1 = q''_1,$$

$$e'_1 = (e_1 \mathbin{\widetilde{/}} q'_1) \mathbin{\widetilde{\times}} q_2 \ge (e_1 \mathbin{\widetilde{/}} q''_1) \mathbin{\widetilde{\times}} q_2 = e''_1 \ge 0,$$

hence $q_2 \mathbin{\widetilde{-}} e'_1 \le q_2 \mathbin{\widetilde{-}} e''_1$, and therefore

$$q'_2 = ((q_2 \mathbin{\widetilde{-}} e'_1) \mathbin{\widetilde{-}} v_1) \mathbin{\widetilde{+}} e_2 \le ((q_2 \mathbin{\widetilde{-}} e''_1) \mathbin{\widetilde{-}} v_2) \mathbin{\widetilde{+}} e_2 = q''_2,$$

etc., until $q''_n \ge q'_n$; q.e.d.

Remark. We must allow here $e''_k = 0$, even though it was assumed that $e'_k > 0$. It is also possible that for all k, $q'_k = q''_k$, $e'_k = e''_k$.

We can state, therefore, that with increasing v the q'_k do not become larger and the e'_k do not become smaller, and this also in numerical computation.

Furthermore, Theorem A11 also is persistent. Here, the upper bound for λ_n given in (A2, 6), however, has to be rewritten in the form (cf. (A2, 7))

$$\lambda_n \le \min_{k} (q'_k - e_k) = \min_{k} (q_k - e'_{k-1}), \tag{4}$$

where q'_k, e'_k are the elements of the row obtained from Z by $Z \xrightarrow{\;0\;} Z'$.

Theorem A21. *If one obtains from the positive row Z numerically by $Z \xrightarrow{\;0\;} Z'$ the row Z', and if*

$$v_2 = \min_{k} (q_k \simeq e'_{k-1}),$$

then the numerically executed qd-step $Z \xrightarrow{\;v_2\;} Z''$ cannot produce a positive row Z''.

Proof. Let $v_2 = q_p \simeq e'_{p-1}$, and suppose that $Z'' > 0$. Then from the proof of Theorem A18 and from Theorem A14 there follows $v_2 \ge 0$. Furthermore,

$$q''_p = ((q_p \simeq e''_{p-1}) \simeq (q_p \simeq e'_{p-1})) \mp e_p.$$

Since by Theorem A20 (with $0 = v_1 \le v_2$) $e''_{p-1} \ge e'_{p-1}$ (for $p = 1$ one has to put $e'_0 = e''_0 = 0$), there follows $q''_p \le e_p$ by Axiom VI_3 and Theorem A14, which for $p = n$ leads to $q''_n \le 0$ and for $p < n$ contradicts Theorem A19; q.e.d.

§A4.2. Coincidence

In the proof of Theorem A3 it is shown that in a step $Z \xrightarrow{\;0\;} Z'$ (we call this a *zero step*) the following inequalities hold,

$$\begin{aligned} q'_k &> e_k & (k = 1, \ldots, n), \\ e'_k &< q_{k+1} & (k = 1, \ldots, n-1), \end{aligned} \tag{5}$$

provided $Z > 0$. This property (5), however, is not persistent; rather, in numerical computation, also the equality sign must be permitted, as can be seen from the proof of Theorem A18. But when in the course of the process $Z \xrightarrow{\hspace{0.2em}0\hspace{0.2em}} Z'$ the inaccuracies of the numerical computation once cause $q'_k = e_k$ to occur, then on the basis of the algorithm (A1, 3) one necessarily has also $e'_k = q_{k+1}$, $q'_{k+1} = e_{k+1}, \ldots, q'_n = e_n = 0$. This event is called *coincidence*; in such a case, the e'_k, q'_{k+1}, $e'_{k+1}, \ldots$ expediently are no longer computed at all, but are simply copied.

Example. Let $Z = \{1, 10^4, 1, 10^4, 1, 1, 1\}$. In 5-digit computation the first two rows of the qd-scheme are (the arrows mean ''copy''):

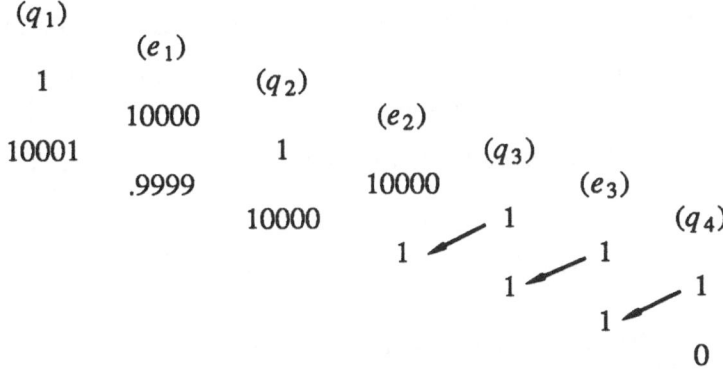

Evidently, one obtains necessarily a semipositive row, and with it also an eigenvalue. Such an incident, therefore, is quite welcome, although it requires a modification of the computing algorithm:

$$q'_1 := q_1 + e_1;$$

```
for k := 2 step 1 until n do
begin
      if q'_{k-1} = e_{k-1} then
      begin comment coincidence;
            for l := k step 1 until n do                    (6)
            begin
                  e'_{l-1} := q_l;
                  q'_l := e_l;
            end for l;
            goto ex;
      end if q;
      e'_{k-1} := (e_{k-1}/q'_{k-1}) × q_k;
```

$$q'_k := (q_k - e'_{k-1}) + e_k;$$
$$\textbf{end } for \ k;$$

$$ex:$$

It should not be overlooked, however, that a coincidence permits such a simplification only in a zero step; in the case $v \neq 0$ the occurrence of $q'_k = e_k$ (with $k < n$) means a failure of the step.

This modified computing algorithm at the same time removes the danger of divisions by zero, which even for a positive initial row Z would be possible, in principle. While it is true that the success of the first step $Z \xrightarrow{\ 0\ } Z'$ is guaranteed by Theorem A18, the row Z' need no longer be semipositive, since individual e'-values can become 0 through underflow. If this happens, there is no longer any guarantee that the next step $Z' \xrightarrow{\ 0\ } Z''$ will succeed, as the following example demonstrates: Let

$$Z = \{10{-}30, \ 10{-}30, \ 1, \ 10{-}30, \ 10{-}30, \ 1, \ 1\}, \qquad (7)$$

and assume that $\mathfrak{u} = \{x \mid |x| < 10{-}50\}$ is the underflow domain and $\theta = 10{-}10$. Then, with two zero steps of the type (A1, 3), one obtains:

```
10-30
            10-30
2 10-30                 1
            .5                   10-30
 .5                  .5                   10-30
            .5               0                   1
            0                    1                    1
                                      1
                                           0
```

Here, $q''_2 = 0$, so that e''_2, according to (A1, 3), cannot even be computed. With (6), however, one obtains without difficulty the row $Z'' = \{.5, .5, 0, 1, 1, 0, 0\}$, since, as the proof of Theorem A18 shows, $q'_{k-1} \geq e_{k-1}$ is still satisfied and therefore the vanishing of q'_{k-1} implies that of e_{k-1}, so that a coincidence necessarily occurs, whereupon the $e'_{k-1}, q'_k, e'_k, q'_{k+1}, \ldots$ are simply copied. (Note that copying is admissible also in the case

$q'_k = e_k = 0$, since the only thing that matters is the similarity of the matrices **A** in (A1, 1) and **B** in (A1, 5), which, in view of $\mathbf{A} = \mathbf{XY}$, $\mathbf{B} = \mathbf{YX}$ is guaranteed in any case.)

§A4.3. The differential form of the progressive qd-algorithm

A zero step carried out with coincidence according to (6) produces the value $q'_n = 0$, while the theoretically exact value by Theorem A3 would be different from 0; we thus have an error of 100%. But the algorithm (A1, 3) can now be modified in such a way that the differences $q'_k - e_k$ appear as independent quantities d_k (cf. (A2, 7)). These, namely, satisfy the recursion formula (A2, 9a):

$$d_k = \frac{q_k}{q'_{k-1}} d_{k-1} \,,$$

where $q'_{k-1} = e_{k-1} + d_{k-1} > d_{k-1}$. The following algorithm then results:

```
d₁ := q₁;
q'₁ := e₁ + d₁;
for k := 2 step 1 until n do
begin
        if d_{k-1} = 0 then
        begin comment coincidence;
                for ℓ := k step 1 until n do
                begin
                        e'_{ℓ-1} := q_ℓ;
                        q_ℓ' := e_ℓ;
                end for ℓ;
                goto ex;
        end if d;
        e'_{k-1} := (e_{k-1}/q'_{k-1}) × q_k;
        d_k := (d_{k-1}/q'_{k-1}) × q_k;
        q'_k := e_k + d_k;
end for k;
ex:
```

This form yields more accurate q-values; coincidence, and thus $q'_n = 0$, is now only possible when a d_k has become 0 through underflow.

Examples. Upon application to the row (7) one obtains (under the conditions stated above):

$$d_1 = {}_{10}{-}30, \qquad q'_1 = 2_{10}{-}30, \qquad e'_1 = .5,$$
$$d_2 = .5, \qquad q'_2 = .5, \qquad e'_2 = 0,$$
$$d_3 = {}_{10}{-}30, \qquad q'_3 = 1, \qquad e'_3 = 1,$$
$$d_4 = {}_{10}{-}30, \qquad q'_4 = {}_{10}{-}30,$$

that is, the row

$$Z' = \{2_{10}{-}30,\ .5,\ .5,\ 0,\ 1,\ 1,\ {}_{10}{-}30\}.$$

Likewise, from $Z = \{1, 10^4, 1, 10^4, 1, 1, 1\}$ one obtains in this way $\{10001, .9999, 10000, 1, 1, 1, .9999_{10}{-}8\}$.

This differential form, in general, produces more accurate results in those cases where the q- and e-values have markedly different orders of magnitude.

§A4.4. The influence of rounding errors on convergence

Naturally, during the numerical execution of the algorithm (A1, 3), the eigenvalues of a qd-row are perturbed, and in the algorithm (A1, 17) they do not decrease exactly by v. It can even happen that the eigenvalues increase upon execution of a step $Z \xrightarrow{\;v\;} Z'$ with positive shift. Consider, for example,

$$Z = \{10, 1, 10^4, 10^6, 10^4, 10^6, 10^4\}.$$

The smallest eigenvalue here is $\lambda_4 \approx .884$, but with $Z \xrightarrow{\;.01\;} Z'$ one obtains in 5-digit floating-point arithmetic

$$Z' = \{10.990, \ 909.92, \ 1009100, \ 9909.8, \ 1000100, \ 9999, \ .99\},$$

with $\lambda_4' \approx .971$.

This hesitation in the convergence of λ_n can have very unpleasant consequences if – as indicated in the example of A2.2 – a large number of steps are necessary, with some shifts extremely small, until q_n finally becomes small. It may indeed happen that in spite of continually positive shifts, λ_n again and again runs away from 0, so that for a very long time no semipositive row is forthcoming. Such a delay in convergence then also gives rise to large errors in the eigenvalues.

This situation – markedly different q-values and large e-values – may also develop only after a while, as the process unfolds([1]). It is precisely this phenomenon which for the reasons stated above leads to difficulties which can only be removed by a special variant of the qd-algorithm.

---

[1] The situation usually occurs, for example, when the q_k initially are incorrectly ordered. (Editors' remark)

CHAPTER A5

Stationary Form of the qd-Algorithm

§A5.1. Development of the algorithm

Suppose one starts with a positive qd-row $\bar{Z}$ and twice carries out a progressive qd step, namely:

$$\bar{Z} \xrightarrow{\;0\;} Z, \; \bar{Z} \xrightarrow{\;v\;} Z^*. \tag{1}$$

Then the following relations hold:

$$q_k = \bar{q}_k + \bar{e}_k - e_{k-1}, \quad q_k^* = \bar{q}_k + \bar{e}_k - e_{k-1}^* - v \quad (k = 1, \ldots, n)$$

(where $e_0 = e_0^* = \bar{e}_n = 0$), and

$$e_k = (\bar{e}_k/q_k)q_{k+1}; \quad e_k^* = (\bar{e}_k/q_k^*)q_{k+1} \quad (k = 1, \ldots, n-1).$$

From these one obtains by elimination of the $\bar{q}_k, \bar{e}_k$

$$q_k^* = q_k + e_{k-1} - e_{k-1}^* - v,$$
$$e_k^* = q_k e_k / q_k^*, \tag{2}$$

and can thus construct the algorithm

$$q_1^* := q_1 - v;$$
for $k := 2$ **step** 1 **until** n **do**
begin   (3)
$$\quad e_{k-1}^* := e_{k-1} \times (q_{k-1}/q_{k-1}^*);$$
$$\quad q_k^* := q_k + (e_{k-1} - e_{k-1}^*) - v;$$
end *for* $k;$

The operation (3) is called a *stationary qd-step* and is formally written as

$$Z \overset{v}{\longrightarrow} Z^*. \tag{4}$$

It is to be noted that a stationary step with $v = 0$ has no effect (in contrast to a progressive zero step).

The adjunct ''stationary'' derives from the behavior of the generating function: by (1) and (A1, 16),

$$f(z) = \frac{z\overline{f}(z) - 1}{\overline{q}_1}, \quad f^*(z - v) = \frac{z\overline{f}(z) - 1}{\overline{q}_1},$$

and thus

$$f^*(z - v) = f(z). \tag{5}$$

The step (3), therefore, does not involve a multiplication of the generating function by z, which is an essential feature in a progressive iteration.

§A5.2. The differential form of the stationary qd-algorithm

Our goal of reducing the influence of rounding errors is only partly achieved by the stationary algorithm; for example, the quantities e_{k-1} and e_{k-1}^* occurring in the assignment statement

$$q_k^* := q_k + (e_{k-1} - e_{k-1}^*) - v$$

are often nearly equal, but much larger than q_k^*. By introducing the differences $t_k = q_k - q_k^*$, however, one obtains

$$t_k = q_k - q_k^*$$
$$= v - e_{k-1} + e_{k-1}^*$$
$$= v - e_{k-1} + e_{k-1}(q_{k-1}/q_{k-1}^*)$$
$$= v + e_{k-1}(q_{k-1} - q_{k-1}^*)/q_{k-1}^*,$$

and thus

$$t_k = v + (e_{k-1}/q_{k-1}^*)t_{k-1}. \tag{6}$$

One can therefore replace (3) by:

$$
\begin{aligned}
&t_1 := v; \\
&q_1^* := q_1 - t_1; \\
&\textbf{for } k := 2 \textbf{ step } 1 \textbf{ until } n \textbf{ do} \\
&\textbf{begin} \\
&\qquad s := e_{k-1}/q_{k-1}^*; \\
&\qquad e_{k-1}^* := s \times q_{k-1}; \\
&\qquad t_k := v + s \times t_{k-1}; \\
&\qquad q_k^* := q_k - t_k; \\
&\textbf{end } for\ k;
\end{aligned}
\tag{7}
$$

Evidently, the q_k^* in this algorithm are computed theoretically as follows:

$$q_k^* = q_k - vG_k^*, \tag{8a}$$

where

$$G_1^* = 1,$$
$$G_k^* = \left[1 + \frac{e_{k-1}}{q_{k-1}^*}\left[1 + \frac{e_{k-2}}{q_{k-2}^*}\left[\cdots\left[1 + \frac{e_1}{q_1^*}\right]\cdots\right]\right]\right], \tag{8b}$$
$$k = 2, \ldots, n,$$

so that only the true reduction νG_k^* is now subtracted from q_k. *We shall henceforth carry out the stationary form of the qd-algorithm only in this differential form*

§A5.3. **Properties of the stationary qd-algorithm**

Many properties of the stationary step $Z \xrightarrow{\nu} Z^*$ (where always $Z > 0$) are the same as for the progressive step $Z \xrightarrow{\nu} Z'$. First a few facts that hold only in exact computation([1]):

a) The step $Z \xrightarrow{\nu} Z^*$ diminishes all eigenvalues by ν:

$$\lambda_k^* = \lambda_k - \nu \quad (k = 1, \ldots, n).$$

b) $Z^* > 0$ precisely if $\nu < \lambda_n$.
c) With $\nu = \lambda_n$, Z^* becomes semipositive.

Certain properties, however, are quite different from those valid for the progressive algorithm (still under the assumption of exact arithmetic):

d) $Z \xrightarrow{\nu} Z^*$, $Z^* \xrightarrow{\nu_1} Z^{**}$ $\;=>\;$ $Z \xrightarrow{\nu + \nu_1} Z^{**}$.

e) If $0 < \nu < \lambda_n$ then

$$\begin{aligned}
q_k^* &< q_k \quad (k = 1, \ldots, n), \\
e_k^* &> e_k \quad (k = 1, \ldots, n - 1).
\end{aligned} \tag{9a}$$

Some of the statements have counterparts in numerical computation:

---

[1] The proofs of statements a) to f) are simple. (Editors' remark)

f) As long as none of the quantities q_k^* becomes ≤ 0 during $Z \xrightarrow{\ \nu\ } Z^*$ with $\nu > 0$, one has in numerical computation,

$$q_k^* \leq q_k \quad (k = 1, \ldots, n),$$
$$e_k^* \geq e_k \quad (k = 1, 2, \ldots, n - 1). \tag{9b}$$

From f) and (A2, 5) there obviously follows

Theorem A22. *If a stationary qd-step* $Z \xrightarrow{\ \nu\ } Z^*$ *(with* $Z > 0$, $Z^* > 0$ *or* $Z^* \geq 0$, $\nu > 0$*) is carried out numerically, then for the quantities* $\widetilde{F}_k$, $\widetilde{G}_k^*$, $\widetilde{F}_k^*$ *computed numerically from* Z *and* Z^* *(where* $\widetilde{F}_k^*$ *is formed analogously to* $\widetilde{F}_k$, *but with* Z^* *instead of* Z*) one has*

$$\widetilde{F}_k \leq \widetilde{G}_k^* \leq \widetilde{F}_k^* \quad (k = 1, \ldots, n). \tag{10}$$

The assertion follows immediately from the observation that, on the basis of (8b) and (9b), when going from F_k to G_k^*, only the denominators in (A2, 5) change, and in fact become smaller (not greater). When going from G_k^* to F_k^*, on the other hand, only the numerators change, and become larger (not smaller).

An important persistent property of the stationary qd-algorithm is the decrease of the smallest eigenvalue when the shift ν is positive. In contrast to the progressive form, the phenomenon of "running away" observed in the example of §A4.4 does no longer occur. We first prove:

Theorem A23. *When in a positive qd-row* Z *at least one q-element is decreased, or at least one e-element increased, then the smallest eigenvalue of* Z *must decrease (so long as* Z *remains positive or semipositive).*

Proof. λ_n is characterized by the fact that with $\nu = \lambda_n$ the step $Z \xrightarrow{\ \nu\ } Z'$ (in exact arithmetic) yields a semipositive row Z'; then $q_k' > 0$, $e_k' > 0$ $(k = 1, \ldots, n - 1)$, $q_n' = 0$. If for a modified row $Z + \delta Z$ one can show that with the same shift it yields a row with some negative q-element, then it is shown that the smallest eigenvalue of the perturbed row is smaller than λ_n.

a) Decrease of a q-element: If exactly one q_k is replaced by $q_k - \varepsilon$, then by (A1, 17) the new elements q_1', e_1', q_2' ,...., q_{k-1}' are not changed, but $e_{k-1}' = (e_{k-1}/q_{k-1}')q_k$ is decreased by $\varepsilon(e_{k-1}/q_{k-1}')$; with this, $q_k' = q_k - e_{k-1}' - \nu + e_k$ is decreased by $\varepsilon - \varepsilon(e_{k-1}/q_{k-1}') = \varepsilon(q_{k-1}' - e_{k-1})/q_{k-1}'$. Subsequently, e_k' becomes larger, q_{k+1}' smaller, etc. (cf. the proof of Theorem A7), and thus finally $q_n' < 0$, if a negative q'-element has not occurred already before.

b) Increase of an e-element: If e_k is replaced by $e_k + \varepsilon$, then by (A1, 17) again q_1', e_1', q_2', ..., q_{k-1}', e_{k-1}' are not changed, but q_k' increases by ε. Therefore, in

$$e_k' = \frac{e_k}{q_k'}\, q_{k+1}$$

both numerator and denominator are increased by ε, but since the numerator by Theorem A8 is smaller, e_k' is in fact increased. Subsequently, q_{k+1}' becomes smaller, e_{k+1}' larger, etc. until a q-element becomes negative, q.e.d.

When carrying out a stationary qd-step $Z \overset{\nu}{\longrightarrow} Z^*$, it may happen that $Z^* = Z$, even though $\nu > 0$. This occurs whenever ν is so small that $q_k \simeq \nu \widetilde{\times} \widetilde{G}_k^* = q_k$ for all k, which by Axiom VII_2 (§A3.4) is certainly the case – since then also $\widetilde{G}_k^* = \widetilde{F}_k$ – when

$$\nu \widetilde{\times} \widetilde{F}_k < \vartheta\, q_k \quad (k = 1,2,....,n). \tag{11}$$

If, on the other hand, $Z^* \neq Z$, then at least one q_k^* is smaller than the corresponding q_k, or at least one e_k^* larger than e_k; therefore:

Theorem A24. *If a stationary step $Z \overset{\nu}{\longrightarrow} Z^*$ (with $Z > 0$, $Z^* > 0$ or $Z^* \geq 0$, $\nu > 0$) is carried out numerically, then for the smallest eigenvalue one has $\lambda_n^* \leq \lambda_n$, where equality sign can hold only in the case $Z = Z^*$.*

Remark. In exact computation, of course, one would have $\lambda_n^* = \lambda_n - \nu$, but this cannot be guaranteed in numerical computation. In view of §A4.4, however, Theorem A24 is already a significant improvement. Indeed, we will succeed in making the smallest eigenvalue of a qd-row as small as we like, something that cannot be guaranteed with progressive steps.

§A5.4. Safe qd-steps

The determination of eigenvalues – whether by means of progressive or stationary qd-steps – always boils down to trying to achieve a semipositive row, and then to proceed as in §A1.7. In "normal" cases, a semipositive row can essentially be obtained by the algorithm of §A2.4, but in critical cases, one must work with the stationary variant.

The primary problem, then, lies in the choice of the shifts v. For this, unfortunately, one has, on the whole, only negative information available, particularly when the numerical realization is to be a matter of concern (see, for example, Theorem A21 for the progressive algorithm). The objective of guaranteeing that the algorithm (7) with a suitable choice of v leads to a row $Z^* > 0$, however, can be achieved only with positive information, but the statement (A2, 8) most relevant in this connection is precisely one that is not persistent.

In the following, F_k and *sup* will denote the quantities computed *exactly* from q_k and e_k by (A2, 5), (A2, 6), while d_k, t_k, q_k^* are the quantities computed *numerically* by

$$
\begin{aligned}
d_1 &= q_1, \\
d_k &= q_k \, \widetilde{/} \, (1 \mp e_{k-1} \, \widetilde{/} \, d_{k-1})
\end{aligned}
\tag{12}
$$

and (7), respectively.   Then the following can be proved:

Theorem A25. *A stationary qd-step* $Z \xrightarrow{\;v\;} Z^*$ *with* $Z > 0$ *and*

$$
0 < v \le \left\{ \frac{1}{n\,(1+4\theta)^n} - 4\theta \right\} (1-4\theta)^n \min_{1 \le k \le n} d_k
\tag{13}
$$

must yield a $Z^* > 0$ *also in numerical computation.*

Proof. In the first place, we show: For the quantities d_k computed numerically by (12), one has[1]

$$d_k < \frac{q_k}{F_k(1 - 4\theta)^k} \ . \tag{14}$$

Because of $d_1 = q_1$, $F_1 = 1$, this is true for $k = 1$, and from

$$d_{k-1} < \frac{q_{k-1}}{F_{k-1}(1 - 4\theta)^{k-1}}$$

there follows

$$d_k = q_k \,\tilde{/}\, (1 \,\widetilde{\mp}\, e_{k-1} \,\tilde{/}\, d_{k-1}) \le \frac{1}{1 - 4\theta} \ \frac{q_k}{1 + \dfrac{e_{k-1}}{d_{k-1}}} \ .$$

(The factor $1 - 4\theta$ takes into account one addition[2], multiplication and division, each, with relative errors 2θ, θ, θ (cf. Theorem A17).) Of course, d_k could underflow to 0; then (14) would be true also. Otherwise,

$$d_k < \frac{1}{1 - 4\theta} \ \frac{q_k}{1 + \dfrac{e_{k-1}}{q_{k-1}} F_{k-1}(1 - 4\theta)^{k-1}}$$

$$< \frac{1}{(1 - 4\theta)^k} \ \frac{q_k}{1 + \dfrac{e_{k-1}}{q_{k-1}} F_{k-1}} = \frac{1}{(1 - 4\theta)^k} \ \frac{q_k}{F_k} \ .$$

---

[1] Terms of second order in θ are neglected here and in the sequel. By applying this consistently, some of the subsequent formulae indeed could be simplified. (Editors' remark)

[2] For the addition of two positive numbers one has in first approximation: $|(a \,\widetilde{\mp}\, b) - (a + b)| \le 2\theta(a + b) \approx 2\theta(a \,\widetilde{\mp}\, b)$. (Editors' remark)

Consequently, the numerical bound

$$\tilde{sup} = \min_{1 \le k \le n} d_k,$$

multiplied by $(1 - 4\theta)^n$, is still below the exact bound *sup*.

   We now choose $0 < v < \alpha \, sup$, where

$$\alpha = \frac{1}{n(1 + 4\theta)^n} - 4\theta,$$

and claim([3]):

$$t_k < \frac{(1 + 4\theta)^k}{1 - (k - 1)(\alpha + 4\theta)(1 + 4\theta)^{k-1}} \, v \, F_k, \tag{15}$$

$$q_k^* > \frac{1 - k(\alpha + 4\theta)(1 + 4\theta)^k}{1 - (k - 1)(\alpha + 4\theta)(1 + 4\theta)^{k-1}} \, q_k. \tag{16}$$

For $k = 1$ one has $t_1 = v$, $F_1 = 1$, hence (15) is satisfied; $q_1^* = q_1 \simeq v \ge q_1 (1 - \theta) - v(1 + \theta) - \theta q_1^*$ (see Theorem A17), thus $q_1^* > q_1 (1 - 2\theta - \alpha)$, since $v < \alpha d_1 = \alpha q_1$; hence also (16) is satisfied for $k = 1$. It remains to establish the induction step from $k - 1$ to k:

$$t_k = v \,\tilde{+}\, (e_{k-1} \,\tilde{/}\, q_{k-1}^*) \,\tilde{\times}\, t_{k-1}$$

$$\le (1 + 4\theta)[v + (e_{k-1}/q_{k-1}^*)t_{k-1}].$$

(The term $1 + 4\theta$ again accounts for one addition, multiplication and division, each.) Thus,

---

[3] Note that from the way α was chosen, the denominators in (15) and (16) are positive. The numerator in (16) is positive for $k < n$ and 0 for $k = n$. (Editors' remark)

$$t_k < (1 + 4\theta) \left[v + \frac{e_{k-1}}{q_{k-1}} \frac{1 - (k-2)(\alpha + 4\theta)(1 + 4\theta)^{k-2}}{1 - (k-1)(\alpha + 4\theta)(1 + 4\theta)^{k-1}} \right.$$

$$\left. \times \frac{(1 + 4\theta)^{k-1} v F_{k-1}}{1 - (k-2)(\alpha + 4\theta)(1 + 4\theta)^{k-2}} \right]$$

$$= (1 + 4\theta) \left[v + v \frac{e_{k-1}}{q_{k-1}} F_{k-1} \frac{(1 + 4\theta)^{k-1}}{1 - (k-1)(\alpha + 4\theta)(1 + 4\theta)^{k-1}} \right].$$

Since the last fraction in brackets is larger than 1, one also has

$$t_k < \frac{(1 + 4\theta)^k}{1 - (k-1)(\alpha + 4\theta)(1 + 4\theta)^{k-1}} v \left[1 + \frac{e_{k-1}}{q_{k-1}} F_{k-1} \right],$$

and this is (15). Furthermore,

$$q_k^* = q_k \overset{\sim}{=} t_k \geq q_k (1 - 4\theta) - t_k.$$

(Here the term 4θ collects all contributions from rounding errors under the assumption that $0 \leq t_k \leq q_k$.) Therefore,

$$q_k^* > q_k(1 - 4\theta) - \frac{(1 + 4\theta)^k}{1 - (k-1)(\alpha + 4\theta)(1 + 4\theta)^{k-1}} v F_k \, ;$$

but since $v < \alpha \ sup \leq \alpha q_k / F_k$,

$$q_k^* > q_k \left[1 - 4\theta - \frac{(1 + 4\theta)^k \alpha}{1 - (k-1)(\alpha + 4\theta)(1 + 4\theta)^{k-1}} \right]$$

$$> q_k \left[1 - \frac{(\alpha + 4\theta)(1 + 4\theta)^k}{1 - (k-1)(\alpha + 4\theta)(1 + 4\theta)^{k-1}} \right],$$

from which there follows also (16). In particular, $q_k^* > 0$ $(k = 1, \ldots, n)$, hence together with (9b) also $Z^* > 0$, provided (as is assumed in (13))

$$0 < v \leq \alpha(1 - 4\theta)^n \; s\tilde{u}p < \alpha \; sup; \quad \text{q.e.d.}$$

It is to be noted, though, that $s\tilde{u}p$ may become 0 owing to underflow; then the shift computed according to (13) also necessarily becomes $v = 0$, and therefore $Z = Z^*$, in which case the theorem becomes trivial. As we shall see, however, the case $s\tilde{u}p = 0$ (because of underflow) need not be seriously considered.

It is now our purpose, however, to show that a qd-step $Z \xrightarrow{v} Z^*$ can be constructed for which not only Z^* becomes positive, but also the quantity sup actually decreases. This, above all, requires that the occurrence of $Z^* = Z$ be avoided, which is something that (13) alone cannot yet exclude.

Analogously to the formula (14) in the proof of Theorem A25, it can first be shown that

$$d_k \geq \frac{q_k}{F_k(1 + 4\theta)^{k-1}} . \tag{17}$$

Then one obtains – even simpler than in (15) – the bound

$$t_k \geq (1 - 4\theta)^{k-1} v F_k ; \tag{18}$$

for $k = 1$, indeed, $t_1 = v$, $F_1 = 1$; furthermore, from $t_{k-1} \geq (1 - 4\theta)^{k-2} v \cdot F_{k-1}$, there follows (since $q_k^* \leq q_k$, see (9b)):

$$t_k = v \mp (e_{k-1} \tilde{/} q_{k-1}^*) \tilde{\times} t_{k-1}$$

$$\geq (1 - 4\theta)[v + (e_{k-1}/q_{k-1}^*)t_{k-1}]$$

$$\geq (1 - 4\theta)[v + (e_{k-1}/q_{k-1})t_{k-1}]$$

$$\geq (1 - 4\theta)[v + (e_{k-1}/q_{k-1})(1 - 4\theta)^{k-2}vF_{k-1}]$$

$$> (1 - 4\theta)^{k-1}[v + (e_{k-1}/q_{k-1})vF_{k-1}] = (1 - 4\theta)^{k-1}vF_k.$$

To proceed, we need a relationship between the quantities d_k computed numerically by (12) and the quantities d_k^* computed analogously from the q_k^*, e_k^* after the step $Z \xrightarrow{\ v\ } Z*$:

First, $d_1^* = q_1^* = q_1 \cong v \leq q_1 = d_1$. Then, from the induction hypothesis $d_{k-1}^* \leq d_{k-1}$ there follows, by (9b),

$$d_k^* = q_k^* \,\tilde{/}\, (1 \mp e_{k-1}^* \,\tilde{/}\, d_{k-1}^*) \leq q_k \,\tilde{/}\, (1 \mp e_{k-1} \,\tilde{/}\, d_{k-1}) = d_k;$$

thus,

$$d_k^* \leq d_k \quad (k = 1, 2, \ldots, n). \tag{19}$$

More precisely, one has $d_k^* \leq q_k^* \,\tilde{/}\, (1 \mp e_{k-1} \,\tilde{/}\, d_{k-1})$, or

$$d_k^* \leq \frac{1+\theta}{1-\theta} \frac{q_k^*}{q_k} \{q_k \,\tilde{/}\, (1 \mp e_{k-1} \,\tilde{/}\, d_{k-1})\},$$

$$d_k^* \leq \frac{1+\theta}{1-\theta} \frac{q_k^*}{q_k} d_k. \tag{20}$$

Theorem A26. *Let $Z \xrightarrow{\ v\ } Z*$, where $Z > 0$, $v > 0$ and $Z* > 0$ (or $Z* \geq 0$), and let*

$$s\tilde{u}p = \min_{1 \leq k \leq n} d_k, \quad s\tilde{u}p^* = \min_{1 \leq k \leq n} d_k^*$$

be the numerically computed upper bounds for the smallest eigenvalue $\lambda_{\min}$ of Z and the smallest eigenvalue $\lambda_{\min}^*$ of $Z*$. Then

$$\tilde{sup}^* \leq \left[\frac{1+\theta}{1-\theta} \right]^2 \tilde{sup} - \frac{1+\theta}{1-\theta} \left[\frac{1-4\theta}{1+4\theta} \right]^{n-1} v. \tag{21}$$

Proof. By (7), $q_k^* = q_k \simeq t_k$. By Theorem A17, however,

$$q_k \simeq t_k \leq q_k - t_k + \theta q_k + \theta t_k + \theta(q_k \simeq t_k),$$

thus,

$$(1-\theta)(q_k \simeq t_k) \leq q_k - t_k + \theta(q_k + t_k),$$

$$q_k^* \leq \frac{1+\theta}{1-\theta} q_k - t_k. \tag{22}$$

There exists a $k = p$ with $\tilde{sup} = d_p$. On the basis of (17) we certainly have

$$F_p \geq \frac{q_p}{(1+4\theta)^{p-1} d_p}$$

and therefore, by (18),

$$q_p^* \leq \frac{1+\theta}{1-\theta} q_p - (1-4\theta)^{p-1} v F_p$$

$$\leq \frac{1+\theta}{1-\theta} q_p - (1-4\theta)^{p-1} v \frac{q_p}{(1+4\theta)^{p-1} d_p},$$

that is,

$$q_p^* \leq \left[\frac{1+\theta}{1-\theta} - \left[\frac{1-4\theta}{1+4\theta} \right]^{p-1} \frac{v}{d_p} \right] q_p. \tag{23}$$

By (20), finally,

$$s\tilde{u}p^* \le d_p^* \le \frac{1+\theta}{1-\theta} \left[\frac{1+\theta}{1-\theta} - \left(\frac{1-4\theta}{1+4\theta} \right)^{n-1} \frac{v}{d_p} \right] d_p,$$

from which, because of $d_p = s\tilde{u}p$, the assertion follows; q.e.d.

With this, a strict decrease of at least one q_k and of the quantity $s\tilde{u}p$ as well, is guaranteed, at least so long as, say, $n \le 1/100\theta$. In fact, by Theorem A25,

$$v \le \left[\frac{1}{n\left[1 + \frac{1}{25n} \right]^n} - \frac{1}{25n} \right] \left[1 - \frac{1}{25n} \right]^n s\tilde{u}p$$

is then sufficient for the success of the step $Z \xrightarrow{\ v\ } Z^*$; this means, however (in first approximation),

$$v \le \frac{.88}{n} s\tilde{u}p. \tag{24}$$

Choosing v so large that equality holds in (24), it follows from (21) that

$$s\tilde{u}p^* \le s\tilde{u}p \left[1 - \frac{.77}{n} \right]. \tag{25}$$

We thus have linear convergence so long as v does not become 0 by underflow; *sup* can therefore practically be made as small as one likes.

Bibliography to the Appendix

[1] HENRICI P.: The quotient-difference algorithm, *Appl. Math. Series* **49**, 23–46 (1958). National Bureau of Standards, Washington, D.C.

[2] HENRICI P.: Some applications of the quotient-difference algorithm, *Proc. Symp. Appl. Math.* **15**, 159–183 (1963). Amer. Math. Soc., Providence, R.I.

[3] HENRICI P.: Quotient-difference algorithms. *Mathematical Methods for Digital Computers,* Vol. 2a, A. Ralston and H.S. Wilf (eds.), Wiley, New York 1967.

[4] HENRICI P.: *Applied and Computational Complex Analysis,* Vol. 1, Wiley, New York 1974. Chapter 7.

[5] HOUSEHOLDER A. S.: *The Numerical Treatment of a Single Non-Linear Equation,* McGraw-Hill, New York 1970.

[6] PERRON O.: *Die Lehre von den Kettenbrüchen,* Teubner, Leipzig 1929.

[7] REINSCH C., BAUER F. L.: Rational QR transformation with Newton shift for symmetric tridiagonal matrices, *Numer. Math.* **11**, 264–272 (1968).

[8] RUTISHAUSER H.: Der Quotienten-Differenzen-Algorithmus, *Z. Angew. Math. Phys.* **5**, 233–251 (1954).

[9] RUTISHAUSER H.: Anwendungen des Quotienten-Differenzen-Algorithmus, *Z. Angew. Math. Phys.* **5**, 496–508 (1954).

[10] RUTISHAUSER H.: Ein infinitesimales Analogon zum Quotienten-Differenzen-Algorithmus, *Arch. Math.* **5**, 132–137 (1954).

[11] RUTISHAUSER H.: Bestimmung der Eigenwerte und Eigenvektoren einer Matrix mit Hilfe des Quotienten-Differenzen-Algorithmus, *Z. Angew. Math. Phys.* **6**, 387–401 (1955).

[12] RUTISHAUSER H.: Une méthode pour la détermination des valeurs propres d'une matrice, *C. R. Acad. Sci. Paris* **240**, 34–36 (1955).

[13] RUTISHAUSER H.: Eine Formel von Wronski und ihre Bedeutung für den Quotienten-Differenzen-Algorithmus, *Z. Angew. Math. Phys.* **7**, 164–169 (1956).

[14] RUTISHAUSER H.: *Der Quotienten-Differenzen-Algorithmus,* Mitt. Inst. f. angew. Math. ETH Zürich, Nr. 7, Birkhäuser Verlag, Basel 1957.

[15] RUTISHAUSER H.: Solution of eigenvalue problems with the LR-transformation, *Appl. Math. Series* **49**, 47–81 (1958). National Bureau of Standards, Washington, D.C.

[16] RUTIHAUSER H.: Zur Bestimmung der Eigenwerte einer schief-symmetrischen Matrix, *Z. Angew. Math. Phys.* **9b**, 586–590 (1958).

[17] RUTISHAUSER, H.: Uber eine kubisch konvergente Variante der LR-Transformation, *Z. Angew. Math. Mech.* **40**, 49–54 (1960).

[18] RUTISHAUSER H.: On a modification of the QD-algorithm with Graeffe-type convergence, *Z. Angew. Math. Phys.* **13**, 493–496 (1962).

[19] RUTISHAUSER H.: Algorithm 125 – WEIGHTCOEFF, *Comm. ACM* **5**, 510–511 (1962).

[20] RUTISHAUSER H.: Stabile Sonderfälle des Quotienten-Differenzen-Algorithmus, *Numer. Math.* **5**, 94–112 (1963).

[21] RUTISHAUSER H.: Les propriétés numériques de l'algorithme quotient-différence. Rapport EUR 4083f, Communauté Européenne de l'Energie Atomique–EURATOM, Luxembourg 1968.

[22] RUTISHAUSER H.: Exponential interpolation with QD-algorithm, Mimeographed Report, ca. 1965.

[23] RUTISHAUSER H., BAUER F. L.: Détermination des vecteurs propres d'une matrice par une méthode itérative avec convergence quadratique, *C.R. Acad. Sci. Paris* **240**, 1680–1681 (1955).

[24] SCHWARZ H. R., RUTISHAUSER H., STIEFEL E.: *Numerical Analysis of Symmetric Matrices*, Prentice-Hall, Englewood Cliffs, N.J. 1973.

[25] STEWART G. W.: On a companion operator for analytic functions, *Numer. Math.* **18**, 26–43 (1971)

[26] STIEFEL E.: Zur Interpolation von tabellierten Funktionen durch Exponentialsummen und zur Berechnung von Eigenwerten aus den Schwarzschen Konstanten, *Z. Angew. Math. Mech.* **33**, 260–262 (1953).

[27] STIEFEL E.: Kernel polynomials in linear algebra and their numerical applications, *Appl. Math. Series* **49**, 1–22 (1958). National Bureau of Standards, Washington, D.C.

Author Index

Subject Index

A

abscissas 294, 359, 393
action integral 390
Adams predictor-corrector formulae 274
– , computer program for 274
Adams-Bashforth method 250, 251, 257
– , case study of stability for 261–262
– , order of 251
– , stability of 258
Adams-Moulton method 249, 251
– , order of 249
– , stability of 258
admissible points 46
affine, locally 282
ALCOR users group 6
algebraic equation 77, 88, 448
– , Wilkinson's example of an 100
algebraic stability 274
algorithm
– , backward recurrence 205
– , formal 1, 486
– , – , for the determination of eigenvalues
        486
– , Lanczos' 438
– , of Gauss 12, 18, 19
– , qd- 463
– , quotient-difference 463
– , Remez 199, 201
– , – , for rational approximation 205
– , – , most common variant of 201
alternation 195
alternation property 205
alternation theorem 194
– , proof of 195
amplification factor 235, 236, 237, 241
amplitude errors 389
analytic function, vector-valued 444
angle of rotation 409, 411, 415
approximate quadrature 163
approximation 175
– , best polynomial 182
– , best rational 205
– , polynomial 175
approximation polynomial 175

B

arithmetic
– , finite 1, 489
– , – , effects of 1
– , floating-point 1
– , interval 8
– , monotonicity of 492
– , optimal 497
– , precision of 495
– , properties of 491
– , unbalanced 495
arithmetic operations 496
– , rounding errors in 496, 497
ASA-FORTRAN 2
A-stability 275
A-stability property 274
A-stable 273, 274
A(α)-stable 273, 274, 275
autonomous form, initial value problems in 271

back substitution 14, 19, 21, 29, 66, 80, 116,
    117, 362
backward deflation 100
backward differentiation formulae 274, 275
– , program for 274
backward error analysis 8, 50
backward recurrence algorithm 205
band matrix 422, 437
– , storage of 422
– , symmetric positive definite 417
bandwidth 417, 418, 422
barycentric formula 133
– , computations for the 171
– , first form of 133
– , for trigonometric interpolation 171
– , operations count for 171
– , programming of 134
– , second (proper) form of 134
basis, normal 444
– , general position relative to a 444, 448
BC-scheme 17, 27, 29
beam clamped at one end 392, 396
– , eigenfrequencies of a 395
– , eigenoscillations of a 417

Y

Z